EDWARDS, Charles Henry. The historical development of the
    calculus. Springer-Verlag, 1979. 351p ill bibl index 79-15461.
    28.00 ISBN 0-387-90436-0. CIP
Edwards has done a notable job in making discovery and development in
mathematics (specifically, of course, in calculus) come alive. The earlier
chapters give an appreciation of the scope and power of early Greek methods
in geometry. The author stimulates one's curiosity about mathematical
investigation and provides an eye-opener concerning the "unexpected nature
and sequence of mathematical discovery" and the "crucial distinction
between rigorous proof and the process of discovery that must precede it."
The many problems included constitute a challenge to the mathematical
initiative and ability of the reader; however, the author is quick to point out
that the reader need not solve them to appreciate their role in the elucidation
of the text. The references at the chapter ends furnish an excellent
bibliography for anyone wanting to delve further into this area. Altogether
this is a delightful trip from earliest beginnings through the 19th-century
work of Cauchy, Riemann, and Weierstrass, with a brief postscript on
developments in the 20th century. Upper-division undergraduate level.

C. H. Edwards, Jr.

The Historical Development of the Calculus

With 150 Illustrations

Springer-Verlag

New York   Heidelberg   Berlin

C. H. Edwards, Jr.
Department of Mathematics
University of Georgia
Athens, GA 30602
USA

AMS Subject Classification (1980): 01A45, 01A50, 26-03, 26A06

Library of Congress Cataloging in Publication Data

Edwards, Charles Henry, 1937-
  The historical development of the calculus.
  Bibliography: p.
  Includes index.
  1. Calculus—History. I. Title.
QA303.E224   515'.09   79–15461

ISBN 0-387-90436-0 Springer-Verlag New York
ISBN 3-540-90436-0 Springer-Verlag Berlin Heidelberg

Preface

The calculus has served for three centuries as the principal quantitative language of Western science. In the course of its genesis and evolution some of the most fundamental problems of mathematics were first confronted and, through the persistent labors of successive generations, finally resolved. Therefore, the historical development of the calculus holds a special interest for anyone who appreciates the value of a historical perspective in teaching, learning, and enjoying mathematics and its applications. My goal in writing this book was to present an account of this development that is accessible, not solely to students of the history of mathematics, but to the wider mathematical community for which my exposition is more specifically intended, including those who study, teach, and use calculus.

The scope of this account can be delineated partly by comparison with previous works in the same general area. M. E. Baron's *The Origins of the Infinitesimal Calculus* (1969) provides an informative and reliable treatment of the precalculus period up to, but not including (in any detail), the time of Newton and Leibniz, just when the interest and pace of the story begin to quicken and intensify. C. B. Boyer's well-known book (1949, 1959 reprint) met well the goals its author set for it, but it was more appropriately titled in its original edition—*The Concepts of the Calculus*—than in its reprinting. Boyer gives an excellent account of the historical development of the concepts that lie at the foundations of the calculus (as opposed to the evolution of the calculus itself as a computational discipline); his essentially verbal exposition was well adapted to his emphasis on what used to be called the "metaphysics" of the calculus.

However, the calculus as a distinct mathematical discipline is not solely an abstract body of fundamental concepts. It is, above all, the calculating instrument *par excellence*; its greatest successes have always been answers to the question "How does one actually compute it?" The solution of specific problems has, in the development of the calculus, often played a role not unlike that of the *experimentum crucis* in natural science. Typically, a particular problem solution yields, by inference, a general technique or procedure which confronts first the question of what new problems it can solve, and in turn raises conceptual questions regarding the range of its applicability, the answers to which may finally illuminate the original problem. In this book I have tried to mirror this complex historical process by anchoring my account of the development of fundamental concepts and general methods in the computational paradigms that I feel have played the central role in the development of the calculus.

Our connected narrative makes its way (if sometimes unevenly, as dictated by the unsteady course of history) from the measurement of land area in antiquity to the nonstandard analysis of the twentieth century. The first two chapters detail those aspects of Greek mathematics that provided the foundation for the development of the calculus. Chapter 3 outlines the absorption and transmission of the Greek legacy by the Arab hegemony, the medieval scholastic speculations that contributed to an environment sympathetic to infinitesimal investigations, and the eventual renaissance of mathematical progress in Western Europe. Chapters 4–7 deal with several ingredients (logarithmic computations, infinitesimal area and tangent methods, and infinite series techniques) of the fertile amalgam that fueled the mathematical explosion of the later seventeenth century.

The centerpiece of any history of the calculus will inevitably be its treatment of the contributions of Newton and Leibniz. In Chapter 8 I have mined the riches of D. T. Whiteside's monumental edition of Newton's mathematical papers to outline (as would not previously have been possible) his calculus researches over a quarter-century period, beginning with the plague years 1665–66 that were "the prime of his age for invention." Leibniz seems to me to have been ill served by most English-language accounts of his mathematical work—I hope that Chapter 9 will help to promote a wider and more accurate understanding of the origin and distinct motivation of his approach to the calculus.

Chapter 10 deals with the onrushing technical progress of the eighteenth century, exemplified by Euler's work, and with controversies about the meaning of it all. Chapter 11 gives the nineteenth century's answers to the questions of the preceding two centuries. Chapter 12 discusses two twentieth-century developments that serve to round out our story.

A principal feature of this book is the inclusion of exercises interspersed throughout the text as an integral part of the exposition. The history of mathematics, like mathematics itself, is best learned not by passive reading, but with pen in hand. Moreover, the solution of problems typical of a particular historical period, using the tools of that time, enables the reader to share (if only at a distance) in the excitement of first discovery. And what better way to penetrate the thought of Archimedes and Newton than to work some of their problems using their own methods? Even so, I should point out that the exercises are annotated in a way that permits them simply to be read (rather than be worked) like footnotes or further remarks. Thus I have adopted the systematic use of annotated exercises as a convenient device for the inclusion of additional insights and supplementary material without a surplus of technical detail.

I hope that this book encourages further study by raising more questions than it answers. Each chapter is provided with its own bibliography, to which references for additional reading are made (within square brackets) throughout the chapter. Indeed, a book on the history of mathematics can serve no finer purpose than to guide its readers to the original sources that are the shrines of our subject.

Although the study of the history of mathematics has an intrinsic appeal of its own, its chief *raison d'etre* is surely the illumination of mathematics itself. For example, the gradual unfolding of the integral concept—from the volume computations of Archimedes to the intuitive integrals of Newton and Leibniz and finally the definitions of Cauchy, Riemann, and Lebesgue—cannot fail to promote a more mature appreciation of modern theories of integration. Because of the wide range of elementary mathematical topics that have contributed to the development of the calculus, I have found the sequence of topics covered in this book suitable for an introductory history of mathematics course. Moreover, in the context of a continuous exposition, I have included throughout the book examples and units of material that should be convenient for inclusion in a wide range of courses—introductory and advanced calculus, the general history of mathematics, and certain precalculus courses.

I owe a special debt to my wife Alice for actively sharing my enthusiasm and interest in this project. In addition to her reading and occasional criticism of the manuscript, her constant support and encouragement entitles her to a full share of satisfaction at its completion.

C. H. Edwards, Jr.

Contents

1   Area, Number, and Limit Concepts in Antiquity   **1**

Babylonian and Egyptian Geometry   1
Early Greek Geometry   5
Incommensurable Magnitudes and Geometric Algebra   10
Eudoxus and Geometric Proportions   12
Area and the Method of Exhaustion   16
Volumes of Cones and Pyramids   19
Volumes of Spheres   24
References   28

2   Archimedes   **29**

Introduction   29
The Measurement of a Circle   31
The Quadrature of the Parabola   35
The Area of an Ellipse   40
The Volume and Surface Area of a Sphere   42
The Method of Compression   54
The Archimedean Spiral   54
Solids of Revolution   62
The Method of Discovery   68
Archimedes and Calculus?   74
References   75

3   Twilight, Darkness, and Dawn   **77**

Introduction   77
The Decline of Greek Mathematics   78
Mathematics in the Dark Ages   80

The Arab Connection   81
Medieval Speculations on Motion and Variability   86
Medieval Infinite Series Summations   91
The Analytic Art of Viète   93
The Analytic Geometry of Descartes and Fermat   95
References   97

4   **Early Indivisibles and Infinitesimal Techniques**   **98**

Introduction   98
Johann Kepler (1571–1630)   99
Cavalieri's Indivisibles   104
Arithmetical Quadratures   109
The Integration of Fractional Powers   113
The First Rectification of a Curve   118
Summary   120
References   121

5   **Early Tangent Constructions**   **122**

Introduction   122
Fermat's Pseudo-equality Methods   122
Descartes' Circle Method   125
The Rules of Hudde and Sluse   127
Infinitesimal Tangent Methods   132
Composition of Instantaneous Motions   134
The Relationship Between Quadratures and Tangents   138
References   141

6   **Napier's Wonderful Logarithms**   **142**

John Napier (1550–1617)   142
The Original Motivation   143
Napier's Curious Definition   148
Arithmetic and Geometric Progressions   151
The Introduction of Common Logarithms   153
Logarithms and Hyperbolic Areas   154
Newton's Logarithmic Computations   158
Mercator's Series for the Logarithm   161
References   164

7   **The Arithmetic of the Infinite**   **166**

Introduction   166
Wallis' Interpolation Scheme and Infinite Product   170

Quadrature of the Cissoid   176
The Discovery of the Binomial Series   178
References   187

8   The Calculus According to Newton   189

The Discovery of the Calculus   189
Isaac Newton (1642–1727)   190
The Introduction of Fluxions   191
The Fundamental Theorem of Calculus   194
The Chain Rule and Integration by Substitution   196
Applications of Infinite Series   200
Newton's Method   201
The Reversion of Series   204
Discovery of the Sine and Cosine Series   205
Methods of Series and Fluxions   209
Applications of Integration by Substitution   210
Newton's Integral Tables   212
Arclength Computations   217
The Newton–Leibniz Correspondence   222
The Calculus and the *Principia Mathematica*   224
Newton's Final Work on the Calculus   226
References   230

9   The Calculus According to Leibniz   231

Gottfried Wilhelm Leibniz (1646–1716)   231
The Beginning—Sums and Differences   234
The Characteristic Triangle   239
Transmutation and the Arithmetical Quadrature of
    the Circle   245
The Invention of the Analytical Calculus   252
The First Publication of the Calculus   258
Higher-Order Differentials   260
The Meaning of Leibniz' Infinitesimals   264
Leibniz and Newton   265
References   267

10   The Age of Euler   268

Leonhard Euler (1707–1783)   268
The Concept of a Function   270
Euler's Exponential and Logarithmic Functions   272
Euler's Trigonometric Functions and Expansions   275
Differentials of Elementary Functions *à la* Euler   277

Interpolation and Numerical Integration   281
Taylor's Series   287
Fundamental Concepts in the Eighteenth Century   292
References   299

11   The Calculus According to Cauchy, Riemann, and
      Weierstrass   301

      Functions and Continuity at the Turn of the Century 301
      Fourier and Discontinuity   304
      Bolzano, Cauchy, and Continuity   308
      Cauchy's Differential Calculus   312
      The Cauchy Integral   317
      The Riemann Integral and Its Reformulations   322
      The Arithmetization of Analysis   329
      References   333

12   Postscript: The Twentieth Century   335

      The Lebesgue Integral and the Fundamental Theorem of
          Calculus   335
      Non-standard Analysis—The Vindication of Euler?   341
      References   346

      Index   347

Area, Number, and Limit Concepts in Antiquity 1

Babylonian and Egyptian Geometry

The historical origins of what we now call mathematical concepts—those that deal with number, magnitude, and form—can be traced to the rise of civilizations in the fertile river valleys of China, Egypt, India, and Mesopotamia. In particular, fairly detailed and reliable information is now available concerning the highly organized cultures of the peoples who lived along the Nile in Egypt and in the "fertile crescent" of the Tigris and Euphrates rivers in Mesopotamia in the early centuries of the second millenium B.C.

The Greeks, whose geometrical investigations provided the foundations for the development of much of modern mathematics (including the calculus), generally assumed that geometry had its origin in Egypt. For example, the Greek historian Herodotus (fifth century B.C.) wrote that agricultural plots along the Nile were taxed according to area, so that when the annual flooding of the river swept away part of a plot and its owner applied for a corresponding reduction in his taxes, it was necessary for surveyors to determine how much land had been lost. Obviously, this would have required the invention of elementary techniques of geometrical measurement.

More direct information is provided by the Egyptian papyri that have been rediscovered in modern times. In regard to Egyptian mathematics, the most important of these is the *Rhind Papyrus* which was copied in about 1650 B.C. by a scribe named Ahmes who states that it derives from a prototype from the "middle kingdom" of about 2000 to 1800 B.C. This papyrus consists mainly of a list of problems and their solutions, about twenty of which relate to the areas of fields and volumes of granaries. Each

1

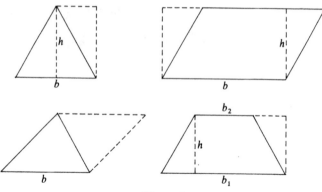

Figure 1

problem is stated in terms of particular numbers (rather than literal variables), and its solution carried out in recipe fashion, without explicitly specifying either the general formula (if any) used or the source or derivation of the method.

Apparently it is taken for granted that the area of a rectangle is the product of its base and height. The area of a triangle is calculated by multiplying half of its base times its height. In one problem the area of an isosceles trapezoid, with bases 4 and 6 and height 20, is calculated by taking half the sum of the bases, "so as to make a rectangle," and multiplying this times the height to obtain the correct area of 100. This and similar examples suggest that Egyptian prescriptions for area computations may have stemmed from elementary *dissection methods* involving the idea of cutting a rectilinear figure into triangles and then rearranging the parts so as to obtain a rectangle.

EXERCISE 1. Use the dissections suggested by Figure 1 to derive the familiar formulas for the areas of triangles ($\frac{1}{2}bh$), parallelograms (bh), and trapezoids ($\frac{1}{2}(b_1 + b_2)h$).

EXERCISE 2. A later papyrus calculates the area of a quadrilateral (4-sided polygon) by multiplying half the sum of two opposite sides times half the sum of the other two sides. Does this give the correct result for a trapezoid or parallelogram that is not a rectangle?

EXERCISE 3. (a) In one of the Rhind papyrus problems the area of a circle is calculated by squaring 8/9 of its diameter. Compare this method with the area formula $A = \pi r^2$ to obtain the Egyptian approximation $\pi \cong 3.16$.

(b) This very good approximation to π may have been obtained as follows. Trisect each side of the square circumscribed about a circle of diameter d, and cut off its 4 corners as indicated in Figure 2. Show that the area of the resulting octagon is

$$A = \frac{7}{9}d^2 = \frac{63}{81}d^2 \cong \frac{64}{81}d^2 = \left(\frac{8}{9}d\right)^2.$$

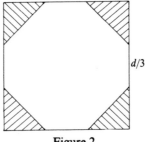

Figure 2

During the past half-century a great many mathematical cuneiform tablets, dating from the Old Babylonian age of the Hammurabi dynasty (ca. 1800–1600 B.C.), have been unearthed and deciphered. It now appears that Babylonian mathematics was considerably more advanced than Egyptian mathematics. For example, the Babylonians were adept at the solution of algebraic problems involving quadratic equations or pairs of equations, either two linear equations in two unknowns or one linear and one quadratic equation. They computed accurate numerical answers using a positional sexagesimal (base 60) system of numeration. For example, they calculated $\sqrt{2}$ as

$$1 + \frac{24}{60} + \frac{51}{60^2} + \frac{10}{60^3} = 1.414213,$$

which differs by less than 0.000001 from the true value.

In regard to geometry, the Babylonians correctly calculated the areas of triangles and trapezoids, and the volumes of cylinders and prisms (as the area of the base times the height). They were also familiar (at least on an empirical basis), well over a millenium before the time of Pythagoras, with the so-called Pythagorean theorem to the effect that the sum of the squares of the legs of a right triangle is equal to the square of its hypotenuse. In a typical Babylonian problem to be solved using this result, a ladder of given length would be standing against a wall, and it would be asked how far the bottom of the ladder slides away from the wall, if its top is lowered by a given distance.

Just as in the case of the Egyptian papyri, the Babylonian tablets mainly present problems solved by means of prescriptions that do not provide the basis for their methods. However, the following exercise presents a derivation of the Pythagorean theorem that would have been well within their range, because they were familiar with the formula $(a + b)^2 = a^2 + 2ab + b^2$.

EXERCISE 4. Four copies of a right triangle with legs a and b and hypotenuse c, together with a square of edge c, are assembled as in Figure 3 to form a square of edge $a + b$. Explain why the assembled figure *is* a square, and derive the Pythagorean relation by computing its area in two different ways.

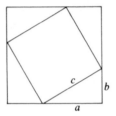

Figure 3

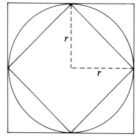

Figure 4

EXERCISE 5. The Babylonians generally used $3r^2$ for the area of a circle of radius r, corresponding to the poor approximation $\pi \cong 3$. Show that this approximation could have been obtained by averaging the areas of the inscribed and circumscribed squares in Figure 4.

EXERCISE 6. The Babylonians generally calculated the volume of a frustum of a cone or pyramid by means of the plausible (?) formula $V = \frac{1}{2}(A_1 + A_2)h$, where h is its height and A_1, A_2 the areas of its top and bottom. Show that this formula is incorrect by calculating the volume of a frustum of height 2, cut from a cone of height 4 and base radius 2 (Fig. 5). Use the (correct) formula $V = \frac{1}{3}\pi r^2 h$ for the volume of a cone.

In summary, the Egyptians and especially the Babylonians acquired a significant accumulation of elementary geometrical facts that they used to solve particular numerical problems. However, their surviving texts include

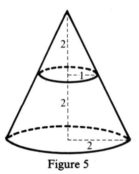

Figure 5

few if any explicit statements of general rules or methods of procedure. They made no clearcut distinctions between exact and approximate results. There is no indication of any emphasis on logical proofs or derivations in Egyptian and Babylonian thought. Thus their mathematics, despite its notable accomplishments, seems not to have been organized into any deductive system of investigation.

More complete accounts of Egyptian and Babylonian mathematics may be found in the books by Boyer [1], Neugebauer [8], and van der Waerden [12] cited in the references at the end of this chapter. Detailed discussions of ancient approximations and computations of the number π may be found in the articles by Seidenberg [9] and Smeur [10].

Early Greek Geometry

The Babylonian and Egyptian lore of number and geometry was assimilated by the Greeks, who contributed to mathematics the consciously logical and explicitly deductive approach that is now its distinguishing feature. The history of Greek mathematics begins in the sixth century B.C. with Thales and Pythagoras, both of whom are said to have traveled to Babylonia and Egypt to acquire the knowledge of those lands.

Thales lived in the first half of the sixth century B.C. On the basis of a late fourth century B.C. history of Greek mathematics that is now lost, the fourth century A.D. philosopher Proclus (in his commentary on the first book of Euclid's *Elements*) states that Thales proved the following theorems:

1. A diameter of a circle divides it into two equal parts.
2. The base angles of an isosceles triangle are equal.
3. The vertical angles formed by two intersecting straight lines are equal.
4. The angle-side-angle congruence theorem for triangles.

In addition, the fact that an angle inscribed in a semi-circle is a right angle is still known as the "theorem of Thales." Whether or not the tools for actual proofs of such theorems existed as early as Thales, it is significant that he is the first human being to whom proofs of specific mathematical results have even been attributed. Proclus adds (as quoted by van der Waerden [12], p. 90) that Thales

made many discoveries himself, in many other things he showed his successors the road to the principles. Sometimes he treated questions in a more general manner, sometimes in a more intuitive way ... Pythagoras, who came after him, transformed this science into a free form of education; he examined this discipline from its first principles and he endeavored to study the propositions, without concrete representation, by purely logical thinking.

Pythagoras is thought to have died about 500 B.C. He established a secret society or cult with distinctly mystical aspects that continued after his death. However, the Pythagoreans were actively engaged in the pursuit of learning, including mathematics (which orginally meant "that which is learned"). At their hands the subject gradually assumed an abstract character that distinguished it from the empirical and pragmatic mathematics of the Babylonians and Egyptians. Before the end of the fifth century B.C. they had formulated and proved on a rational basis the common theorems dealing with relations between triangles and other rectilinear plane figures and their areas.

"All is number" is quoted as the motto of the Pythagorean school. The Greeks used the word number to mean a "whole" number, a positive integer. In Greek theoretical mathematics (as distinguished from practical or commercial arithmetic) a fraction that we would write as a/b was not regarded as a number, as a single entity, but as a relationship or ratio $a : b$ between the (whole) numbers a and b. Thus the ratio $a : b$ was, in modern terms, simply an ordered pair, rather than a rational number.

Two ratios were said to be *proportional*, $a : b = c : d$, if (with the obvious meaning) a is the same part or parts or multiple of b as c is of d. For example, $6 : 9 = 10 : 15$ because 6 is two of the three parts of 9, as 10 is two of the three parts of 15. More formally, $a : b = c : d$ provided that there exist integers p, q, m, n such that $a = mp, b = mq, c = np, d = nq$ (so a/b and c/d are both integral multiples of p/q). On this basis the early Pythagoreans developed an elementary theory of proportionality.

EXERCISE 7. Establish the following implications.

(i) $a : b = c : d \implies a : c = b : d \implies ad = bc$
(ii) $a : c = b : c \implies a = b$
(iii) $a : b = c : d \implies (a + b) : b = (c + d) : d$
(iv) $a : b = c : d \implies (a - b) : b = (c - d) : d$ if $a > b$.

This *discrete* view of number or size was also applied to geometrical magnitudes—lengths, areas, and volumes. In particular, it was believed by the early Pythagoreans that any two line segments are *commensurable*, that is, are multiples of a common unit. On this assumption, the theory of integer ratios and proportions readily extends so as to apply to lengths and

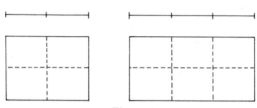

Figure 6

areas of simple figures such as line segments and rectangles. For example, the ratio $a : b$ of the lengths of the two line segments in Figure 6 is equal to the ratio $2 : 3$ of integers, while the ratio $A : B$ of areas of the two rectangles is equal to $4 : 6$. Thus we can talk about proportions $a : b = A : B = 2 : 3$ between ratios of magnitudes of different types—numbers, lengths, and areas.

For simple geometric figures with commensurable dimensions, the usual results involving area relationships are then easily established. For example, given two rectangles R and S with commensurable bases a and b and equal height h, the ratio $A : B$ of their areas is equal to the ratio $a : b$ of their bases. For if $a = mc$ and $b = nc$ where m and n are integers, then R consists of m subrectangles with base c and height h, while S consists of n such subrectangles. Hence $A : B = m : n = a : b$.

EXERCISE 8. Suppose that two rectangles are *similar*, meaning that the ratio of their bases is proportional to the ratio of their heights. If their bases and heights are commensurable, prove that the ratio of their areas is proportional to the ratio of the *squares* of (or on) their bases. By taking halves, the same result obtains for similar triangles (why?).

EXERCISE 9. A *regular* polygon is one with equal sides and equal angles. Define similarity for regular polygons. Then prove that the ratio of the areas of two similar regular polygons (with commensurable sides) is proportional to the ratio of the squares of their respective sides. *Hint:* By joining its vertices to its center, any regular polygon can be dissected into congruent isosceles triangles.

According to a fragment from the lost history of Eudemus that was allegedly copied verbatim in the sixth century A.D. by the Aristotelian commentator Simplicius, Hippocrates of Chios (ca. 430 B.C.; not to be confused with the physician Hippocrates of Cos) proved that the ratio of the areas of two circles is equal to the ratio of the squares of their diameters (or radii). Presumably he deduced (if not rigorously proved) this result by inscribing in the two circles similar regular polygons, and then "exhausting" the areas of the circles by increasing indefinitely the number of sides of the polygons (Fig. 7). Since, at each stage, the ratio of the areas of the two inscribed polygons is equal to the ratio of the squares of the radii of the two circles (as a consequence of Exercise 9), it would seem to follow "in the limit" that the same is true of the areas of the circles. However, Hippocrates probably had no limit concept sufficient to "clinch" this essentially infinitesimal argument.

Although it appears that the area of a circle can be approximated arbitrarily closely by the area of an inscribed regular polygon with sufficiently many sides, the area of the circle is not precisely equal to that of any inscribed polygon. The quadrature or "squaring of the circle"—the problem of finding a square with area precisely equal to that of a given circle—was one of the three classical problems of antiquity (together with

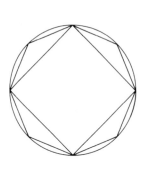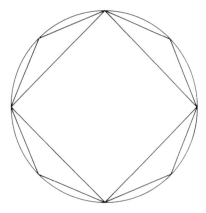

Figure 7

the duplication of a cube and the trisection of an angle). This is an example of a problem, involving a distinction between approximate and exact computation, that is unlike any considered by the Babylonians and Egyptians.

EXERCISE 10. Hippocrates applied his result on areas of circles to obtain the quadrature of a certain "lune." Consider a semicircle circumscribed about an isosceles right triangle ABC (Fig. 8). Let $ADBE$ be a circular segment on the base (hypotenuse) that is similar to the circular segments on the legs of the right triangle. Use the fact that similar circular segments are in area as the squares of their bases (why?), and the Pythagorean theorem applied to the right triangle ABC, to show that the area of the lune $ADBC$ between the circular arcs is equal to the area of the triangle ABC, and hence to half of the area of the square on AB.

According to the introduction to Archimedes' treatise *The Method*, Democritus (ca. 460 B.C.–ca. 370 B.C.) was the first to discover the fact that the volume of a pyramid (or cone) is one-third that of a prism (or cylinder) with the same base and height, but he did not rigorously prove it. A possible indication of Democritus' approach is indicated by the following question attributed to him by Plutarch (quoted by van der Waerden [12], p. 138):

> If a cone is cut by surfaces [i.e. planes] parallel to the base, then how are the sections, equal or unequal? If they were unequal then [i.e. thinking of

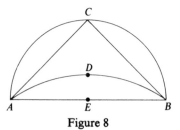

Figure 8

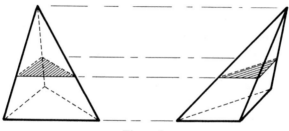

Figure 9

the slices as cylinders] the cone would have the shape of a staircase; but if they were equal, then all sections will be equal, and the cone will look like a cylinder, made up of equal circles; but this is entirely nonsensical.

Here Democritus is thinking of a solid as being composed of sections parallel to its base. From this idea it is plausible to conclude that two solids composed of equal parallel sections at equal distances from their bases should have equal volumes (Fig. 9). This fact was exploited extensively by Cavalieri in the early seventeenth century, and now bears his name. It implies that triangular pyramids with the same height and bases of equal areas will have equal volumes.

EXERCISE 11. If the bases of the two pyramids in Figure 9 have equal areas, why does it follow that corresponding sections parallel to their bases have equal areas?

Given a triangular pyramid *ABCE*, it can be "completed" to form a prism of the same base and height (Fig. 10). But then the pyramids *ABCE*, *DEFC*, *ADEC* have equal volumes, because the first and second have equal bases and heights, as do the first and third. Consequently the volume of the pyramid *ABCE* is one-third that of the prism. Since any pyramid with polygonal base can be dissected into triangular pyramids, the same result obtains for arbitrary polygonal pyramids.

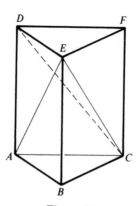

Figure 10

Just as Hippocrates exhausted a circle with inscribed regular polygons, a circular cone can similarly be exhausted with pyramids over regular polygons inscribed in its circular base. It seems to follow "in the limit" that the volume of the cone will be one-third that of the cylinder with the same base and height.

We will see that these infinitesimal plausibility arguments of the late fifth century B.C. were, about a half century later, converted into rigorous proofs by Eudoxus.

Incommensurable Magnitudes and Geometric Algebra

The Pythagorean geometry of the fifth century B.C. was based on the discrete number concept and theory of proportions discussed in the previous section. During the latter part of that century it was discovered that there exist pairs of line segments, such as the edge and diagonal of a square, that are not commensurable—they cannot be subdivided as integral multiples of segments of the same length, and hence the ratio of their lengths is not equal to the ratio of two integers. For example, the Pythagorean theorem says that the square on the diagonal of the unit square has area 2, whereas (in modern terms) $\sqrt{2}$ is not a rational number. The chronology of this discovery is discussed in detail by Knorr [6], Chapter II.

The existence of incommensurable geometric magnitudes (lengths, areas, volumes) necessitated a thorough reexamination and recasting of the foundations of mathematics, a task that occupied much of the fourth century B.C. During this period Greek algebra and geometry assumed the highly organized and rigorously deductive form that is set forth in the 13 books of the *Elements* that Euclid wrote about 300 B.C. This systematic exposition of the Greek mathematical accomplishments of the preceding three centuries is the earliest major Greek mathematical text that is now available to us (due perhaps to the extent to which the *Elements* subsumed previous expositions).

Today we simply say that $\sqrt{2}$ is an irrational number. However, for the Greeks, the discovery of incommensurability meant that there existed geometric magnitudes that could not be measured by numbers! For, as we have seen, their conception of numbers as integers alone was *discrete* in character, whereas the phenomenon of incommensurable lengths implied that geometric magnitudes have some sort of inherently (and unavoidable) *continuous* character. It followed that geometric magnitudes could not be manipulated without hesitation in algebraic computations just as though they were numbers. Although it was obvious that lengths or areas could be added by taking unions of sets, what, for example, would be meant by the product or quotient of two lengths or areas?

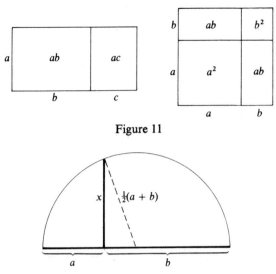

Figure 11

Figure 12

The Greek answer to such fundamental questions is presented in the *geometric algebra* of Books II and VI of Euclid's *Elements*. The product of two lengths a and b is not a third length, but rather the area of a rectangle with sides a and b. Algebraic identities, such as $a(b+c) = ab + ac$ and $(a+b)^2 = a^2 + 2ab + b^2$, are interpreted as the geometric propositions whose proofs are indicated in Figure 11.

Whereas such a simple equation as $x^2 = 2$ has no solution in the domain of (Greek, rational) numbers, the equation $x^2 = ab$ where a and b are given lengths can be solved geometrically by constructing a square with edge x whose area is equal to that of the rectangle with sides a and b. This is the real point (not always understood) to the "ruler and compass" constructions of the *Elements*—the solution of algebraic equations in terms of geometric magnitudes.

EXERCISE 12. If x is the chord in Figure 12 of a semicircle of diameter $a+b$, apply the Pythagorean theorem to show that $x^2 = ab$.

The principal Greek technique for the geometric solution of algebraic equations was based on the "application of areas." For example, given a segment AB of length a, the construction in Proposition I.44 (Prop. 44 of Book I) of the *Elements*, of a rectangle with base AB and area equal to that of a given square of edge b (Fig. 13), provides a solution of the equation $ax = b^2$. This corresponds to geometric division, and we say that the given area b^2 has been *applied* to the given segment AB.

Proposition VI.28 of the *Elements* shows how to apply a given area b^2 to a given segment of length a, but "deficient" by a square. That is, a

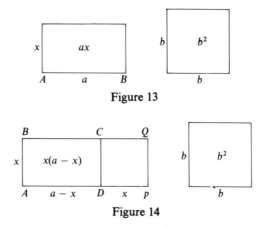

Figure 13

Figure 14

rectangle *ABCD* of area b^2 is constructed with its base lying along the given segment *AP* (of length *a*), but falling short (of the rectangle on the whole segment *AP*) by a square (*CDPQ*). This construction (Fig. 14) provides a geometric solution of the quadratic equation $ax - x^2 = b^2$.

EXERCISE 13. Proposition VI.29 of Euclid's *Elements* shows how to apply a given area b^2 to a line segment of length *a*, but "in excess" by a square. That is, the base *AB* of the constructed rectangle with area b^2 extends beyond the given segment *AP* of length *a*, with the "excess" part of this rectangle being a square. Draw the indicated figure, and interpret the construction as a geometric solution of the quadratic equation $ax + x^2 = b^2$.

   The Greeks used these admittedly cumbersome techniques of geometric algebra to handle with power and precision the staple fare of today's high school algebra, but without assuming the existence of irrational numbers. They were well aware of the existence of geometric magnitudes that we call "irrational," but simply did not think of them as numbers. This was not a lack of sophistication on their part, but rather a direct result of their unyielding insistence on logical rigor. In this connection, it is instructive to examine Book X of Euclid's *Elements*, which devotes 115 propositions and over 250 pages (in Heath's annotated translation [2]) to a comprehensive classification of irrational magnitudes of the forms $a \pm \sqrt{b}$, $\sqrt{a} \pm \sqrt{b}$, $\sqrt{a \pm \sqrt{b}}$, and $\sqrt{\sqrt{a} \pm \sqrt{b}}$, where *a* and *b* are commensurable lengths.

Eudoxus and Geometric Proportions

Any two line segments can be compared (by ruler and compass methods if one insists) to determine which has the greater length, and two lengths can be added by placing two line segments end-to-end to form a third one. The application of areas technique made possible the same operations with areas, because any rectilinear plane figure could be transformed into a rectangle with the same area and with a preassigned height. The areas of two rectangles with the same height could then be compared by comparing the lengths of their bases, and could be added by placing the two rectangles side-by-side to form a third one. By repeated addition, a geometric magnitude (length, area, or volume) can be multiplied by a positive integer.

However, the discovery of incommensurables made the Pythagorean theory of integral proportions useless for the comparison of ratios of geometric magnitudes, and thereby invalidated those geometric proofs that had utilized proportionality concepts. This crisis in the foundations of geometry was resolved by Eudoxus of Cnidus (408?–355? B.C.), a student at Plato's Academy in Athens who became the greatest mathematician of the fourth century B.C.

The key to Eudoxus' accomplishment was (as often happens in mathematics) the proper formulation of a definition—in this case, the definition of proportionality of ratios of geometric magnitudes. Let a and b be geometric magnitudes of the same type (both lengths or areas or volumes). Let c and d be a second pair of geometric magnitudes, both of the same type (but not necessarily the same type as the first pair). Then Eudoxus defines the ratios $a : b$ and $c : d$ to be *proportional*, $a : b = c : d$, provided that, given any two positive integers m and n, it follows that either

$$na > mb \quad \text{and} \quad nc > md, \tag{1}$$

or

$$na = mb \quad \text{and} \quad nc = md, \tag{2}$$

or

$$na < mb \quad \text{and} \quad nc < md. \tag{3}$$

EXERCISE 14. Show that Eudoxus' definition generalizes the familiar notion of proportionality (or equality) of ratios of integers. In particular, if a, b, c, d are integers such that $a/b = c/d$, and m and n are two positive integers, show that (1), (2), or (3) holds, depending on whether m/n is less than, equal to, or greater than $a/b = c/d$.

Thus Eudoxus' definition of proportionality for geometric ratios is simply a necessarily ponderous way of saying what is essentially obvious in the case of proportional ratios of numbers. In addition, it may be noted that, given incommensurable magnitudes a and b, this definition effectively splits the field of rational numbers m/n into two disjoint sets: the set L of those for which (1) holds, or $m : n < a : b$, and the set U of those for which (3) holds, or $m : n > a : b$. A separation of the rational numbers into two disjoint subsets L and U, such that every element of L is less than every element of U, is now called a "Dedekind cut," after Richard Dedekind, who in the nineteenth century defined a *real* number to be precisely such a "cut" of the *rational* numbers. Dedekind thereby established a firm foundation for the real number system by retracing some of Eudoxus' steps of over two thousand years earlier.

The general theory of proportionality that Eudoxus erected on the basis of the above definition is presented in Book V of Euclid's *Elements*. A critical assumption is innocuously included in Definition 4 of Book V, which states that two geometric magnitudes a and b "are said to have a ratio to one another which are capable, when multiplied, of exceeding one another," that is, if there exists an integer n such that $na > b$. The assumption that, given two comparable geometric magnitudes a and b, there exists an integer n such that $na > b$, was first stated explicitly as an axiom by Archimedes, with whose name it is therefore usually associated. We prefer, however, to call it here the "axiom of Eudoxus."

The critical role of this axiom of Eudoxus is illustrated by the proof that

$$a : c = b : c \quad \text{implies} \quad a = b. \tag{4}$$

Suppose to the contrary that $a > b$. Then there exists an integer n such that

$$n(a - b) > c. \tag{5}$$

Let mc be the smallest multiple of c that exceeds nb, so

$$mc > nb \geqslant (m-1)c. \tag{6}$$

Addition of (5) and (6) then gives

$$na > mc, \quad \text{while} \quad nb < mc,$$

which contradicts the definition of the proportionality $a : c = b : c$. It follows that $a = b$, as desired.

As a further example of the extreme care with which Eudoxus framed his theory of proportions, let us apply (4) to show that

$$a : b = c : d \quad \text{implies} \quad ad = bc. \tag{7}$$

First note that

$$a : b = ad : bd$$

because $na > mb$ implies $nad > mbd$, etc. Similarly

$$c : d = bc : bd,$$

so it follows that

$$ad : bd = bc : bd,$$

which by (4) implies that $ad = bc$, as desired.

EXERCISE 15. (Euclid V.16) Show that $a : b = c : d$ implies $a : c = b : d$. *Hints*: First apply the definition of proportionality to show that

$$a : b = c : d \quad \text{implies} \quad na : nb = mc : md$$

for any two integers m and n. Then apply (4) and/or its proof to show that $na > mc$, $na = mc$, $na < mc$ imply $nb > md$, $nb = md$, $nb < md$ respectively. Finally apply the definition of proportionality again.

EXERCISE 16. Consider two rectangles with bases a and b, areas A and B, that have the same height. Apply Eudoxus' definition of proportionality to show that $A : B = a : b$.

The proofs of (4) and (7) above indicate the manner in which the "usual" properties of proportions are demonstrated in Book V of the *Elements*, to an extent that enabled the Greeks to work with ratios of geometric magnitudes in much the same way, and to the same ends, that we today carry out arithmetical computations with real numbers. On this basis Eudoxus proceeded to give rigorous proofs of the results of Hippocrates and Democritus on areas of circles and volumes of pyramids and cones.

These area and volume computations form the content of Book XII of Euclid's *Elements*, and of the remainder of this chapter. In order to spare the reader a heavy burden of geometric algebra and Eudoxian proportions, our exposition will make free use of real numbers and modern algebraic notation. However, in order to preserve the original flavor and spirit as carefully as possible, we will follow closely both the geometrical constructions and the logical sequence of the proofs presented by Euclid.

Before proceeding in this fashion, however, it may be instructive to interpret in quite modern terms the Greek view of geometric magnitudes, taking the case of area of plane figures as an example. Say that two *polygonal* figures "have the same area" if by application of areas techniques they can be transformed to the same rectangle. This is an equivalence relation that separates the class of all polygonal plane figures into a set $\mathcal{C}$ of equivalence classes. Given a polygonal figure P, denote by $a(P) \in \mathcal{C}$ the equivalence class containing P, and call $a(P)$ the *area* of P. Then the set $\mathcal{C}$ of areas is what we might call a "Eudoxian semigroup"— an ordered commutative semigroup satisfying the axiom of Eudoxus. That is, given $a, b, c \in \mathcal{C}$, it follows that

1. (Associativity) $a + (b + c) = (a + b) + c$
2. (Commutativity) $a + b = b + a$
3. $a > b$ implies $a + c > b + c$
4. There exists an integer n such that $na > b$.

This interpretation emphasizes the fact that it is not necessary to think of areas as numbers (as the Greeks did not).

Area and the Method of Exhaustion

The Greeks assumed on intuitive grounds that simple curvilinear figures, such as circles or ellipses, have areas that are geometric magnitudes of the same type as areas of polygonal figures, and that these areas enjoy the following two natural properties.

(i) (Monotonicity) If S is contained in T, then $a(S) \leqslant a(T)$.
(ii) (Additivity) If R is the union of the non-overlapping figures S and T, then $a(R) = a(S) + a(T)$.

Given a curvilinear figure S, they attempted to determine its area $a(S)$ by means of a sequence $P_1, P_2, P_3, \ldots$, of polygons that fill up or "exhaust" S, analogous to Hippocrates' sequence of regular polygons inscribed in a circle. The so-called method of exhaustion was devised, apparently by Eudoxus, to provide a rigorous alternative to merely taking a vague and unexplained limit of $a(P_n)$ as $n \to \infty$. Indeed, the Greeks studiously avoided "taking the limit" explicitly, and this virtual "horror of the infinite" is probably responsible for the logical clarity of the method of exhaustion.

In any event, the crux of the matter consists of showing that the area $a(S - P_n)$, of the difference between the figure S and the inscribed polygon P_n, can be made as small as desired by choosing n sufficiently large. For this purpose the following consequence (Euclid X.1) of the Archimedes-Eudoxus axiom is repeatedly applied.

> Two unequal magnitudes being set out, if from the greater there be subtracted a magnitude greater than its half, and from that which is left a magnitude greater than its half, and if this process be repeated continually, there will be left some magnitude which will be less than the lesser magnitude set out.

This result, which we will call "Eudoxus' principle," may be phrased as follows. Let M_0 and ϵ be the two given magnitudes, and $M_1, M_2, M_3, \ldots$, a sequence such that $M_1 < \frac{1}{2}M_0$, $M_2 < \frac{1}{2}M_1$, $M_3 < \frac{1}{2}M_2$, etc. Then we want to conclude that $M_n < \epsilon$ for some n. To see that this is so, choose an integer N such that

$$(N+1)\epsilon > M_0.$$

Then ϵ is at most half of $(N+1)\epsilon$, so it follows that

$$N\epsilon \geqslant \tfrac{1}{2}M_0 > M_1.$$

Similarly, ϵ is at most half of $N\epsilon$, so

$$(N-1)\epsilon \geqslant \tfrac{1}{2}M_1 > M_2.$$

Proceeding in this way, at each step subtracting ϵ (which is at most half) from the left-hand-side and halving the right-hand-side, we arrive in N steps at the desired inequality

$$\epsilon > M_N. \qquad \square$$

EXERCISE 17. Conclude from Eudoxus' principle that, if M, ϵ, and $r \leq \frac{1}{2}$ are given positive numbers, then $Mr^n < \epsilon$ for n sufficiently large. Is it necessary that r be at most $\frac{1}{2}$?

We first apply Eudoxus' principle to describe precisely the manner in which the area of a circle can be exhausted by means of inscribed polygons.

Given a circle C and a number $\epsilon > 0$, there exists a regular polygon P inscribed in C such that

$$a(C) - a(P) < \epsilon. \qquad (8)$$

PROOF. We start with a square $P_0 = EFGH$ inscribed in the circle C, and write $M_0 = a(C) - a(P_0)$. Doubling the number of sides, we obtain a regular octagon P_1 inscribed in C (Fig. 15).

Continuing in this fashion, we obtain a sequence $P_0, P_1, P_2, \ldots,$ $P_n, \ldots,$ with P_n having 2^{n+2} sides. Writing

$$M_n = a(C) - a(P_n),$$

we want to show that

$$M_n - M_{n+1} > \tfrac{1}{2} M_n. \qquad (9)$$

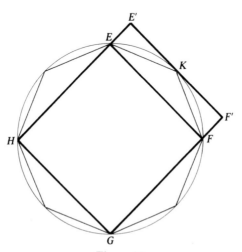

Figure 15

It will then follow from Eudoxus' principle that $M_n < \epsilon$ for n sufficiently large, and we will be finished.

The proof of (9) is essentially the same for all n, so we consider the case $n = 0$ illustrated by Fig. 15. Then

$$
\begin{aligned}
M_0 - M_1 &= a(P_1) - a(P_0) \\
&= 4a(\Delta EFK) \\
&= 2a(EFF'E') \\
&> 2a(\overparen{EKF}) \\
&= \tfrac{1}{2} \cdot 4a(\overparen{EKF}) \\
&= \tfrac{1}{2}\left[a(C) - a(P_0) \right]
\end{aligned}
$$

$$ M_0 - M_1 > \tfrac{1}{2} M_0, $$

where we denote by $\overparen{EFK}$ the circular segment cut off the circle by the side EF of the square P_0. In the general case, we obtain

$$
\begin{aligned}
M_n - M_{n+1} &= a(P_{n+1}) - a(P_n) \\
&> \tfrac{1}{2}\left[a(C) - a(P_n) \right] = \tfrac{1}{2} M_n,
\end{aligned}
$$

where $a(C) - a(P_n)$ is the sum of the areas of the 2^{n+1} circular segments cut off by the edges of P_n. $\square$

The above lemma provides the basis for a rigorous proof of the theorem on areas of circles (Euclid XII.2).

If C_1 and C_2 are circles with radii r_1 and r_2, then

$$ \frac{a(C_1)}{a(C_2)} = \frac{r_1^2}{r_2^2}. \tag{10} $$

PROOF. If $A_1 = a(C_1)$, $A_2 = a(C_2)$, then either

$$ \frac{A_1}{A_2} = \frac{r_1^2}{r_2^2} \quad \text{or} \quad \frac{A_1}{A_2} < \frac{r_1^2}{r_2^2} \quad \text{or} \quad \frac{A_1}{A_2} > \frac{r_1^2}{r_2^2}. $$

The proof is a double *reductio ad absurdum* argument, characteristic of Greek geometry, in which we show that the assumption of either of the inequalities leads to a contradiction.

Suppose first that

$$ \frac{A_1}{A_2} < \frac{r_1^2}{r_2^2}, \quad \text{or} \quad A_2 > \frac{A_1 r_2^2}{r_1^2} = S, $$

and let $\epsilon = A_2 - S > 0$. Then, by the lemma, there exists a polygon P_2 inscribed in C_2 such that

$$ A_2 - a(P_2) < \epsilon = A_2 - S, $$

so $a(P_2) > S$. But

$$\frac{a(P_1)}{a(P_2)} = \frac{r_1^2}{r_2^2} = \frac{A_1}{S}, \quad \text{(Exercise 9)}$$

where P_1 is the similar regular polygon inscribed in C_1. It follows that

$$\frac{S}{a(P_2)} = \frac{A_1}{a(P_1)} = \frac{a(C_1)}{a(P_1)} > 1$$

so $S > a(P_2)$, which is a contradiction. Hence the assumption $A_1/A_2 < r_1^2/r_2^2$ is false.

By interchanging the roles of the two circles, we find similarly that the assumption

$$\frac{A_1}{A_2} > \frac{r_1^2}{r_2^2} \quad \text{or} \quad \frac{A_2}{A_1} < \frac{r_2^2}{r_1^2}$$

is also false. We therefore conclude that (10) holds, as desired.   □

If we rewrite (10) as

$$\frac{a(C_1)}{r_1^2} = \frac{a(C_2)}{r_2^2}, \tag{11}$$

and denote by π the common value of these two ratios, then we obtain the familiar formula $A = \pi r^2$ for the area of a circle. In fact, however, the Greeks could not do this, because for them (11) was a proportion between ratios of areas, rather than a numerical equality. Hence the number π does not appear in this connection in Greek mathematics.

EXERCISE 18. Apply the lemma on the exhaustion of a circle by inscribed polygons, together with the fact that the volume of a prism is the product of its height and the area of its base, to give a double *reductio ad absurdum* proof that the volume of a circular cylinder is equal to the product of its height and the area of its base. Given a polygon P inscribed in the base circle, consider the prism Q with base P and height equal to that of the cylinder. Then the cylinder can be exhausted by prisms like Q.

Volumes of Cones and Pyramids

If P is either a triangular pyramid or a circular cone, then its volume is given by

$$v(P) = \tfrac{1}{3}Ah, \tag{12}$$

where h is its height and A the area of its base. According to Archimedes, the two results described by this formula were discovered by Democritus,

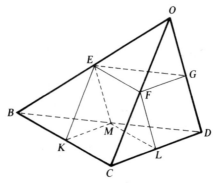

Figure 16

but were first proved by Eudoxus. In this section we discuss their treatment by Euclid in Book XII of the *Elements*.

The calculation of the volume of a pyramid is based on the dissection of an arbitrary pyramid with triangular base into two prisms and two similar pyramids, as indicated in Figure 16. The points *E, F, G, K, L, M* are the midpoints of the six edges of the pyramid *OBCD*. It is clear that the pyramids *OEFG* and *EBKM* are similar to *OBCD* and are congruent to each other. The crucial fact about this dissection is that the sum of the volumes of the two prisms

$$EKMFCL \quad \text{and} \quad MLDEFG$$

is *greater than half* the volume of the original pyramid *OBCD*. This is true because

$$v(OEFG) = v(FKCL) < v(EKMFCL)$$

and

$$v(EBKM) = v(GMLD) < v(MLDEFG).$$

If we denote by *h* the height and by *A* the area of the base *BCD* of the pyramid *OBCD*, then

$$v(MLDEFG) = \tfrac{1}{8}Ah$$

because the height of this prism is $\tfrac{1}{2}h$ and the area of its base *MLD* is $\tfrac{1}{4}A$. Also,

$$v(EKMFCL) = \tfrac{1}{8}Ah$$

because the area of the parallelogram *KCLM* is $\tfrac{1}{2}A$, and the prism *EKMFCL* is half of a parallelopiped with base *KCLM* and height $\tfrac{1}{2}h$. Consequently the sum of the volumes of the two smaller prisms is $\tfrac{1}{4}Ah$.

Now let us similarly dissect each of the two pyramids *OEFG* and *EBKM* into two smaller pyramids and two prisms. The sum of the volumes of the four resulting smaller prisms is then greater than half of the sum of the volumes of the pyramids *OEFG* and *EBKM*. Because these two latter pyramids both have height $h/2$ and base area $A/4$, it follows that the sum of the volumes of the four smaller prisms is

$$2 \cdot \frac{1}{4} \cdot \frac{A}{4} \cdot \frac{h}{2} = \frac{Ah}{4^2}.$$

After n steps of this sort, we obtain an *n-step-dissection* of the original pyramid. At the kth step we have 2^k subdivided small pyramids, and hence 2^k pairs of smaller prisms. Each of the 2^k small pyramids has height $h/2^k$ and base area $A/4^k$, so the sum of the volumes of the 2^k pairs of smaller prisms is

$$2^k \cdot \frac{1}{4} \cdot \frac{A}{4^k} \cdot \frac{h}{2^k} = \frac{Ah}{4^{k+1}}.$$

Finally, if P denotes the union of all the prisms obtained in all the steps of this n-step-dissection, it follows that

$$v(P) = Ah\left(\frac{1}{4} + \frac{1}{4^2} + \cdots + \frac{1}{4^{n+1}}\right). \qquad (13)$$

Furthermore, because at each step the sum of the volumes of the prisms is greater than half the sum of the pyramids obtained in the previous step, Eudoxus' principle implies that, given $\epsilon > 0$,

$$V - v(P) < \epsilon \qquad (14)$$

if n is sufficiently large, and $V = v(OBCD)$. This construction is the basis for Euclid's proof of Proposition XII.5.

Given two triangular pyramids with the same height and with base areas A_1 and A_2, the ratio of their volumes V_1 and V_2 is equal to that of their base areas,

$$\frac{V_1}{V_2} = \frac{A_1}{A_2}. \qquad (15)$$

PROOF. The demonstration of (15) is a double *reductio ad absurdum* argument almost identical to that used in the proof of the theorem on areas of circles. Suppose first that

$$\frac{V_1}{V_2} < \frac{A_1}{A_2}, \quad \text{or} \quad V_2 > \frac{V_1 A_2}{A_1} = S,$$

and let $\epsilon = V_2 - S$. Denote by P_2 the union of all the prisms obtained in an n-step-dissection of the second pyramid, with n sufficiently large that

$$V_2 - v(P_2) < \epsilon = V_2 - S,$$

so $v(P_2) > S$. It then follows from (13) that, if P_1 is the similar union of prisms obtained in an n-step-dissection of the first pyramid, then

$$\frac{v(P_1)}{v(P_2)} = \frac{A_1}{A_2} = \frac{V_1}{S}.$$

Hence

$$\frac{S}{v(P_2)} = \frac{V_1}{v(P_1)} > 1$$

because P_1 is properly contained in the first pyramid. But $S > v(P_2)$ is a contradiction, so the assumption $V_1/V_2 < A_1/A_2$ is false.

By interchanging the roles of the two pyramids, we find that the assumption $V_1/V_2 > A_1/A_2$ is also false. It therefore follows that $V_1/V_2 = A_1/A_2$, as desired. $\square$

We have previously seen that the formula $V = \frac{1}{3}Ah$, for the volume of a triangular pyramid, follows from the fact that two pyramids with equal heights and base areas must have the same volumes. For any given pyramid is one of three pyramids with equal volumes, whose union is a prism with height and base area equal to those of the given pyramid (see Figure 10). We assume here (as in the above construction) the elementary fact that the volume of a prism is the product of its height and its base area.

Alternatively, it is interesting to derive the formula $V = \frac{1}{3}Ah$ directly by using the sum of the geometric series

$$\sum_{n=0}^{\infty} \frac{1}{4^n} = \frac{4}{3}.$$

Given $\epsilon > 0$, we see from (13) that

$$V - Ah\left(\frac{1}{4} + \frac{1}{4^2} + \cdots + \frac{1}{4^{n+1}}\right) < \epsilon$$

if n is sufficiently large. It follows that the volume of the pyramid is

$$V = \frac{Ah}{4} \sum_{n=0}^{\infty} \frac{1}{4^n} = \frac{Ah}{4} \cdot \frac{4}{3} = \frac{1}{3}Ah.$$

Although the Greeks knew how to sum a finite geometric progression, they used *reductio ad absurdum* arguments to avoid the formal summation of an infinite series.

EXERCISE 19. Show that the volume formula $V = \frac{1}{3}Ah$ holds for a pyramid whose base is an arbitrary convex polygon (that can be dissected into triangles, thereby dissecting the pyramid into triangular pyramids for which the formula is already known).

EXERCISE 20. Show that the ratio of the volumes of two similar pyramids is equal to the ratio of the cubes of corresponding edges.

In Proposition XII.10 Euclid uses inscribed pyramids to exhaust a circular cone so as to establish the volume formula $V = \frac{1}{3} Ah$ for cones. To outline this proof, let T be a cone with vertex O, base circle C, and height h. Let

$$P_0, P_1, P_2, \ldots, P_n, \ldots$$

be the sequence of inscribed regular polygons previously used to exhaust the circle, with P_n having 2^{n+2} sides. If T_n denotes the pyramid with vertex O and base P_n, then

$$T_0, T_1, T_2, \ldots, T_n, \ldots$$

is a sequence of pyramids inscribed in the cone T, and $v(T_n) = \frac{1}{3} a(P_n) h$, where h is the height of T.

Recall that we proved that, if $M_n = a(C) - a(P_n)$, then $M_n - M_{n+1} > \frac{1}{2} M_n$. By joining every polygon involved with the vertex O, we can similarly prove that, if

$$\overline{M_n} = v(T) - v(T_n),$$

then

$$\overline{M_n} - \overline{M_{n+1}} > \tfrac{1}{2} \overline{M_n}.$$

Eudoxus' principle therefore implies that, given $\epsilon > 0$,

$$\overline{M_n} = v(T) - v(T_n) < \epsilon \tag{16}$$

if n is sufficiently large. Also, if Q is the cylinder with base C and height h, and Q_n is the inscribed prism with base P_n and height h, then

$$v(Q) - v(Q_n) < \epsilon$$

for n sufficiently large (why?).

We are now ready for the *reductio ad absurdum* proof that

$$v(T) = \tfrac{1}{3} v(Q) = \tfrac{1}{3} Ah. \tag{17}$$

Otherwise, either $v(T) < \frac{1}{3} v(Q)$ or $v(T) > \frac{1}{3} v(Q)$.

Assuming that $v(T) < \frac{1}{3} v(Q)$, choose n sufficiently large that

$$v(Q) - v(Q_n) < v(Q) - 3v(T).$$

Then $v(Q_n) > 3v(T) > 3v(T_n)$ because the pyramid T_n is inscribed in the cone T. But the conclusion that $v(T_n) < \frac{1}{3} v(Q_n)$ is a contradiction, because the pyramid T_n and the prism Q_n have the same base and height, so we know (Exercise 19) that $v(T_n) = \frac{1}{3} v(Q_n)$.

Assuming that $v(T) > \frac{1}{3}v(Q)$, choose n sufficiently large that

$$v(T) - v(T_n) < v(T) - \tfrac{1}{3}v(Q).$$

Then $v(T_n) > \frac{1}{3}v(Q) > \frac{1}{3}v(Q_n)$ because the prism Q_n is inscribed in the cylinder Q. But this is a contradiction for the same reason as before, so we conclude that $v(T) = \frac{1}{3}v(Q)$ as desired.   □

Volumes of Spheres

The final result in Book XII of the *Elements* is Proposition 18, to the effect that the volume of a sphere is proportional to the cube of its radius. Euclid proves this in the following form.

If S_1 and S_2 are two spheres with radii r_1 and r_2 and volumes V_1 and V_2, then

$$\frac{V_1}{V_2} = \frac{r_1^3}{r_2^3}. \tag{18}$$

As a preliminary lemma (XII.17) he shows that, given two concentric spheres S and S' with S' interior to S, there exists a polyhedral solid P inscribed in S that contains S' in its interior. The polyhedral solid P is a union of finitely many pyramids, each of which has the common center O of the two spheres as its vertex, with its base being a polygon inscribed in the outer sphere S (Fig. 17).

In his proof of Proposition 18, Euclid assumes without proof that, given a sphere S with volume V and $V' < V$, there exists a concentric sphere S' with $v(S') = V'$. We will repair this minor gap by using the slightly simpler

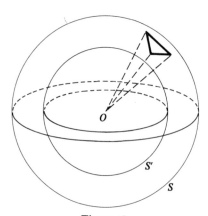

Figure 17

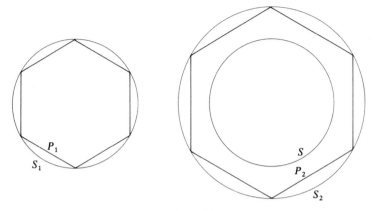

Figure 18

fact that there exists a concentric sphere S' with $V' < v(S') < V$ (Exercise 21 below).

Assuming that $V_1/V_2 < r_1^3/r_2^3$, let

$$\epsilon = V_2 - \frac{r_2^3 V_1}{r_1^3},$$

and let S be a sphere interior to and concentric with S_2 (see Fig. 18) such that

$$v(S) = V > V_2 - \epsilon = \frac{r_2^3 V_1}{r_1^3}. \tag{19}$$

Now let P_2 be a polyhedral solid inscribed in S_2 that contains S in its interior. If P_1 is the similar polyhedral solid inscribed in S_1, then

$$\frac{v(P_1)}{v(P_2)} = \frac{V_1'}{V_2'} = \frac{r_1^3}{r_2^3} \tag{20}$$

by Exercise 20, because P_1 and P_2 are made up of pairwise similar pyramids with corresponding edges r_1 and r_2. Hence

$$V_2' > V > \frac{r_2^3 V_1}{r_1^3}$$

by (19), so

$$\frac{V_1}{V_2'} < \frac{r_1^3}{r_2^3} = \frac{V_1'}{V_2'}$$

by (20). Thus $v(S_1) = V_1 < V_1' = v(P_1)$. But this is a contradiction, because S_1 contains P_1.

Interchanging the roles of the two spheres S_1 and S_2, the assumption that $V_1/V_2 > r_1^3/r_2^3$ leads similarly to a contradiction. Consequently we conclude that $V_1/V_2 = r_1^3/r_2^3$, as desired. $\square$

EXERCISE 21. Let S be a sphere of radius r and equatorial circle C, and $\epsilon > 0$. Show as follows, without using the formula for the volume of a sphere, that there is a sphere $\bar{S}$ with $v(S) - \epsilon < v(\bar{S}) < v(S)$. First choose $\delta > 0$ such that

$$\pi(r+\delta)^2 - \pi(r-\delta)^2 = 4\pi r\delta < \frac{\epsilon}{4\pi r}.$$

If $r - \delta < \bar{r} < r$, then the annular ring A bounded by C and the concentric circle $\bar{C}$ of radius $\bar{r}$ can be covered by non-overlapping rectangles $R_1, R_2, \ldots, R_n$ such that

$$\sum_{i=1}^{n} a(R_i) < \frac{\epsilon}{4\pi r}. \qquad \text{(Why?)}$$

If T_i is the cylinder-with-hole obtained by revolving the rectangle R_i about a horizontal axis through the center of the circles, then the sets $T_1, T_2, \ldots, T_n$ cover the spherical shell between the sphere S and the sphere $\bar{S}$ of radius $\bar{r}$. Now apply the formula for the volume of a cylinder to show that

$$\sum_{i=1}^{n} v(T_i) < \epsilon.$$

Why does this imply that $v(S) - \epsilon < v(\bar{S}) < v(S)$?

Let S_1 be an arbitrary sphere with radius r and volume V, and denote by α the volume of a sphere S_2 with unit radius. Then Equation (18) yields the volume formula

$$V = \alpha r^3, \qquad (21)$$

according to which the volume of a sphere is proportional to the cube of its radius. There is no indication that Euclid or his predecessors knew the relationship between α and π; it was Archimedes who discovered that $\alpha = 4\pi/3$ (see Chapter 2).

It is instructive to examine the common pattern of the proofs of the five basic results from Euclid XII that we have discussed in this and the preceding two sections. Each of these theorems compares the areas or volumes of two sets A and B that are either

1. two circles,
2. two cylinders with the same height,
3. two pyramids with the same height,
4. a cone and a cylinder with the same height, or
5. two spheres.

In particular, we want to prove that

$$v(B) = kv(A), \qquad (22)$$

where the proportionality constant k is equal in the five cases, respectively, to

1. the ratio of the squares of their radii.
2. the ratio of the squares of their radii.
3. the ratio of their base areas.
4. one-third.
5. the ratio of the cubes of their radii.

The first step in each proof is the construction of two sequences of polygonal or polyhedral figures, $\{P_n\}_1^\infty$ inscribed in A and $\{Q_n\}_1^\infty$ inscribed in B, such that

$$v(Q_n) = kv(P_n)$$

for all n. Eudoxus' principle is applied to the construction to show that, given $\epsilon > 0$,

$$v(A) - v(P_n) < \epsilon \quad \text{and} \quad v(B) - v(Q_n) < \epsilon$$

if n is sufficiently large.

In terms of the modern limit concept we would complete the proof by simply noting that

$$v(B) = \lim_{n \to \infty} v(Q_n)$$
$$= \lim_{n \to \infty} kv(P_n)$$
$$= k \lim_{n \to \infty} v(P_n)$$
$$v(B) = kv(A).$$

In essence, the Greeks avoided this explicit use of limits by completing the proof by means of the double *reductio ad absurdum* argument. Assuming that

$$v(B) > kv(A),$$

and writing $\epsilon = v(B) - kv(A)$, we choose inscribed figures P in A and Q in B such that

$$v(Q) = kv(P) \quad \text{and} \quad v(Q) > v(B) - \epsilon = kv(A).$$

But this is a contradiction, since $v(P) \leqslant v(A)$ because A contains P. Reversing the roles of A and B, the assumption $v(A) > v(B)/k$ leads similarly to a contradiction, so we must conclude that $v(B) = kv(A)$ as desired. □

A logically complete indirect proof is thereby obtained without explicit reference to limits. The mystery which the Greeks attached to the infinite and, in particular, to what we call the limit concept, is absorbed (if not obviated) in Eudoxus' principle. In this connection, Aristotle remarked that mathematicians make no use of magnitudes infinitely large or small, but content themselves with magnitudes that can be made as large or as small as they please (quoted by Heath [3], p. 272).

References

[1] C. B. Boyer, *A History of Mathematics*. New York: Wiley, 1968.

[2] T. L. Heath, *The Thirteen Books of Euclid's Elements*. Cambridge University Press, 1908. (Dover reprint, 1956).

[3] T. L. Heath, *A History of Greek Mathematics*, Vol. I. Oxford University Press, 1921.

[4] T. L. Heath, Greek geometry with special reference to infinitesimals. *Math Gaz* **11**, 248–259, 1922–23.

[5] J. Hjelmslev, Eudoxus' axiom and Archimedes' lemma. *Centaurus* **1**, 2–11, 1950.

[6] W. R. Knorr, *The Evolution of the Euclidean Elements*. Dordrecht: Reidel, 1975.

[7] G. R. Morrow, *Proclus, A Commentary on the First Book of Euclid's Elements*. Princeton, NJ: Princeton University Press, 1970.

[8] O. Neugebauer, *The Exact Sciences in Antiquity*. Brown University Press, 1957, 2nd ed.

[9] A. Seidenberg, On the area of a semi-circle. *Arch Hist Exact Sci* **9**, 171–211, 1972–73.

[10] A. J. E. M. Smeur, On the value equivalent to π in ancient mathematical texts. *Arch Hist Exact Sci* **6**, 249–270, 1969–70.

[11] I. Thomas, *Greek Mathematical Works*, 2 vols. Cambridge, MA: Harvard University Press, 1951.

[12] B. L. van der Waerden, *Science Awakening*. Oxford University Press, 1961, 2nd ed.

[13] K. von Fritz, The discovery of incommensurability by Hippasus of Metapontum, *Ann Math* (2) **46**, 242–264, 1945.

Archimedes 2

Introduction

Archimedes of Syracuse (287–212 B.C.) was the greatest mathematician of ancient times, and twenty-two centuries have not diminished the brilliance or importance of his work. Another mathematician of comparable power and creativity was not seen before Newton in the seventeenth century, nor one with similar clarity and elegance of mathematical thought before Gauss in the nineteenth century.

He was famous in his own time for his mechanical inventions—the so-called Archimedean screw for pumping water, lever and pulley devices ("give me a place to stand and I can move the earth"), a planetarium that duplicated the motions of heavenly bodies with such accuracy as to show eclipses of the sun and moon, machines of war that terrified Roman soldiers in the siege of Syracuse (which, however, resulted in Archimedes' death). For Archimedes himself these inventions were merely the "diversions of geometry at play," and the writings that he left behind are devoted entirely to mathematical investigations. These treatises have been described by Heath (editor of the standard English edition of the works of Archimedes [5]) as

> without exception, models of mathematical exposition; the gradual unfolding of the plan of attack, the masterly ordering of the propositions, the stern elimination of everything not immediately relevant, the perfect finish of the whole, combine to produce a deep impression, almost a feeling of awe, in the mind of the reader. There is here, as in all the great Greek mathematical masterpieces, no hint as to the kind of analysis by which the results were first arrived at; for it is clear that they were not discovered by the steps which lead up to them in the finished treatise. If

29

the geometrical treatises had stood alone, Archimedes might seem, as Wallis said, "as it were of set purpose to have covered up the traces of his investigations, as if he has grudged posterity the secret of his method of inquiry, while he wished to extort from them assent to his results" ([7], p. 281).

In this chapter we discuss those of Archimedes' extant works that deal primarily with area, length, and volume computations, in the following order:

1. *Measurement of a Circle*
2. *Quadrature of the Parabola*
3. *On the Sphere and Cylinder*
4. *On Spirals*
5. *On Conoids and Spheroids*
6. *The Method*

The first five of these develop the method of exhaustion into a technique of remarkable power which Archimedes applied to a wide range of problems that today are typical applications of the integral calculus, and which provided the starting point for the modern development of the calculus. Treatise (6), which was unknown until its rediscovery in 1906, describes the heuristic infinitesimal method by which Archimedes first discovered many of his results.

Throughout this chapter, as in the later sections of Chapter 1, modern algebraic symbolism is used to palatably translate verbal statements and arguments that were originally presented in the cumbersome language of classical geometric algebra. Recall that Greek mathematics did not represent geometric magnitudes in terms of real numbers. Consequently, in order to specify the area of a given figure represented geometrically as the product of two linear factors, the Greek geometer had to introduce a plane figure with area equal to that of the given figure. For example, Archimedes would say that the surface area (A) of a right circular cylinder (excluding its bases) is equal to the area of a circle whose radius is the mean proportional between the height (h) of the cylinder and the diameter (d) of its base. For us, A is the product of the height and the circumference of the base, and we simply write $A = \pi dh$.

It must be recognized that this concession to ease of understanding on the part of the modern reader entails a loss of certain characteristic features of geometric algebra that are important for a full understanding and appreciation of classical Greek mathematics. However, we will adhere closely to Archimedes' basic geometric constructions, and thereby attempt to preserve those features that seem most important for an understanding of the historical development of the calculus. The best comprehensive analysis of Archimedes' works, one that faithfully preserves the flavor of antiquity, is that of E. J. Dijksterhuis [3].

The Measurement of a Circle

As we saw in Chapter 1, the awareness (on some level) that the area of a circle is proportional to the square of its radius, $A = \pi_1 r^2$ for some constant π_1, dates back to earliest times. Similarly, the proportionality between a circle's circumference and diameter, $C = \pi_2 d$ for some constant π_2, is an ancient one. However, it is not clear when it was first realized that the two proportionality constants are the same, $\pi_1 = \pi_2 = \pi$. In the *Measurement of a Circle*, Archimedes provided the first rigorous proof of this fact by showing that the area of a circle is equal to that of a triangle with base equal to its circumference and height equal to its radius,

$$A = \tfrac{1}{2} rC. \tag{1}$$

To see that (1) implies that $\pi_1 = \pi_2$, simply substitute $A = \pi_1 r^2$ and $C = 2\pi_2 r$. He then showed that

$$3\tfrac{10}{71} < \pi < 3\tfrac{1}{7} \tag{2}$$

by explicitly establishing this inequality for the ratio of the circumference of a circle to its diameter.

Formula (1) was certainly known before Archimedes, and it was probably deduced by regarding the circle as the union of indefinitely many isosceles triangles with the center as their common vertex, and with their bases forming an inscribed regular polygon with each of its indefinitely many sides almost coinciding with a small arc of the circle. Since the height of each triangle will virtually equal the radius of the circle, and the sum of their bases will virtually equal its circumference, this picture makes the truth of (1) seem evident.

This heuristic derivation supplies the motivation for Archimedes' rigorous proof. In it he extends the method of exhaustion to what has been termed the "method of compression." Instead of dealing only with inscribed polygons, he employs both inscribed and circumscribed polygons. The area of the circle is then "compressed" between the areas of inscribed and circumscribed polygons that closely approximate the circle (Fig. 1). The following two exercises are preliminaries to the proof.

EXERCISE 1. Consider a circle with circumference C, and let P and Q be inscribed and circumscribed polygons, respectively. Show that the perimeter of P is less than C, while the perimeter of Q is greater than C. Use the facts that $\sin \theta < \theta < \tan \theta$ if θ is an angle less than $\pi/2$ radians, and that a central angle of θ radians subtends an arc of length $r\theta$ in a circle of radius r.

EXERCISE 2. In Chapter 1 we saw that, given a circle with area A and $\epsilon > 0$, there is an inscribed regular polygon P with $a(P) > A - \epsilon$. Show similarly that there is a circumscribed regular polygon Q with $a(Q) < A + \epsilon$.

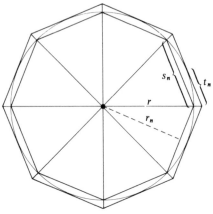

Figure 1

The proof of (1) is a typical *reductio ad absurdum* argument. Assuming that $A > \frac{1}{2}rC$, let $\epsilon = A - \frac{1}{2}rC$, and choose a regular n-sided polygon P inscribed in the circle such that

$$a(P) > A - \epsilon = \tfrac{1}{2}rC.$$

If s_n is the length of a side of P, and r_n is the length of a perpendicular from the center to a side of P, then

$$r_n < r \quad \text{and} \quad ns_n < C$$

by Exercise 1. Because P is the union of n isosceles triangles with base s_n and height r_n, it follows that

$$a(P) = n \cdot \tfrac{1}{2}r_n s_n = \tfrac{1}{2}r_n(ns_n) < \tfrac{1}{2}rC.$$

But this is a contradiction, so A is not greater than $\frac{1}{2}rC$.

Assuming that $A < \frac{1}{2}rC$, let $\epsilon = \frac{1}{2}rC - A$, and choose a regular n-sided circumscribed polygon Q such that

$$a(Q) < A + \epsilon = \tfrac{1}{2}rC.$$

For the purpose of subsequent computations, let t_n denote *half* the length of a side of Q. Then

$$a(Q) = n \cdot \tfrac{1}{2}r(2t_n) = \tfrac{1}{2}r(2nt_n) > \tfrac{1}{2}rC,$$

because the perimeter $2nt_n$ of Q is greater than C (Exercise 1). This contradiction completes the proof of (1).   □

EXERCISE 3. Let A_n and C_n denote the area and perimeter, respectively, of a regular polygon with n sides inscribed in a circle of radius r. Show that

$$A_n = nr^2 \sin \frac{\pi}{n} \cos \frac{\pi}{n} \quad \text{and} \quad C_n = 2nr \sin \frac{\pi}{n}.$$

Deduce that $A = \frac{1}{2}rC$ by taking the limit of A_n/C_n as $n \to \infty$.

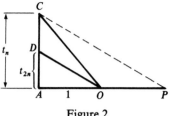

Figure 2

EXERCISE 4. With A_n and C_n as in Exercise 3, and r_n, s_n as in Figure 1, write $A_n = \frac{1}{2}nr_ns_n$ and $C_n = ns_n$. Deduce that $A = \frac{1}{2}rC$ without using trigonometric functions, by taking the limit of A_n/C_n as $n \to \infty$. What must you assume to be "obvious"?

In order to obtain the approximation (2) for π, Archimedes began with regular hexagons inscribed in and circumscribed about a circle of radius one. By successively doubling the number of sides he obtained pairs of inscribed and circumscribed regular polygons with 12, 24, 48, and 96 sides, and calculated their perimeters to find upper and lower bounds for π.

Consider first the circumscribed polygons. If t_n denotes *half* the length of a side of a regular circumscribed polygon with n sides, then the relationship between t_n and t_{2n} is indicated in Figure 2, where O is the center of the circle, and OD bisects angle AOC. If CP is parallel to OD, it is easily seen that $OP = CO$. Since triangles ADO and ACP are similar, it follows that

$$\frac{AD}{AO} = \frac{AC}{AO + OP} = \frac{AC}{AO + OC},$$

or

$$t_{2n} = \frac{t_n}{1 + \sqrt{1 + t_n^2}}. \tag{3}$$

Now consider the inscribed polygons. If s_n denotes the side of the regular inscribed polygon with n sides, the relationship between s_n and s_{2n} is indicated by Figure 3, where $s_n = BC$, $s_{2n} = BD$, and AD bisects angle BAC (Why?). It is easily checked that the triangles ABD, BPD, and APC

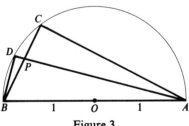

Figure 3

are similar. Hence

$$\frac{AB}{AD} = \frac{BP}{BD} \quad \text{and} \quad \frac{AC}{AD} = \frac{PC}{BD},$$

so

$$\frac{AB + AC}{AD} = \frac{BP + PC}{BD} = \frac{BC}{BD},$$

or

$$\frac{2 + \sqrt{4 - s_n^2}}{\sqrt{4 - s_{2n}^2}} = \frac{s_n}{s_{2n}}.$$

After cross-multiplying and squaring, the resulting equation yields

$$s_{2n}^2 = \frac{s_n^2}{2 + \sqrt{4 - s_n^2}}. \tag{4}$$

EXERCISE 5. Observe from the familiar geometry of the regular hexagon that $s_6 = 1$, $t_6 = 1/\sqrt{3}$. Apply formulas (3) and (4) to compute $s_{12}, s_{24}, s_{48}, s_{96}$ and $t_{12}, t_{24}, t_{48}, t_{96}$ recursively (using a hand calculator) so as to obtain

$$s_{96} = 0.065438 \quad \text{and} \quad t_{96} = 0.032737.$$

Since $48s_{96} < \pi < 96t_{96}$, conclude that

$$3\tfrac{10}{71} < \pi < 3\tfrac{1}{7}.$$

Hence $\pi = 3.14$, rounded off to two decimal places.

Of course, Archimedes did not have a hand calculator available. He started with the approximation

$$\tfrac{265}{153} < \sqrt{3} < \tfrac{1351}{780}$$

and proceeded manually, carefully rounding down in calculating the s_n's and rounding up in calculating the t_n's, finally obtaining

$$3\frac{10}{71} < \frac{6336}{2017\tfrac{1}{4}} < \pi < \frac{14688}{4673\tfrac{1}{2}} < 3\frac{1}{7}.$$

It is generally believed that the extant *Measurement of a Circle* is only a fragment of Archimedes' original and more comprehensive treatment of the circle. In a recent article W. R. Knorr [10] argues persuasively that, to obtain a more accurate approximation to π, Archimedes started with inscribed and circumscribed decagons (regular 10-sided polygons) and successively doubled sides six times to obtain inscribed and circumscribed regular polygons with 640 sides.

EXERCISE 6. Starting with the fact that the side of a regular decagon inscribed in the unit circle is $s_{10} = 2 \sin 18° = (\sqrt{5} - 1)/2$, apply formulas (3) and (4) to recursively calculate s_{640} and t_{640}, carrying 8 decimal places on a hand calculator. Thence verify that $\pi = 3.1416$, rounded off to 4 decimal places.

EXERCISE 7. Let p_n and P_n denote the perimeters of the inscribed and circumscribed regular n-sided polygons for the unit circle. Noting that

$$s_n = 2 \sin \frac{\pi}{n} \quad \text{and} \quad t_n = \tan \frac{\pi}{n},$$

show that

$$p_{2n} = \sqrt{p_n P_{2n}} \quad \text{and} \quad P_{2n} = \frac{2 p_n P_n}{p_n + P_n}.$$

Starting with $p_4 = 4\sqrt{2}$ and $P_4 = 8$ for inscribed and circumscribed squares, use these recursive formulas to calculate p_{64} and P_{64}. What bounds on π does this computation give?

EXERCISE 8. If a_n and A_n are the areas of the inscribed and circumscribed polygons in the previous exercise, show that

$$a_{2n} = \sqrt{a_n A_n} \quad \text{and} \quad A_{2n} = \frac{2 a_{2n} A_n}{a_{2n} + A_n}.$$

The Quadrature of the Parabola

A *segment* of a convex curve is a region bounded by a straight line and a portion of the given curve (Fig. 4). In the preface to the *Quadrature of the Parabola*, Archimedes remarks that earlier mathematicians had successfully attempted to find the area of a segment of a circle or hyperbola, but that apparently no one had previously attempted the quadrature of a segment of a parabola—precisely the one that can be carried out by the method of exhaustion.

The parabola was originally defined by the Greeks as a conic section. That is, given a circular (double) cone with vertical axis, a parabola is the

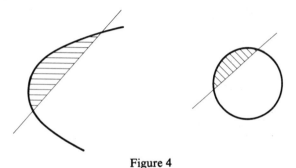

Figure 4

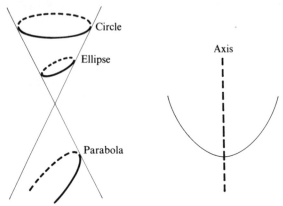

Figure 5

curve of intersection of the cone with a plane that is parallel to a generating element of the cone. Other positions of the plane yield ellipses and hyperbolas. If the plane is horizontal then the section is a circle. The parabola is obviously symmetric with respect to a certain straight line in the plane containing it; this line is called the *axis* of the parabola (Fig. 5).

Given a parabolic segment with *base AB* (Fig. 6), the point *P* of the segment that is farthest from the base is called the *vertex* of the segment,

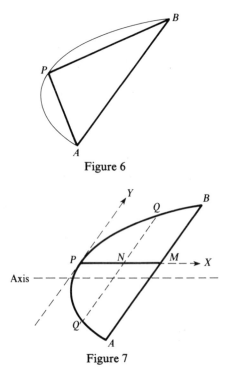

Figure 6

Figure 7

and the (perpendicular) distance from P to AB is its *height*. Archimedes showed that the area of the segment is four-thirds that of the inscribed triangle APB. That is, the area of a segment of a parabola is $4/3$ times the area of a triangle with the same base and height. He gave two separate proofs of the result; we will discuss here the second one.

By the time of Archimedes, the following facts were known concerning an arbitrary parabolic segment APB.

(a) The tangent line at P is parallel to the base AB.
(b) The straight line through P parallel to the axis intersects the base AB in its midpoint M.
(c) Every *chord* QQ' parallel to the base AB is bisected by the *diameter* PM.
(d) With the notation in Figure 7,

$$\frac{PN}{PM} = \frac{NQ^2}{MB^2}. \tag{5}$$

That is, in the pictured oblique xy-coordinate system, the equation of the parabola is of the form $x = ky^2$. Archimedes quotes these facts without proof, referring to earlier treatises on the conics by Euclid and Aristaeus.

There is a natural parallelogram circumscribed about a parabolic segment APB, having AB as a side, and with its base AA' and top BB' parallel to the diameter PM (Fig. 8). Since the area of the inscribed triangle APB is half that of the circumscribed parallelogram, it follows that the area of this triangle is more than half of the area of the parabolic segment APB.

Now consider the two smaller parabolic segments with bases PB and AP; let their vertices be P_1 and P_2, respectively (Fig. 8). In the same way as above, it follows that the areas of the inscribed triangles PP_1B and AP_2P are more than half of the areas of these two segments.

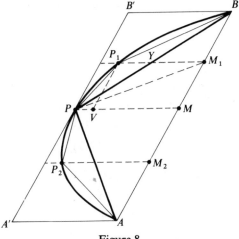

Figure 8

We have begun to exhaust the area of the original parabolic segment APB with inscribed polygons. The triangle APB is our first inscribed polygon, and AP_2PP_1B is the second. We continue in this way, adding at each step the triangles inscribed in the parabolic segments remaining from the previous step. Since the total area of these inscribed triangles is more than half that of the segments, it follows from Eudoxus' principle that, given $\epsilon > 0$, *we obtain after a finite number of steps an inscribed polygon whose area differs from that of the segment APB by less than ϵ.*

Now we want to show that the sum of the areas of the triangles AP_2P and PP_1B is $\frac{1}{4}$ that of $\triangle APB$. Let M_1 be the midpoint of BM, Y the point of intersection of P_1M_1 and PB, and V the intersection with PM of the line through P_1 parallel to AB. Then

$$BM^2 = 4M_1M^2,$$

so it follows from (5) that

$$PM = 4PV \quad \text{or} \quad P_1M = 3PV.$$

But $YM_1 = \frac{1}{2}PM = 2PV$, so that

$$YM_1 = 2P_1Y.$$

It follows from this that

$$a(\triangle PP_1B) = \tfrac{1}{2}a(\triangle PM_1B) = \tfrac{1}{4}a(\triangle PMB),$$

applying twice the fact that the ratio of the areas of two triangles with the same base is equal to the ratio of their heights. Similarly

$$a(\triangle AP_2P) = \tfrac{1}{4}a(\triangle APM),$$

so we find that

$$a(\triangle PP_1B) + a(\triangle AP_2P) = \tfrac{1}{4}a(\triangle PMB) + \tfrac{1}{4}a(\triangle APM)$$
$$= \tfrac{1}{4}a(\triangle APB)$$

as desired.

In the same way it can be proved that the sum of the areas of the inscribed triangles added at each step is equal to $\frac{1}{4}$ of the sum of the areas of the triangles added at the previous step. If we write

$$\alpha = a(\triangle APB),$$

it follows that the polygon $\mathcal{P}_n$ obtained after n steps has area

$$a(\mathcal{P}_n) = \alpha + \frac{\alpha}{4} + \frac{\alpha}{4^2} + \cdots + \frac{\alpha}{4^n}. \tag{6}$$

Consequently, given $\epsilon > 0$, the area $a(APB)$ of the parabolic segment differs from the right-hand side of (6) by less than ϵ if n is sufficiently large.

At this point Archimedes derives the elementary identity

$$1 + \frac{1}{4} + \frac{1}{4^2} + \cdots + \frac{1}{4^n} + \frac{1}{3} \cdot \frac{1}{4^n} = \frac{4}{3}. \tag{7}$$

This follows from the observation that

$$\frac{1}{4^k} + \frac{1}{3} \cdot \frac{1}{4^k} = \frac{4}{3 \cdot 4^k} = \frac{1}{3} \cdot \frac{1}{4^{k-1}},$$

for then

$$1 + \frac{1}{4} + \frac{1}{4^2} + \cdots + \left(\frac{1}{4^n} + \frac{1}{3} \cdot \frac{1}{4^n} \right)$$

$$= 1 + \frac{1}{4} + \frac{1}{4^2} + \cdots + \left(\frac{1}{4^{n-1}} + \frac{1}{3} \cdot \frac{1}{4^{n-1}} \right)$$

$$\vdots$$

$$= 1 + \left(\frac{1}{4} + \frac{1}{3} \cdot \frac{1}{4} \right)$$

$$= \frac{4}{3}.$$

It is tempting to simply sum the geometric series by letting $n \to \infty$ in (7) to obtain

$$1 + \frac{1}{4} + \frac{1}{4^2} + \cdots + \frac{1}{4^n} + \cdots = \frac{4}{3}.$$

We would then conclude that

$$a(APB) = \lim_{n \to \infty} a(\mathcal{P}_n)$$

$$= \lim_{n \to \infty} \alpha \left(1 + \frac{1}{4} + \cdots + \frac{1}{4^n} + \frac{1}{3} \cdot \frac{1}{4^n} \right)$$

$$= \alpha \left(1 + \frac{1}{4} + \cdots + \frac{1}{4^n} + \cdots \right)$$

$$a(APB) = \frac{4}{3} \alpha = \frac{4}{3} a(\triangle APB)$$

as desired. □

No doubt Archimedes intuitively obtained the answer $4/3$ in similar fashion but, rather than taking limits explicitly, he concluded the proof with a typical double *reductio ad absurdum* argument which we leave to the reader.

EXERCISE 9. Supply this concluding argument, using the facts that

(a) Given $\epsilon > 0$, $a(APB)$ and $a(\mathcal{P}_n)$ differ by less than ϵ if n is sufficiently large, and

(b) $a(\mathcal{P}_n) = (4\alpha/3) - (\alpha/3) \cdot (1/4^n)$ (from (6) and (7)).

If Archimedes' result is applied to the segment bounded by the parabola $y = x^2$ and the horizontal line $y = 1$, we find that its area is $4/3$, so it follows that the area under the parabola and over the interval $0 \leqslant x \leqslant 1$ is $\frac{1}{3}$. In modern integral notation this means that

$$\int_0^1 x^2 dx = \tfrac{1}{3}.$$

The Area of an Ellipse

Although Archimedes was unable to compute the area of an arbitrary segment of an ellipse, he did show (in *On Conoids and Spheroids*) that the area of the complete ellipse with major and minor semi-axes a and b is

$$A = \pi ab, \tag{8}$$

a pleasant generalization of the formula for the area of a circle (the circle of radius r being an ellipse with $a = b = r$).

Archimedes' proof of (8) is based on the following characteristic property of an ellipse. The circle of radius a, circumscribed about the ellipse as in Figure 9, is called its *auxiliary circle*. Given a point P on the major (horizontal) axis of the ellipse, let Q be the point on the ellipse and R the point on the circle above P. Then

$$\frac{PQ}{PR} = \frac{b}{a}. \tag{9}$$

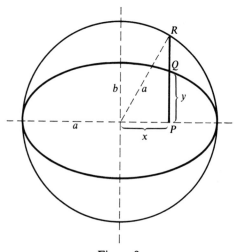

Figure 9

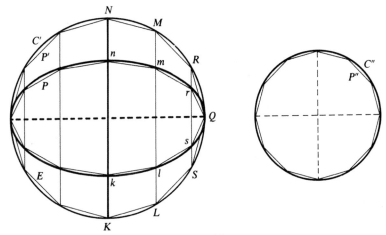

Figure 10

This is obvious from the rectangular coordinates equation

$$\frac{x^2}{a^2} + \frac{y^2}{b^2} = 1$$

for the ellipse, which gives

$$PQ = y$$
$$= \frac{b}{a}\sqrt{a^2 - x^2} = \frac{b}{a}PR.$$

To give the proof of (8), we start with an ellipse E with major and minor semi-axes a and b, and with auxiliary circle C'. Let C'' be a circle of radius $r = \sqrt{ab}$, so $a(C'') = \pi ab$ (see Fig. 10). We want to prove that

$$a(E) = a(C'').$$

Assuming that $a(E) < a(C'')$, let P'' be a regular polygon inscribed in C'', having its number of sides equal to a multiple of 4, and with opposite ends of the horizontal diameter of C'' as vertices, such that

$$a(P'') > a(E). \tag{10}$$

If P' is a similar regular polygon inscribed in the auxiliary circle C', then

$$\frac{a(P'')}{a(P')} = \frac{r^2}{a^2} = \frac{ab}{a^2} = \frac{b}{a}. \tag{11}$$

Now let P be the polygon inscribed in the ellipse E whose vertices are the intersections with E of the perpendiculars from the vertices of P' to the horizontal axis of E. We can consider the polygons P and P' as unions of corresponding pairs of triangles like Qrs and QRS, and corresponding pairs of trapezoids like $klmn$ and $KLMN$. Now the characteristic property

(9) of the ellipse implies that

$$\frac{lm}{LM} = \frac{kn}{KN} = \frac{rs}{RS} = \frac{b}{a},$$

so it follows that

$$\frac{a(klmn)}{a(KLMN)} = \frac{a(Qrs)}{a(QRS)} = \frac{b}{a}.$$

Consequently, by pairwise comparison of the triangles and trapezoids forming P with those forming P', we see that

$$\frac{a(P)}{a(P')} = \frac{b}{a}. \tag{12}$$

But (11) and (12) imply that $a(P) = a(P'')$, which contradicts (10) because P is inscribed in E. Therefore $a(C'')$ is *not greater* than $a(E)$.

Assuming next that $a(C'') < a(E)$, we start with a polygon P like the one above inscribed in the ellipse E, such that

$$a(P) > a(C'').$$

Then let P' be the polygon inscribed in the auxiliary circle C' whose vertices are the intersections with C' of vertical lines through pairs of vertices of P, and let P'' be the similar polygon inscribed in the circle C''. By the same computations as in the first case we find that

$$\frac{a(P)}{a(P')} = \frac{b}{a} = \frac{a(P'')}{a(P')}.$$

But then $a(P'') = a(P) > a(C'')$, which is a contradiction because P'' is inscribed in C''. This completes the double *reductio ad absurdum* proof that

$$a(E) = a(C'') = \pi ab. \qquad \square$$

In essence, Archimedes has simply given a rigorous exhaustion proof of the intuitively evident fact that the area of the ellipse is b/a times the area πa^2 of its auxiliary circle, corresponding to the observation that the circle is transformed into the ellipse by shrinking its vertical dimension by the factor b/a.

The Volume and Surface Area of a Sphere

To the modern reader, the treatise *On the Sphere and Cylinder* probably seems the most elegant and inventive of Archimedes' works. The author himself apparently concurred with this judgment, for he requested that on his tombstone be carved a sphere inscribed in a right circular cylinder whose height equals its diameter. When the Roman orator Cicero was later

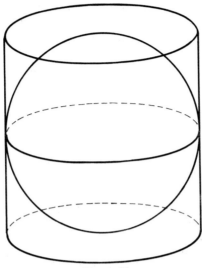

Figure 11

serving as quaestor in Sicily, he found and restored the tomb with this inscription. The Romans had so little interest in pure mathematics that this action by Cicero was probably the greatest single contribution of any Roman to the history of mathematics.

This tombstone carving symbolized the two principal results of Book I (of two) of *On the Sphere and Cylinder*, to the effect that the surface area S and the volume V of a sphere of radius r are given by

$$S = 4\pi r^2 \quad \text{and} \quad V = \tfrac{4}{3}\pi r^3. \tag{13}$$

The connection with the circumscribed cylinder is that S is two-thirds of the total surface area of the cylinder (including its two ends), while V is two-thirds of the volume of the cylinder. Thus the ratio of the surface areas of the sphere and cylinder is the same as the ratio of their volumes!

We saw in Chapter 1 that Euclid (or Eudoxus) proved that the volume of a sphere is proportional to the cube of its radius, $V = \alpha r^3$, but did not discover that $\alpha = 4\pi/3$. The relationship between V and the surface area S is suggested by a heuristic argument similar to the one mentioned earlier in this chapter in the discussion of the area and the circumference of a circle. We regard the sphere as approximately the union of indefinitely many pyramids with the center of the sphere as their common vertex, and with their bases forming a polyhedral surface with indefinitely many faces inscribed in the sphere, each of which almost coincides with a small piece of the sphere. Since the height of each of the pyramids will virtually equal the radius of the sphere, and the volume of a pyramid is one-third of the product of its height and its base, it seems evident that

$$V = \tfrac{1}{3}rS. \tag{14}$$

Assuming formula (14), either of the formulas in (13) may be deduced from the other. Archimedes initially derived $S = 4\pi r^2$ from $V = \frac{4}{3}\pi r^3$, for he states in his treatise *The Method* (discussed later in this chapter) that

> From the theorem that a sphere is four times as great as the cone with a great circle of the sphere as base and with height equal to the radius of the sphere I conceived the notion that the surface of any sphere is four times as great as a great circle in it; for, judging from the fact that any circle is equal to a triangle with base equal to the circumference and height equal to the radius of the circle, I apprehended that, in like manner, any sphere is equal to a cone with base equal to the surface of the sphere and height equal to the radius.

The concept of area is inherently more complicated for curvilinear surfaces in space than it is for plane regions. Before he could proceed to compute surface areas, it was necessary for Archimedes first to limit the class of surfaces to be considered, and then to introduce axioms that serve (in modern terms) to define the concept of surface area. With striking insight he stated the following definitions and axioms at the beginning of *On the Sphere and Cylinder* I.

Let C be a bounded plane curve (i.e., one having two endpoints) that lies on one side of the line L through its endpoints. Then the curve C is called *convex* if, for any two points P and Q of C, the line segment from P to Q is wholly contained in the region that is bounded by the curve C and the segment of L joining the endpoints of C (Fig. 12).

Similarly, let S be a surface bounded by a simple closed curve J in the plane M, such that S lies on one side of M, and denote by Σ the plane region in M that is bounded by J. Then the surface S is called *convex* if, for any two points P and Q of S, the line segment joining P and Q is wholly contained by the region (in space) that is bounded by $S \cup \Sigma$ (Fig. 13).

For *simple* (unbounded) *closed* curves or surfaces the definition of convexity can be stated more simply. A simple closed curve (in a plane) or surface (in space) is *convex* if the region bounded by it wholly contains any line segment joining two points of it.

Archimedes discusses curve length and surface area only for convex curves and surfaces. Realizing that these are new geometric magnitudes

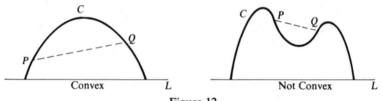

Figure 12

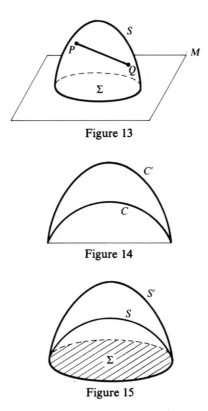

Figure 13

Figure 14

Figure 15

different from any studied by his predecessors, he introduces the following *convexity axioms* for their definition and computation.

I. (For curves). If C and C' are convex curves with the same endpoints, and C is contained in the region bounded by C' and the line segment joining its endpoints, then the length $l(C)$ of C is less than the length of C',

$$l(C) < l(C').$$

That is, if one convex curve is "included within" another (Fig. 14), then the one included is shorter. Also, a straight line segment is the shortest curve joining two given points.

II. (For surfaces). If S and S' are convex surfaces with the same boundary curve which bounds a region Σ in a plane, and S is contained in the region bounded by $S' \cup \Sigma$, then

$$a(S) < a(S').$$

That is, if one convex surface is "included within" another, then the included surface has the lesser area (Fig. 15). Also, of all surfaces bounded by a given plane curve, the region in the plane has the least area.

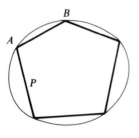 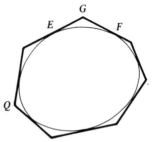

Figure 16

To illustrate the application of the convexity axioms in the method of compression, consider a convex closed curve C with an inscribed polygon P and a circumscribed polygon Q (Fig. 16). Then the convexity axiom for curves implies that the typical line segment $\overline{AB}$ is shorter than the curve segment $\overset{\frown}{AB}$, and that the curve segment $\overset{\frown}{EF}$ is shorter than the portion $\overline{EG} \cup \overline{GF}$ of the circumscribed polygon. Hence

$$l(P) < l(C) < l(Q). \qquad (15)$$

If we can find a sequence $\{P_n\}$ of inscribed polygons and a sequence $\{Q_n\}$ of circumscribed polygons, such that

$$\lim_{n \to \infty} l(P_n) = \lim_{n \to \infty} l(Q_n) = L,$$

where we can compute L, then (15) implies that

$$l(C) = L.$$

In essence, the convexity axiom requires that we *define* $l(C)$ to equal L, this being the only value for the length of C that is consistent with the convexity axiom.

Archimedes begins his investigation of surface area with the proofs of the formulas

$$A = 2\pi rh \qquad (16)$$

for the surface area (excluding the bases) of a cylinder with radius r and height h, and

$$A = \pi rs \qquad (17)$$

for the surface area (excluding the base) of a cone with base radius r and "slant height" $s = \sqrt{r^2 + h^2}$. We shall discuss here the case of the cone, and leave the cylinder to an exercise for the reader.

A heuristic derivation of (17) may be obtained by "unrolling" the cone onto a circular sector (Fig. 17) with radius s and central angle

$$\theta = \frac{2\pi r}{2\pi s} \cdot 2\pi = \frac{2\pi r}{s} \text{ radians.}$$

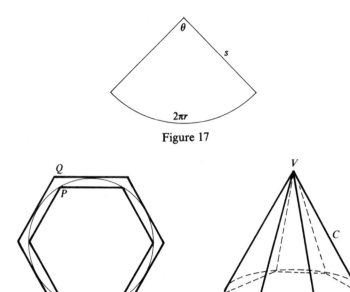

Figure 17

Figure 18

The formula $A = \frac{1}{2}s^2\theta$ for the area of the circular sector then yields (17).

Now let C denote the lateral surface of a right circular cone with vertex V whose base B is a circle of radius r. If P and Q are regular polygons, with P inscribed in B and Q circumscribed about B, then denote by VP and VQ the corresponding pyramids inscribed in and circumscribed about the cone C. We would like to conclude from the convexity axiom for surface area that the lateral surface areas $a(VP)$ and $a(VQ)$ of these inscribed and circumscribed pyramids are, respectively, less than and greater than the lateral surface area $a(C)$ of the cone,

$$a(VP) < a(C) < a(VQ). \tag{18}$$

These inequalities do not follow immediately from the convexity axiom, because the three convex surfaces have different boundary curves, but are established by ingenious arguments in Propositions 9 and 10 of *On the Sphere and Cylinder* I.

Archimedes then proves that $a(C) = \pi rs$ by an interesting ratio version of the method of compression. Let B' be a circle with radius $R = \sqrt{rs}$; we want to show that

$$a(C) = a(B') = \pi R^2 = \pi rs.$$

Assuming first that $a(C) > a(B')$, choose polygons P' and Q' inscribed

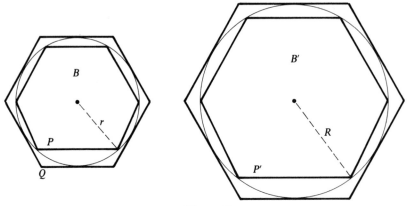

Figure 19

in and circumscribed about B' such that

$$\frac{a(Q')}{a(P')} < \frac{a(C)}{a(B')}.$$

(Archimedes has established in Proposition 3 the rather obvious fact that, given $\alpha > 1$, there exist inscribed and circumscribed polygons P and Q in a circle such that $a(Q)/a(P) < \alpha$). Let P and Q be similar polygons inscribed in and circumscribed about the base B of the cone. Then

$$\frac{a(Q)}{a(Q')} = \frac{r^2}{R^2} = \frac{r^2}{rs} = \frac{r}{s} = \frac{a(Q)}{a(VQ)},$$

so it follows that $a(VQ) = a(Q')$. Hence

$$\frac{a(VQ)}{a(P')} = \frac{a(Q')}{a(P')} < \frac{a(C)}{a(B')}.$$

But this is a contradiction, because $a(VQ) > a(C)$ by (18), while $a(P') < a(B')$ since P' is inscribed in B'. Therefore $a(C)$ is not greater than $a(B')$.

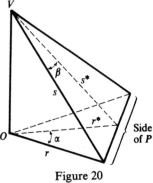

Figure 20

Next suppose that $a(C) < a(B')$, and choose regular polygons P' and Q' inscribed in and circumscribed about B' such that

$$\frac{a(Q')}{a(P')} < \frac{a(B')}{a(C)}.$$

Again let P and Q be similar polygons inscribed in and circumscribed about B. Then

$$\frac{a(P)}{a(P')} = \frac{r^2}{R^2} = \frac{r}{s} > \frac{a(P)}{a(VP)}. \tag{19}$$

This is true because

$$\frac{a(P)}{a(VP)} = \frac{r^*}{s^*},$$

where r^* is a perpendicular from the center of B to a side of P, while s^* is a perpendicular from V to a side of P. But then

$$\frac{r^*}{s^*} = \frac{r \sin \alpha}{r \sin \beta} < \frac{r}{s}$$

because it is evident that $\alpha < \beta$, so $\sin \alpha < \sin \beta$ (Fig. 20).

From (19) we see that $a(VP) > a(P')$, so

$$\frac{a(Q')}{a(VP)} < \frac{a(B')}{a(C)}.$$

But this is a contradiction because $a(VP) < a(C)$ by (18), while $a(Q') > a(B')$ since Q' is circumscribed about B'. Consequently we conclude that

$$a(C) = a(B') = \pi r s$$

as desired.   □

From (17) it is easy to derive the formula for the lateral surface area of a frustum of a cone,

$$A = \pi(r_1 + r_2)s, \tag{20}$$

where r_1 and r_2 are the radii of the bases and s is the slant height of the frustum (Fig. 21).

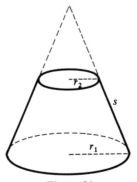

Figure 21

EXERCISE 10. Derive formula (20) for the lateral surface area of a conical frustum.

EXERCISE 11. Give a rigorous proof by the method of compression of the formula $A = 2\pi rh$ for the lateral surface area of a cylinder, following in outline the above proof of the formula $A = \pi rs$ for the lateral surface area of a cone (replacing pyramids with prisms).

The Sphere

In order to investigate the surface area of a sphere S with diameter $d = 2r$, Archimedes inscribes a regular polygon P with $2n$ sides in a great circle of S, and then rotates it about a diameter through a pair of opposite vertices of P. The surface Σ' of the resulting solid of revolution V' then consists of two cones and $n - 2$ frusta of cones, each with slant height s' equal to the length of a side of P. The radii $a_1, a_2, \ldots, a_{n-1}$ of their bases are semi-chords of the great circle, drawn through the vertices of P perpendicular to the axis AC of revolution (Fig. 22).

Applying the formulas for the surface of a cone and of a frustum of a cone (Eq. (20)) to compute the surface area of the inscribed surface Σ', we

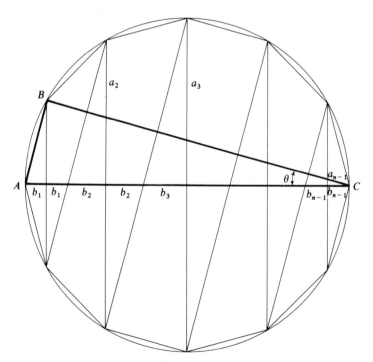

Figure 22

obtain

$$a(\Sigma') = \pi a_1 s' + \pi(a_1 + a_2)s' + \cdots + \pi(a_{n-2} + a_{n-1})s' + \pi a_{n-1}s'$$

$$= 2\pi s' \sum_{i=1}^{n-1} a_i. \tag{21}$$

Now we divide the diameter AC into segments of lengths $b_1, b_1, b_2,$ $b_2, \ldots, b_{n-1}, b_{n-1}$ as indicated in Figure 22. Then from the obvious similar triangles we see that

$$\frac{a_1}{b_1} = \frac{a_2}{b_2} = \cdots = \frac{a_{n-1}}{b_{n-1}} = \frac{BC}{s'}.$$

Since $2(b_1 + b_2 + \cdots + b_{n-1}) = 2r = d$, it follows by addition of these ratios that

$$\frac{2(a_1 + a_2 + \cdots + a_{n-1})}{d} = \frac{BC}{s'}$$

so

$$2s' \sum_{i=1}^{n-1} a_i = d \cdot BC = 4r^2 \cos\theta,$$

where θ is the angle ACB (and, to avoid anachronism in describing Archimedes' computation, $\cos\theta$ is simply an abbreviation for the ratio BC/AC). Consequently (21) becomes

$$a(\Sigma') = 4\pi r^2 \cos\theta. \tag{22}$$

In particular, we see that *the area of the inscribed surface Σ' is less than* $4\pi r^2$. Also, it follows from the convexity axiom that $a(\Sigma') < a(S)$.

Next we want to show that the surface area of a similar but circumscribed surface is greater than $4\pi r^2$. Let Q be a regular polygon with sides of length s'', similar to P but circumscribed about the great circle with diameter AC (Fig. 23). Upon rotating it about this diameter, we obtain a solid of revolution V'' with surface Σ''.

Since Σ'' may be regarded as a surface similar to Σ', but inscribed in a slightly larger sphere with diameter (see Fig. 23)

$$d' = A'C' = 2r \sec\theta,$$

the above calculation immediately yields

$$a(\Sigma'') = d' \cdot B'C' = 4\pi r^2 \sec\theta, \tag{23}$$

since $B'C' = 2r$ by similar triangles (Fig. 23). Because $\sec\theta = AC/BC > 1$, we see that *the area of the circumscribed surface Σ'' is greater than $4\pi r^2$.* Also, $a(\Sigma'') > a(S)$ by the convexity axiom.

Equations (22) and (23) together give

$$4\pi r^2 \cos\theta < a(S) < 4\pi r^2 \sec\theta,$$

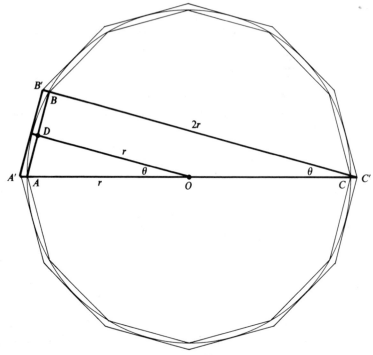

Figure 23

from which the limit as $n \rightarrow \infty$, $\theta \rightarrow 0$ obviously yields $a(S) = 4\pi r^2$. Of course Archimedes concludes the proof with the usual double *reductio ad absurdum* argument.

From the triangles ADO and $A'B'C'$ in Fig. 23 we see that

$$\sin \theta = \frac{\frac{1}{2}s'}{r} = \frac{s'}{d} \quad \text{and} \quad \tan \theta = \frac{s''}{d},$$

so

$$\sec \theta = \frac{\cos \theta}{\sin \theta} = \frac{s''}{s'}.$$

Consequently (22) and (23) imply that

$$\frac{a(\Sigma'')}{a(\Sigma')} = \sec^2 \theta = \left(\frac{s''}{s'} \right)^2, \tag{24}$$

the ratio of the squares of the sides of the polygons Q and P. We are now ready for the proof that $a(S) = 4\pi r^2$ by the ratio form of the compression method.

Assuming that $a(S) < 4\pi r^2$, choose the inscribed and circumscribed polygons P and Q such that

$$\frac{s''}{s'} < \sqrt{\frac{a(S)}{4\pi r^2}} \, .$$

Then equation (24) gives

$$\frac{a(\Sigma'')}{a(\Sigma')} = \left(\frac{s''}{s'}\right)^2 < \frac{a(S)}{4\pi r^2} \, .$$

But this is impossible because $a(\Sigma'') > a(S)$ while $a(\Sigma') < 4\pi r^2$.
  Assuming that $a(S) > 4\pi r^2$, choose P and Q such that

$$\frac{s''}{s'} < \sqrt{\frac{4\pi r^2}{a(S)}} \, .$$

Then equation (24) gives

$$\frac{a(\Sigma'')}{a(\Sigma')} = \left(\frac{s''}{s'}\right)^2 < \frac{4\pi r^2}{a(S)} \, .$$

But this is impossible because $a(\Sigma'') > 4\pi r^2$ while $a(\Sigma') < a(S)$. It finally
follows that $a(S) = 4\pi r^2$ as desired.   □

  The same geometric construction yields a proof that the volume of the
ball V bounded by the sphere S is $v(V) = 4\pi r^3/3$. Archimedes proves that
the volume of the inscribed solid V' is equal to that of a cone whose base
has area $a(\Sigma')$ and whose height is equal to the length p of a perpendicular
from the center O to a side of the polygon P. Hence

$$v(V') = \tfrac{1}{3} p a(\Sigma') < \tfrac{4}{3}\pi r^3. \tag{25}$$

Similarly, the volume of the circumscribed solid V'' is equal to that of a
cone with height r whose base has area $a(\Sigma'')$. Hence

$$v(V'') = \tfrac{1}{3} r a(\Sigma'') > \tfrac{4}{3}\pi r^3. \tag{26}$$

The ratio of these volumes is

$$\frac{v(V'')}{v(V')} = \frac{r a(\Sigma'')}{p a(\Sigma')} = \left(\frac{s''}{s'}\right)^3 \tag{27}$$

because $r/p = \sec\theta = s''/s'$. Finally, equations (25), (26), and (27) provide
the ingredients for a compression proof of $v(V) = 4\pi r^3/3$ that is virtually
identical to the preceding proof that $a(S) = 4\pi r^2$.

EXERCISE 12. Supply the details for this proof that $v(V) = 4\pi r^3/3$ by the ratio form
of the method of compression.

Archimedes also introduced the refinement of the above construction that is necessary to compute the volume and (curved) surface area of a segment Σ of a sphere of radius r that is cut off by a plane at a distance $a < r$ from the center. In terms of the height $h = r - a$ and the base radius $\rho = \sqrt{r^2 - a^2}$ of the segment, these are

$$a(\Sigma) = \pi(\rho^2 + h^2), \qquad v(\Sigma) = \frac{1}{3}\pi\rho^2 h\left(\frac{3r - h}{2r - h}\right).$$

The Method of Compression

Archimedes' method of compression, for proving that a geometric magnitude S (length, area, or volume) is equal to a given magnitude C, may be described in quite general terms as follows. On the basis of the geometry of the figure whose length, area, or volume is sought, two sequences $\{L_n\}$ and $\{U_n\}$ are constructed such that

$$L_n < S < U_n \quad \text{and} \quad L_n < C < U_n \quad \text{for all } n. \tag{28}$$

In the "difference form" of the method it is proved that, given $\epsilon > 0$,

$$U_n - L_n < \epsilon \tag{29}$$

for n sufficiently large. In the "ratio form" of the method it is proved that, given $\alpha > 1$,

$$\frac{U_n}{L_n} < \alpha \tag{30}$$

for n sufficiently large. In either case a double *reductio ad absurdum* proof finally establishes that $S = C$ as desired.

EXERCISE 13. (a) Use (28) and (29) to prove that $S = C$. (b) Use (28) and (30) to prove that $S = C$.

The Archimedean Spiral

Greek mathematicians suffered from a paucity of curves available to serve as objects of their study. Because their algebra was geometric and rhetorical rather than numerical and symbolic, the introduction of new curves by means of equations (as in analytic geometry) was not feasible. In addition, Greek geometry was essentially static rather than dynamic in character. Consequently they could, for the most part, define curves only in terms of simple locus conditions (e.g., the circle as the locus of points equidistant from a given fixed point) or as intersections of given surfaces fixed in position (e.g., a conic section as the intersection of a plane and a cone).

The static character of Greek geometry is a reflection of the very limited role of motion and variability concepts in Greek science generally. Only the cases of *uniform* motion—either rectilinear or circular—were studied in any detail. Other motions (such as those of the planets in Greek astronomical models) could only be analyzed in terms of uniform linear and circular motions.

It was in these terms that Archimedes defined his famous spiral—as the composition of a uniform linear motion and a uniform circular motion.

> If a straight line drawn in a plane revolve at a uniform rate about one extremity which remains fixed and return to the position from which it started, and if, at the same time as the line revolves, a point move at a uniform rate along the straight line beginning from the extremity which remains fixed, the point will describe a *spiral* in the plane (Definition 1 of *On Spirals*).

To describe this curve in modern polar coordinates, let ω (in radians) be the constant angular speed of rotation of the line, and v the constant speed with which the point moves along the line, starting at the origin. Then the polar coordinates of the moving point at time t are $r = vt$ and $\theta = \omega t$, so the polar coordinates equation of the spiral is

$$r = a\theta \tag{31}$$

where $a = v/\omega$.

The first twenty propositions of the treatise *On Spirals* are devoted mainly to the determination of the tangent line to the spiral at a given point of it. Here, as elsewhere in Greek geometry, a static conception of a tangent line to a curve is employed—it is a straight line that "touches" the curve at a single point without crossing it. Although no dynamic considerations appear in the finished exposition, it has been conjectured that Archimedes *discovered* the tangent line to the spiral by means of a parallelogram of velocities determined by the two component motions generating the spiral (see the Appendix to Heath [6]). If so, this would be a rare (if not unique) instance of *differential* calculus methods in antiquity.

At any rate, he shows that the tangent line at the point P to the spiral OPS intersects the perpendicular OQ to OP in a point T such that the line segment OT is equal in length to the circular arc PR intercepted between the polar axis and the radius vector OP (Fig. 24). The proof is a double *reductio ad absurdum* argument showing that the assumptions $OT > PR$ and $OT < PR$ lead to contradictions.

EXERCISE 14. Use a parallelogram of velocities argument to verify Archimedes' construction of the tangent line to the spiral, for the case when P is the point $(0, a\pi/2)$ on the y-axis resulting from one quarter-turn of the radius vector. At this

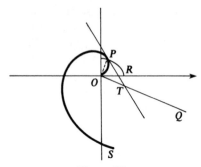

Figure 24

instant P has a vertical velocity component v and a horizontal velocity component $\omega a\pi/2$ (why?).

EXERCISE 15. Suppose the spiral and its tangent line at the point P of the preceding exercise are given. Then explain how on this basis to "square the circle"—to construct by ruler and compass methods a square whose area is equal to that of the circle with radius OP. Recall the first proposition of *On the Measurement of the Circle*.

EXERCISE 16. Show that the Archimedean spiral can be used to trisect an arbitrary angle, given that a line segment can easily be trisected (how?) See Figure 25.

The final eight propositions of *On Spirals* are devoted to area computations. For example, Archimedes proves that the area of the region S, bounded by one turn of the spiral and the line segment joining its initial and final points, is one-third that of the circle C centered at the initial point and passing through the final point (Fig. 26). That is,

$$a(S) = \tfrac{1}{3}\pi(2\pi a)^2. \tag{32}$$

The proof of (32) makes use of the familiar (then as now) formulas for

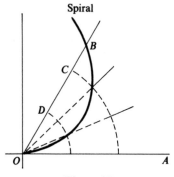

Figure 25

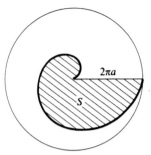

Figure 26

the sum of the terms of an arithmetic progression, and of their squares:

$$1 + 2 + \cdots + n = \frac{n}{2}(n+1) \qquad (33)$$

and

$$1^2 + 2^2 + \cdots + n^2 = \frac{n}{6}(n+1)(2n+1). \qquad (34)$$

Writing (34) in the form

$$1^2 + 2^2 + \cdots + n^2 = \frac{n^3}{3} + \frac{n^2}{2} + \frac{n}{6},$$

it is clear that

$$1^2 + 2^2 + \cdots + (n-1)^2 < \frac{n^3}{3} < 2^2 + 3^2 + \cdots + n^2, \qquad (35)$$

which is what Archimedes actually needed.

EXERCISE 17. Prove by mathematical induction that formulas (33) and (34) hold for all n. That is, note that each of these formulas holds for $n = 1$, and show that its truth for $n = k$ implies its truth for $n = k + 1$. Why does this imply that it holds for all n?

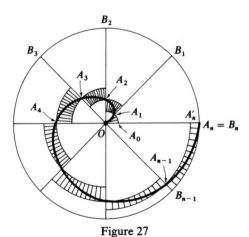

Figure 27

Archimedes' geometric construction for the proof of (32) is very similar to that which would be employed to set up a modern definite integral in polar coordinates for this area computation—see Figure 27. We divide the circle into n equal sectors as shown, with their bounding radii intersecting the spiral in the points $O, A_1, A_2, \ldots, A_n$. If we write $OA_1 = b$, then

$$OA_1 = b, \qquad OA_2 = 2b, \qquad \ldots, \qquad OA_n = nb.$$

Consequently we see that the spiral region S contains a region P consisting of circular sectors with radii

$$0, b, \ldots, (n-1)b,$$

and in turn is contained in a region Q consisting of circular sectors with radii

$$b, 2b, \ldots, nb.$$

Therefore $a(Q) - a(P)$ is equal to the area of a single sector of the circle C, and this can be made as small as we please by choosing n sufficiently large. Thus we have the necessary ingredients for a proof by compression.

If $a(S) < \frac{1}{3}a(C)$, we choose n sufficiently large that

$$a(Q) - a(P) < \frac{1}{3}a(C) - a(S),$$

so it follows that

$$a(Q) < \frac{1}{3}a(C).$$

But since the ratio of the areas of similar circular sectors is equal to the ratio of the squares of their radii, we find that

$$\frac{a(Q)}{a(C)} = \frac{b^2 + (2b)^2 + \cdots + (nb)^2}{n(nb)^2}$$

$$= \frac{1^2 + 2^2 + \cdots + n^2}{n^3} > \frac{1}{3}$$

by (35). This contradiction shows that $a(S)$ is not less than $\frac{1}{3}a(C)$.

Assuming that $a(S) > \frac{1}{3}a(C)$, we choose n sufficiently large that

$$a(Q) - a(P) < a(S) - \frac{1}{3}a(C),$$

so it follows that

$$a(P) > \frac{1}{3}a(C).$$

But then we find that

$$\frac{a(P)}{a(C)} = \frac{b^2 + (2b)^2 + \cdots + [(n-1)b]^2}{n(nb)^2}$$

$$= \frac{1^2 + 2^2 + \cdots + (n-1)^2}{n^3} < \frac{1}{3}$$

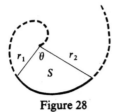

Figure 28

by (35). This contradiction shows that $a(S)$ is not greater than $\frac{1}{3}a(C)$, so we conclude that

$$a(S) = \tfrac{1}{3}a(C) = \tfrac{1}{3}\pi(2\pi a)^2 \qquad \square$$

as desired.

Archimedes went on to calculate the area of a spiral sector S (Fig. 28). If the radii to the initial and final points on the spiral are r_1 and r_2, and the central angle is θ (in radians), then

$$a(S) = \frac{\theta}{2}\left[r_1 r_2 + \tfrac{1}{3}(r_2 - r_1)^2 \right]. \tag{36}$$

Although (36) may not be of great importance in itself, its proof seems worthy of inclusion as a demonstration of the computational mastery that Archimedes exhibited, despite the fact that he had to express his calculations entirely in verbal and geometric language.

The proof of (36) requires the formula for the sum of the squares of the terms of an arbitrary finite arithmetic progression

$$a_1 = a, \qquad a_2 = a + b, \qquad a_3 = a + 2b, \qquad \ldots, \qquad a_n = a + (n-1)b$$

with n terms, initial term a, and common difference b. We find that

$$
\begin{aligned}
a^2 + (a+b)^2 + \cdots &+ [a + (n-1)b]^2 \\
&= na^2 + 2a[b + 2b + \cdots + (n-1)b] \\
&\quad + [b^2 + (2b)^2 + \cdots + (n-1)^2 b^2] \\
&= na^2 + 2ab[1 + 2 + \cdots + (n-1)] \\
&\quad + b^2[1^2 + 2^2 + \cdots + (n-1)^2] \\
&= na^2 + abn(n-1) + \frac{b^2}{6}(n-1)(n)(2n-1) \\
&= na_1 a_2 + \frac{1}{6}(a_n - a_1)^2 \frac{2n^2 - n}{n-1} \tag{37}
\end{aligned}
$$

because $(n-1)b = a_n - a_1$ and $a^2 + ab(n-1) = a_1 a_n$.

Now what we actually need are the inequalities

$$
\begin{aligned}
a_1^2 + \cdots + a_{n-1}^2 &< (n-1)\left[a_1 a_n + \tfrac{1}{3}(a_n - a_1)^2 \right] \\
&< a_2^2 + \cdots + a_n^2 \tag{38}
\end{aligned}
$$

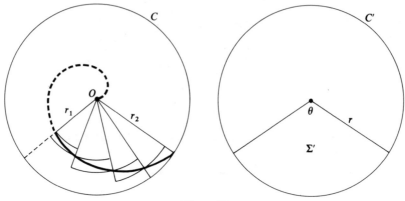

Figure 29

generalizing inequalities (35). In order to prove the right-hand inequality in (38), we need to see that

$$(n-1)\left[a_1a_n + \frac{1}{3}(a_n - a_1)^2\right] < na_1a_n + \frac{1}{6}(a_n - a_1)^2\frac{2n^2 - n}{n - 1} - a_1^2,$$

or

$$(n-1)a_1a_n + a_1^2 + \frac{1}{3}(n-1)(a_n - a_1)^2 < na_1a_n + (a_n - a_1)^2\frac{2n^2 - n}{6(n - 1)}.$$

But it is obvious that

$$(n-1)a_1a_n + a_1^2 < na_1a_n,$$

and

$$\frac{n - 1}{3} < \frac{2n^2 - n}{6(n - 1)}$$

is verified by cross-multiplication. The left-hand inequality in (38) can be established similarly.   □

Now we are ready for the proof of formula (36) for the area of the spiral sector S. Let C be a circle with radius r_2 and center O, and let Σ be the sector of C with central angle θ containing the spiral sector S. Let C' be a circle of radius r such that

$$r^2 = r_1r_2 + \tfrac{1}{3}(r_2 - r_1)^2,$$

and let Σ' be a sector of C' with central angle θ. We want to prove that

$$a(S) = a(\Sigma').$$

We subdivide Σ into $n-1$ equal sectors, each with central angle $\theta/(n-1)$, and obtain figures P and Q inscribed in and circumscribed about S, as in the previous proof. If

$$a = r_1 \quad \text{and} \quad b = \frac{r_2 - r_1}{n - 1},$$

then P consists of sectors with radii

$$a_1 = a, \qquad a_2 = a + b, \qquad \ldots, \qquad a_{n-1} = a + (n-2)b,$$

while Q consists of sectors with radii

$$a_2 = a + b, \qquad a_3 = a + 2b \qquad \ldots, \qquad a_n = a + (n-1)b = r_2.$$

Assuming that $a(S) < a(\Sigma')$, choose n sufficiently large that

$$a(Q) < a(\Sigma').$$

But then

$$\frac{a(Q)}{a(\Sigma)} = \frac{a_2^2 + \cdots + a_n^2}{(n-1)a_n^2} > \frac{a_1 a_n + \frac{1}{3}(a_n - a_1)^2}{a_n^2}, \qquad \text{(by (38))}$$

$$\frac{a(Q)}{a(\Sigma)} > \frac{a(\Sigma')}{a(\Sigma)},$$

which implies that $a(Q) > a(\Sigma')$. This contradiction shows that $a(S)$ is not less than $a(\Sigma')$.

Assuming that $a(S) > a(\Sigma')$, choose n sufficiently large that

$$a(P) > a(\Sigma').$$

But then

$$\frac{a(P)}{a(\Sigma)} = \frac{a_1^2 + \cdots + a_{n-1}^2}{(n-1)a_n^2}$$

$$< \frac{a_1 a_n + \frac{1}{3}(a_n - a_1)^2}{a_n^2} \qquad \text{(by (38))}$$

$$= \frac{a(\Sigma')}{a(\Sigma)},$$

which implies that $a(P) < a(\Sigma')$. This contradiction shows that $a(S)$ is not greater than $a(\Sigma')$, so we conclude that

$$a(S) = a(\Sigma') = \frac{\theta}{2}\left[r_1 r_2 + \frac{1}{3}(r_2 - r_1)^2 \right]$$

as desired.   □

If α and β denote the initial and final angles of the sector S, substitution of $\theta = \beta - \alpha$, $r_1 = a\alpha$, $r_2 = a\beta$ into (36) yields

$$a(S) = \frac{a^2}{6}(\beta^3 - \alpha^3).$$

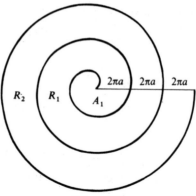

Figure 30

Since integration in polar coordinates gives

$$a(S) = \frac{1}{2} \int_\alpha^\beta r^2 \, d\theta = \frac{a^2}{2} \int_\alpha^\beta \theta^2 \, d\theta,$$

formula (36) is thus equivalent to the fact that

$$\int_\alpha^\beta \theta^2 \, d\theta = \frac{1}{3} (\beta^3 - \alpha^3).$$

EXERCISE 18. Consider in this problem multiple turns of the spiral $r = a\theta$. Let A_n denote the area bounded by the nth turn (for $2\pi(n-1) < \theta < 2\pi n$) and the portion of the polar axis joining its endpoints. For each $n > 2$ let $R_n = A_n - A_{n-1}$ denote the area of the ring between the $(n-1)$st and the nth turns (Fig. 30).
(a) Apply the formula (36) to calculate

$$A_1 = \tfrac{1}{3}\pi(2\pi a)^2 \quad \text{and} \quad A_2 = \tfrac{7}{12}\pi(4\pi a)^2.$$

Conclude that $R_2 = 6A_1$.
(b) For $n > 2$ derive Archimedes' formula

$$R_{n+1} = \frac{nR_n}{n-1} = nR_2.$$

Solids of Revolution

Archimedes' use of the method of compression approached most closely to the construction underlying the modern definite integral in his treatise *On Conoids and Spheroids*. A "conoid" is what we would call a paraboloid or hyperboloid of revolution, and a "spheroid" is an ellipsoid of revolution. He showed that the volume of a paraboloid of revolution P inscribed in a cylinder C with radius R and height H is

$$v(P) = \tfrac{1}{2}\pi R^2 H, \tag{39}$$

one-half of the volume of the circumscribed cylinder (Fig. 31).

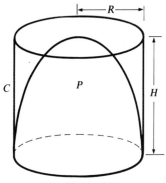

Figure 31

To prove (39) we cut the paraboloid and the cylinder into n slices of equal thickness $h = H/n$ by means of equidistant planes perpendicular to the axis OA of the cylinder. Let $A_1, A_2, \ldots, A_n = A$ be the points of intersection of these planes with the axis OA, and $B_1, \ldots, B_n = B$ their intersections with one side OB of the parabola that generates the paraboloid by revolution about OA (Fig. 32).

Now consider the solid of revolution I inscribed in the paraboloid P that consists of $n-1$ thin cylinders with radii

$$A_1 B_1, \ldots, A_{n-1} B_{n-1},$$

and the circumscribed solid of revolution J that consists of n thin cylinders with radii

$$A_1 B_1, \ldots, A_n B_n.$$

Then

$$v(I) = \pi \left(\sum_{i=1}^{n-1} \overline{A_i B_i}^2 \right) h \quad \text{and} \quad v(J) = \pi \left(\sum_{i=1}^{n} \overline{A_i B_i}^2 \right) h, \qquad (40)$$

so $v(J) - v(I) = \pi \overline{A_n B_n}^2 h$ can be made as small as we please by choosing n sufficiently large, because $A_n B_n = R$ and $h = H/n$.

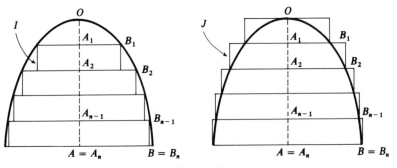

Figure 32

Notice that

$$\frac{\overline{A_i B_i^2}}{R^2} = \frac{OA_i}{OA_n} = \frac{ih}{nh} = \frac{i}{n}$$

by the characteristic property of the parabola. It follows that

$$\frac{v(I)}{v(C)} = \frac{\pi\left(\sum_{i=1}^{n-1} \overline{A_i B_i^2}\right)h}{n\pi R^2 h} = \frac{1 + \cdots + (n-1)}{n^2} < \frac{1}{2}, \qquad (41)$$

while

$$\frac{v(J)}{v(C)} = \frac{\pi\left(\sum_{i=1}^{n} \overline{A_i B_i^2}\right)h}{n\pi R^2 h} = \frac{1 + \cdots + n}{n^2} > \frac{1}{2}, \qquad (42)$$

using the inequalities

$$1 + \cdots + (n-1) < \frac{n^2}{2} < 1 + \cdots + n$$

that follow immediately from the identity

$$1 + \cdots + n = \frac{n}{2}(n+1).$$

If it were true that $v(P) > \frac{1}{2}\pi R^2 H = \frac{1}{2}v(C)$, we could choose n sufficiently large that

$$v(I) > \tfrac{1}{2}v(C)$$

(why?), which contradicts (41). If it were true that $v(P) < \frac{1}{2}\pi R^2 H = \frac{1}{2}v(C)$, we could choose n sufficiently large that

$$v(J) < \tfrac{1}{2}v(C),$$

which contradicts (42). We therefore conclude that

$$v(P) = \tfrac{1}{2}R^2 H = v(C)$$

as desired.   □

In order to interpret this result as an integral, let us turn the parabola on its side (Fig. 33), so it takes the form

$$y^2 = \frac{R^2}{H}x, \qquad 0 \leqslant x \leqslant H.$$

If we subdivide the interval $[0, H]$ into n equal subintervals,

$$0 = x_0 < x_1 < \cdots < x_n = H,$$

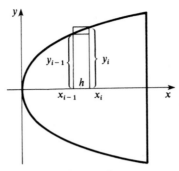

Figure 33

and write $y_i^2 = (R^2/H)x_i$, then we see that

$$v(I) = \pi\left(\sum_{i=1}^{n-1} y_i^2\right)h = \sum_{i=1}^{n-1} \frac{\pi R^2 x_i h}{H}$$

and

$$v(J) = \pi\left(\sum_{i=1}^{n} y_i^2\right)h = \sum_{i=1}^{n} \frac{\pi R^2 x_i h}{H}$$

are simply the lower and upper Riemann sums for the integral

$$\int_0^H \frac{\pi R^2 x\ dx}{H},$$

so Archimedes' analysis shows that

$$\int_0^H \frac{\pi R^2 x}{H}\ dx = \frac{1}{2}\pi R^2 H \quad \text{or} \quad \int_0^H x\ dx = \frac{1}{2}H^2.$$

We have seen that the volume of a sphere is two-thirds that of the circumscribed cylinder. In *On Conoids and Spheroids* Archimedes shows that this relation generalizes to the case of an ellipsoid of revolution inscribed in a cylinder. That is, if the ellipse (Fig. 34)

$$\frac{x^2}{a^2} + \frac{y^2}{b^2} = 1$$

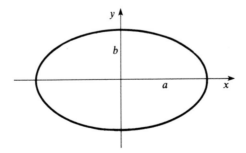

Figure 34

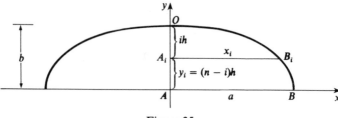

Figure 35

with semi-axes a and b is revolved around the y-axis, then the volume of the ellipsoid E we obtain is

$$v(E) = \tfrac{4}{3}\pi a^2 b.\qquad (43)$$

Archimedes' proof of (43) is based on precisely the same construction as in the case of the paraboloid, applied to the upper half E' of the ellipsoid E, and with the same notation, except that R and H are replaced by a and b respectively. The volumes of the inscribed and circumscribed solids are still given by formulas (40),

$$v(I) = \pi\left(\sum_{i=1}^{n-1}\overline{A_i B_i}^2\right)h \quad \text{and} \quad v(J) = \pi\left(\sum_{i=1}^{n}\overline{A_i B_i}^2\right)h.$$

The difference here is that

$$\frac{\overline{A_i B_i}^2}{a^2} = \frac{x_i^2}{a^2} = 1 - \frac{y_i^2}{b^2}$$

$$= 1 - \frac{(n-i)^2 h^2}{b^2}$$

$$= \frac{b^2 - (n-i)^2 h^2}{b^2}$$

$$= \frac{n^2 - (n-i)^2}{n^2}$$

because $b = nh$. It follows that

$$\frac{v(I)}{v(C)} = \frac{\pi\left(\displaystyle\sum_{i=1}^{n-1}\overline{A_i B_i}^2\right)h}{n\pi a^2 h} = \sum_{i=1}^{n-1}\frac{n^2 - (n-i)^2}{n^3}$$

$$= \frac{(n-1)n^2 - \left(1^2 + 2^2 + \cdots + (n-1)^2\right)}{n^3}$$

$$= 1 - \frac{1^2 + 2^2 + \cdots + n^2}{n^3}$$

$$\frac{v(I)}{v(C)} < \frac{2}{3},$$

because we know from (35) that

$$1^2 + \cdots + (n-1)^2 < \tfrac{1}{3}n^3 < 1^2 + \cdots + n^2.$$

Similarly, we see that

$$
\frac{v(J)}{v(C)} = \frac{\pi \left(\sum\limits_{i=1}^{n} \overline{A_i B_i}^2 \right) h}{n \pi a^2 h}
$$

$$
= \sum_{i=1}^{n} \frac{n^2 - (n-i)^2}{n^3}
$$

$$
= \frac{n^3 - \left(1^2 + \cdots + (n-1)^2 \right)}{n^3}
$$

$$
= 1 - \frac{1^2 + \cdots + (n-1)^2}{n^3}
$$

$$
\frac{v(J)}{v(C)} > \frac{2}{3}.
$$

The facts that

$$v(I) < v(E') < v(J), \qquad v(I) < \tfrac{2}{3}v(C) < v(J)$$

and that $v(J) - v(I)$ can be made arbitrarily small (by choosing n sufficiently large) now imply in the usual way that

$$v(E') = \tfrac{2}{3}v(C)$$

as desired, C being a cylinder with radius a and height b. □

Archimedes applied this "slicing construction" to calculate the volumes of segments of paraboloids, ellipsoids, and hyperboloids of revolution cut off by planes not necessarily perpendicular to the axes of these figures. In these investigations he employed a common general procedure for the solution of several different but similar volume problems. From this point it was a relatively short logical step (albeit one that was not taken for almost nineteen centuries) to the formulation of a general concept of integration.

EXERCISE 19. Apply a slicing construction like that used in *On Conoids and Spheroids* to prove the formulas for the volume of a cone or a pyramid.

EXERCISE 20. Use a slicing construction to prove that the volume of an elliptical cone, with height h and base the ellipse $x^2/a^2 + y^2/b^2 = 1$ is $V = \tfrac{1}{3}\pi abh$. Show first that the area A_z of the horizontal cross-section at a height z above the base is $A_z = \pi ab(h-z)^2/h^2$.

EXERCISE 21. Use a slicing construction to prove that the volume of the ellipsoid

$$\frac{x^2}{a^2} + \frac{y^2}{b^2} + \frac{z^2}{c^2} = 1$$

is $V = \frac{4}{3}\pi abc$. Show first that the area A_z of the horizontal cross-section at height z above the xy-plane is

$$A_z = \pi ab\left(1 - \frac{z^2}{c^2}\right).$$

The Method of Discovery

The works of Archimedes only barely survived the so-called Dark Ages from 500 to 1000 A.D. Of the treatises mentioned previously in this chapter, only two—*Measurement of a Circle* and *On the Sphere and Cylinder*—seem to have been generally known at the time of the Archimedean commentator Eutocius in the sixth century. Almost all modern translations of Archimedes' works stem from a single Greek manuscript that was copied from an earlier original at Constantinople in the ninth or tenth century, was translated into Latin in the thirteenth century, and eventually disappeared without a trace in the sixteenth century. The main exception is a treatise entitled *The Method* that was rediscovered virtually by accident on a palimpsest parchment in Constantinople in 1906 after having been lost since the early centuries of our era. Archimedes' work had been (fortunately) imperfectly erased in about the thirteenth century and replaced with liturgical writings. In this now restored treatise he had detailed his method of discovery, rather than having deliberately concealed it as Wallis and others had speculated.

After stating in the preface to *The Method* a couple of new theorems to be discussed, he adds his intent to "explain in detail in the same book the peculiarity of a certain method," to which he attributes the discovery on heuristic grounds of many of his results, prior to providing them with rigorous proofs (by the method of exhaustion or compression).

> For certain things first became clear to me by a mechanical method, although they had to be demonstrated by geometry afterwards because their investigation by the said method did not furnish an actual demonstration. But it is of course easier, when we have previously acquired, by the method, some knowledge of the questions, to supply the proof than it is to find it without any previous knowledge . . . I am persuaded that it will be of no little service to mathematics; for I apprehend that some, either of my contemporaries or of my successors, will, by means of the method when once established, be able to discover other theorems in addition, which have not yet occurred to me (preface to *The Method* [5]).

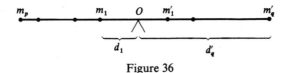

Figure 36

Archimedes' "mechanical method" is based on the *law of the lever*,
according to which a finite system of point masses $m_1, \ldots, m_p$ at distances
$d_1, \ldots, d_p$ from the fulcrum, on one side of a (weightless) lever, balances
another system $m_1', \ldots, m_q'$ on the other side at distances $d_1', \ldots, d_q'$
(Fig. 36) provided that

$$\sum_{i=1}^{p} m_i d_i = \sum_{j=1}^{q} m_j' d_j'. \tag{44}$$

Archimedes had written an earlier treatise *On the Equilibrium of Planes*
devoted to geometric applications and generalizations of the law of the
lever. In it he proved that the centroid of a triangle (the point about which
it balances) is the common point of intersection of the medians from the
vertices to the midpoints of the opposite sides. As is well known from
elementary geometry, this point lies two-thirds of the way from each vertex
to the midpoint of the opposite side.

In its simplest form, the mechanical method for the investigation of
areas and volumes can be described as follows. Suppose that R and S are
two convex regions lying along the same interval of a horizontal axis L
(Fig. 37). Given the area $a(S)$ and the centroid c_S of S, we inquire as to the
area $a(R)$ of R.

We regard the two regions as plane laminas of unit density, each
consisting of an indefinitely large number of elements—line segments or
strips of infinitesimal width—perpendicular to L, and think of L as a lever
with fulcrum O. Suppose there is a constant k such that, for each vertical
line at a distance x from O intersecting the regions R and S in line
segments with lengths l and l' respectively, we can show that

$$k \cdot l = x \cdot l'. \tag{45}$$

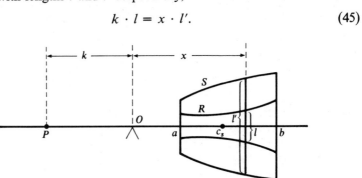

Figure 37

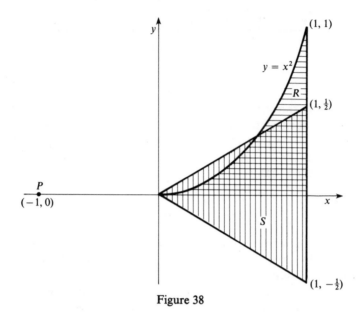

Figure 38

Then the *law of the lever* implies that the segment l, if placed at the point P at a distance k from O, balances the segment l' where it is. It seems to follow that, if the region R is placed with its centroid at P, then it will balance the region S in place where it is, so that

$$a(R) \cdot k = a(S) \cdot \bar{x}_S, \tag{46}$$

where $\bar{x}_S$ is the distance from O to the centroid c_S of S (assuming that each region acts as a point mass placed at its centroid). Since k, $a(S)$, and $\bar{x}_S$ are assumed known, we can solve (46) for $a(R)$. Conversely, if $a(R)$, $a(S)$, and k were known in advance, we could solve (46) to discover the centroid of S.

As a first example, we take R as the region bounded by the parabola $y = x^2$, the x-axis, and the vertical line $x = 1$ (Fig. 38). Let S be the triangle with vertices $(0, 0)$, $(1, \frac{1}{2})$, $(1, -\frac{1}{2})$, whose area is $\frac{1}{2}$ and whose centroid is the point $(\frac{2}{3}, 0)$. Then $l = x^2$ and $l' = x$, so we can take $k = 1$ in (45), i.e. $1 \cdot x^2 = x \cdot x$. Then (46) gives

$$a(R) = a(S) \cdot \bar{x}_S = \frac{1}{2} \cdot \frac{2}{3} = \frac{1}{3}.$$

The computation is similar (although not identical) to Archimedes' mechanical investigation of the area of a segment of a parabola.

As he remarks, "the fact here stated is not actually demonstrated by the argument used; but that argument has given a sort of indication that the conclusion is true." The reason the argument is not rigorous (as it stands) is that a plane region does not consist of a *finite* collection of line segments, whereas we have applied the law of the lever as stated for *finite*

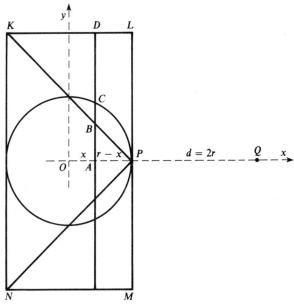

Figure 39

collections of masses. However, the needed fact that (45) implies (46), which Archimedes did not prove, is easily verified using integrals. For if we write $l = w_R(x)$ and $l' = w_S(x)$ for the widths of R and S above x, then

$$a(R) = \int_a^b w_R(x)\, dx, \quad a(S) = \int_a^b w_S(x)\, dx, \quad \bar{x}_S = \frac{1}{a(S)} \int_a^b x w_S(x)\, dx,$$

so it is clear that (46) does indeed follow from (45). In these terms, the effect of "the method" is to express a desired integral (that for $a(R)$) in terms of another integral (that for $a(S)$) which is already known.

If R and S are solid regions, then the widths in the above discussion are replaced by the cross-sectional areas of R and S in the plane perpendicular to L at x. As an example of this case we give a mechanical derivation of Archimedes' favorite result—the formula for the volume of a sphere. We will use his geometric construction (for Proposition 2 of *The Method*), but with Cartesian coordinate computations instead of his arguments in terms of similar triangles.

Starting with the circle $x^2 + y^2 = r^2$ intersecting the positive x-axis at the point $P = (r, 0)$, let $KLMN$ be a rectangle centered at the origin O with base $d = 2r$ and height $2d$, and consider also the triangle KNP (Fig. 39). By revolving these three figures around the x-axis, we generate a sphere S, a cone C, and a cylinder Z. We consider these three solids as being made up of circular disks perpendicular to the x-axis. For example, the plane perpendicular to the x-axis at the point $A = (x, 0)$ intersects the sphere S in a circle S_x with radius $AC = y = \sqrt{r^2 - x^2}$, the cone C in a circle C_x with

radius $AB = r - x$, and the cylinder Z in a circle Z_x with radius $AD = d$.
Now

$$d[a(S_x) + a(C_x)] = \pi d[y^2 + (r - x)^2]$$
$$= \pi d[(r^2 - x^2) + (r^2 - 2rx + x^2)]$$
$$= \pi d(2r^2 - 2rx)$$
$$= \pi d^2(r - x)$$
$$d[a(S_x) + a(C_x)] = (r - x)a(Z_x). \qquad (47)$$

This implies that, if the circles S_x and C_x are placed at the point $Q = (3r, 0)$ a distance d to the right of P, then together they will balance the circle Z_x where it is, considering the x-axis as a lever with fulcrum P. It follows that, if the sphere S and the cone C are placed with their centroids at Q, then together they will balance the cylinder Z where it is. Since by symmetry the centroid of Z is at the origin O, the law of the lever gives

$$2r[v(S) + v(C)] = rv(Z). \qquad (48)$$

Substituting the known volumes $v(C) = \pi d^3/3$ and $v(Z) = \pi d^3$ into (47), we obtain

$$v(S) = \tfrac{1}{6}\pi d^3 = \tfrac{4}{3}\pi r^3.$$

Archimedes indicates that this was his original derivation of the volume of a sphere, from which he inferred (as described previously) the formula for the surface area of the sphere.

EXERCISE 22. Obtain from this same construction the formula

$$V = \frac{1}{3}\pi \rho^2 h\left(\frac{3r - h}{2r - h}\right)$$

for the volume of the smaller spherical segment cut off by the plane $x = a$ (where $0 < a < r$), having base radius $\rho = \sqrt{r^2 - a^2}$ and height $h = r - a$. Use the portions of the cone and cylinder cut off by this same plane.

EXERCISE 23. Let P be the segment of a paraboloid obtained by revolving the parabola $y^2 = x$, $0 \leqslant x \leqslant 1$, about the x-axis. Use Archimedes' mechanical method

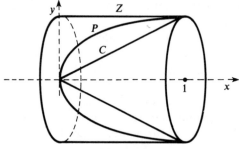

Figure 40

to deduce that the volume of P is one-half that of the circumscribed cylinder Z (Fig. 40).

EXERCISE 24. Balance the paraboloid (where it is) against the inscribed cone C (concentrated at an appropriate point) to show that the centroid of the paraboloid is two-thirds of the way from its vertex to its base.

After applying his mechanical technique to compute the volumes and centroids of segments of ellipsoids, paraboloids, and hyperboloids of revolution, Archimedes concludes *The Method* with computations of the volumes of two special solids that are standard examples in modern calculus textbooks. In Proposition 14 he proves that the volume of the wedge W, cut from a cylinder of diameter d and height $h = d/2$ by a plane through a diameter of one base and a point of the circumference of the other (Fig. 41), is $v(W) = d^3/6$. In Proposition 15 he shows that the volume of the region S^* common to two cylinders with diameter d and perpendicular intersecting axes is $v(S^*) = 2d^3/3$. Figure 42 shows the first octant of the region S^*.

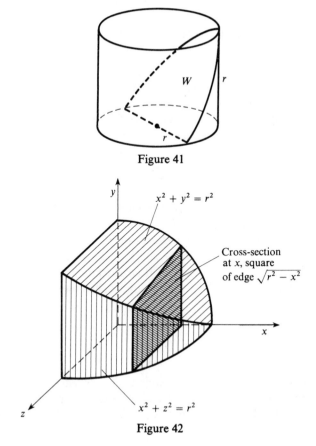

Figure 41

Figure 42

The calculation of the volume of S^* employs the same construction as indicated in Figure 39, except with circular cross-sections replaced by square cross-sections. The cross-section of S^* perpendicular to the x-axis at A is a square S_x^* with edge $2AC = 2\sqrt{r^2 - x^2}$ (see Figures 39 and 42). Let C^* be the pyramid with vertex P and square base with edge $KN = 2d$. Then the cross-section of C^* perpendicular to the x-axis at A is a square C_x^* with edge $2AB = 2(r - x)$. Let Z^* be the rectangular solid with height $KL = d$ and square base with edge $KN = 2d$. Its cross-section Z_x^* perpendicular to the x-axis at A is then a square with area $4d^2$.

Upon adding asterisks and replacing the constant π with 4, the derivation of equation (47) now gives

$$d\left[a(S_x^*) + a(C_x^*) \right] = (r - x)a(Z_x^*). \tag{47*}$$

In the same way that equation (47) implies equation (48), it follows from (47*) that

$$2r\left[v(S^*) + v(C^*) \right] = rv(Z^*). \tag{48*}$$

Substituting $v(Z^*) = 4d^3$ and $v(C^*) = \frac{1}{3}v(Z^*) = \frac{4}{3}d^3$ into (48*), we finally obtain

$$v(S^*) = \tfrac{2}{3}d^3. \tag{49}$$

Since it is evident that S^* is the union of four copies of the wedge W of Figure 41, it follows from (49) that

$$v(W) = \tfrac{1}{6}d^3. \tag{50}$$

However, Archimedes gave a separate mechanical derivation of (50), as well as rigorous compression proofs of both (49) and (50).

In its treatment of plane areas in terms of line elements and volumes in terms of area elements, Archimedes' mechanical method was a precursor to the "indivisibles" techniques that flourished in the early seventeenth century, and which led directly to the rapid development of the calculus.

Archimedes and Calculus?

We have seen in this chapter that Archimedes solved many of the problems that are staples of modern calculus courses, and in particular that his solutions can often be interpreted as computations of definite integrals of the form

$$\int_c^d (ax + bx^2)\, dx.$$

On this basis various authors have credited Archimedes with the original discovery of the calculus.

While it is true that Archimedes' work ultimately (in the seventeenth century) gave birth to the calculus, three indispensable ingredients of the calculus are missing in his methods:

(1) The explicit introduction of limit concepts. Archimedes, at least in his formal proofs if not in his informal analyses, shared the Greek "horror of the infinite." The Greek concept of rigor demanded the cumbersome double *reductio ad absurdum* argument rather than a simple passage to the limit.

(2) A general computational algorithm for the calculation of areas and volumes. A distinctive feature of the calculus is the formulation of general procedures for the exploitation of analogies between different but similar problems to lessen the burden of duplication in their solutions. By contrast, Archimedes (with very few exceptions) started from scratch in each computation, basing the solution of each problem on a construction determined by the special geometric features of that particular problem, and without taking advantage of previous solutions of similar problems. The reliance upon geometric algebra without any simplifying symbolic notation was a substantial impediment to the identification and codification of computational features common to different problems.

(3) A recognition of the inverse relationship between area and tangent problems. The Greek view of tangent lines as merely "touching" lines was inadequate to provide any hint of this relationship or of "rate of change" interpretations.

Nevertheless, the investigations of Archimedes continue to serve as exemplars of originality and precision. His progress with the mathematical tools available to him will always be one of the great landmarks in the history of mathematics.

References

[1] M. Clagett, *Archimedes in the Middle Ages*, Vol. I. University of Wisconsin Press, 1964.
[2] M. Clagett, Archimedes, in *Dictionary of Scientific Biography*. New York: Scribners, 1970, Vol. I, pp. 213–231.
[3] E. J. Dijksterhuis, *Archimedes*. New York: Humanities Press, 1957.
[4] S. H. Gould, The Method of Archimedes, *Am Math Mon* **62**, 473–476, 1955.
[5] T. L. Heath, *The Works of Archimedes*. Cambridge University Press, 1897. (Dover reprint).

[6] T. L. Heath, *A History of Greek Mathematics*, Vol. II. Oxford University Press, 1921.

[7] T. L. Heath, *A Manual of Greek Mathematics*. Oxford University Press, 1931. (Dover reprint 1963).

[8] T. L. Heath, Greek geometry with special reference to infinitesimals. *Math Gaz* 11, 248–259, 1922–23.

[9] J. Hjelmslev, Eudoxus' axiom and Archimedes' lemma. *Centaurus* 1, 2–11, 1950.

[10] W. R. Knorr, Archimedes and the measurement of the circle: a new interpretation. *Arch Hist Exact Sci* 15, 115–140, 1975–76.

[11] W. R. Knorr, Archimedes and the *Elements*: Proposal for a revised chronological ordering of the Archimedean corpus. *Arch Hist Exact Sci* 19, 211–290, 1978.

See also the Chapter 1 references to Seidenberg, Smeur, and van der Waerden.

Twilight, Darkness, and Dawn 3

Introduction

The classical era of Greek mathematical development stretched over a period of approximately ten centuries, from about 600 B.C. to 400 A.D. However, it reached an early climax in the third century B.C. with the work of Archimedes and that of his younger contemporary Apollonius, who elaborated a comprehensive theory of the conic sections. Coincident with the establishment of Roman power in the Mediterranean area during the second century B.C., Hellenistic culture in general, and Greek theoretical mathematics in particular, began a period of decline that produced no new contributions comparable to those of Eudoxus, Euclid, Archimedes, and Apollonius.

With the collapse of the Western Roman Empire in the fifth century A.D., Western Europe entered a long dark age during which the scientific accomplishments of the past were only dimly remembered, when at all. For several centuries after the fall of Rome, the Greek legacy was confined to the beleagured Byzantine or Eastern Roman Empire, where it was preserved but hardly flourished. In 529 A.D. the emperor Justinian closed the Greek schools of "pagan" philosophy at Athens, including Plato's Academy which had survived for nine centuries.

The principal center of surviving Greek learning at Alexandria fell in 641 A.D. to the Moslems, who during the seventh century rapidly conquered many of the territories immediately surrounding the Mediterranean, and established there a stable culture that prospered for at least four centuries. The Moslems eagerly absorbed the available repository of Greek science and mathematics, together with Indian and Babylonian elements, and during the ninth and tenth centuries the works of Euclid, Archimedes, Apollonius, and Ptolemy were translated from Greek into Arabic.

The boundaries of Arab jurisdiction in the west were pushed back in the eleventh century. With the fall of Toledo (Spain) in 1085 and Sicily in 1091, the Greek classics in Arabic were again accessible to Christian Western Europe, and were translated into Latin during the twelfth and thirteenth centuries. The Greek mathematical works thus reacquired, and the medieval scholastic speculations on motion, variability, and the infinite, together with the symbolic algebra and analytic geometry of the late Renaissance, formed the rich amalgam that fueled the seventeenth century explosion of infinitesimal mathematics. In this chapter we outline briefly the significant features of the decline of Greek mathematics, the absorption and transmission of the mathematics of antiquity by the Islamic culture, and the eventual rebirth of mathematical progress in Western Europe.

The Decline of Greek Mathematics

After the golden age of the third and fourth centuries B.C., theoretical progress ceased and Greek mathematics turned toward applications of a sort that failed to stimulate further progress in mathematics itself. The most significant work of the next four centuries was the mathematical astronomy and associated applied trigonometry of Hipparchus (second century B.C.) and Ptolemy (second century A.D.). Scholarly commentaries on the Greek geometrical masterworks were written by Pappus, Proclus, and Eutocius in the fourth, fifth, and sixth centuries A.D., but the earlier level of originality in geometrical analysis was never regained. Although Archimedes had seemingly paved the way for a development of infinitesimal techniques, his work had to wait eighteen centuries for its continuation.

Changing political and social conditions probably played some role in this decline. Greek science, despite its brilliant achievements, was apparently a fragile enterprise, centering on a relatively small number of professional workers concentrated in a few centers (such as Athens and Alexandria), and dependent on both favorable intellectual conditions and royal subsidies for its continued success. Wars and the end of Hellenistic prosperity under Roman domination terminated these favorable conditions. Archimedes was killed in the sack of Syracuse, and a large part of the famous library at Alexandria was burned during the Roman siege of that city. The Romans were an intensely practical people who undertook great construction projects (bridges, highways, viaducts, etc.) on the basis of rule of thumb procedures, but had no interest in and did not support abstract and theoretical studies.

However, these vagaries of the external ancient world were not by themselves responsible for the failure of Greek analysis to advance materially beyond Archimedes. There were also internal mathematical factors

that suffice to explain this failure. These impeding factors centered on the
rigid separation in Greek mathematics between geometry and arithmetic
(or algebra), and a one-sided emphasis on the former. Their analysis dealt
solely with geometrical magnitudes—lengths, areas, volumes—rather than
numerical ones, and their manipulation of these magnitudes was exclusively verbal or rhetorical, rather than symbolic (or algebraic, as we would
say today). As a consequence, their cumbersome computations disguised
the analogies between solutions of similar problems, and thereby prevented
the recognition and codification of general computational algorithms that
could be applied to whole classes of similar problems. In short, the Greeks'
thoroughgoing geometrization of mathematics, to the exclusion of its
algebraic aspect, effectively precluded the growth of an algorithmic tradition based on generally applicable methods.

It is somewhat paradoxical that this principal shortcoming of Greek
mathematics stemmed directly from its principal virtue—the insistence on
absolute logical rigor. The Greeks imposed on themselves standards of
exact thought that prevented them from using and working with concepts
that they could not completely and precisely formulate. For this reason
they rejected irrationals as numbers, and excluded all traces of the infinite,
such as explicit limit concepts, from their mathematics. Although the
Greek bequest of deductive rigor is the distinguishing feature of modern
mathematics, it is arguable that, had all succeeding generations also
refused to use real numbers and limits until they fully understood them,
the calculus might never have been developed, and mathematics might
now be a dead and forgotten science.

The Babylonians had handed down a working body of algebraic (though
still rhetorical) techniques for the solution of problems involving linear and
quadratic equations, based on the uncritical manipulation of numbers and
unquestioning representation of all quantities in terms of numbers, as well
as the free use of convenient approximations. However, the Pythagorean
discovery of incommensurable line segments meant that geometric magnitudes could not be measured by numbers as the Greeks understood them,
and the requirement of exact solutions rendered approximations pointless.

Consequently the handy algebra of Babylonia was converted into a
ponderous geometric algebra based on the technique of application of
areas. It no longer made sense to say that the area of a circle is πr^2; one
had to express the result in terms of Eudoxian proportions between pairs
of comparable areas (e.g., circle to circle as square to square). A magnitude
could not be expressed precisely in terms of a simple symbol like a
number, but only in terms of a line segment constructed in prescribed
fashion on a geometric figure. Algebraic operations became geometric
constructions (the product of two linear magnitudes being a rectangular
area, etc.). What we would call higher degree equations were meaningless
because they did not correspond to geometric magnitudes in three dimensions. Hence complicated computations had to be expressed in terms of
cumbersome geometric transformations of proportions.

Because of these difficulties that are inherent in geometric algebra and the formal theory of proportions, only a gifted Greek mathematician could carry out computations that today can be handled by a school child using elementary algebraic notation. We can understand the formidable sentences of Greek geometric algebra only by transforming them into concise formulas. Van der Waerden has emphasized the importance of a continuous oral tradition to compensate for the thorniness of Greek mathematical exposition: "As long as there was no interruption, as long as each generation could hand over its method to the next, everything went well and the science flourished. But as soon as some external cause brought about an interruption in the oral tradition, and only books remained, it became extremely difficult to assimilate the work of the great precursors and next to impossible to pass beyond it" [14], p. 266. To the evident difficulties of an exclusively written tradition may be added the accounts of recurrent burnings of hundreds of thousands of books at the great Alexandria library by successive waves of military invaders and religious fanatics.

In summary, it was probably a combination of external and internal factors that brought a halt to the great enterprise of Greek theoretical mathematics, and delayed for many centuries the harvest of its eventual fruits.

Mathematics in the Dark Ages

The traditional date for the end of the Roman Empire in the West is 476, when the resident Roman emperor was displaced by a Goth intruder. The fall of Rome was followed by a breakdown of central government and the dissolution of urban life, and continuity of scientific development in Western Europe was lost altogether. For the next several centuries, the remnants of the Western Roman Empire, mainly in the form of the Catholic Church, were occupied with the task of civilizing barbarian tribes so primitive that they had no written language, let alone any science or culture.

Only the Latin encyclopedists preserved any connection, however tenuous, with the intellectual treasures of the past. It had been fashionable for cultured Romans to acquire a nodding acquaintance (though seldom any real understanding) with Greek science and philosophy. In order to meet this need, popularizers incorporated palatable condensations of scientific results into handbooks and manuals that were often superficial "scissors and paste" jobs. Although these books, taken together, presented an unorganized mass of frequently contradictory facts and myths, they constituted virtually the only sources of general scientific information in Western Europe during the early middle ages.

In regard to mathematics, the most important of these Roman writers was Boethius (ca. 480–524), who wrote four elementary textbooks—in arithmetic, geometry, astronomy, and music—that served as the basis for the quadrivium of the medieval monastic schools. His *Arithmetic* was an abridgement of the *Introductio arithmeticae* of Niomachus (ca. 100 A.D.). This latter work was basically a compilation of Pythagorean and Platonic number lore; its level may be judged by the fact that its author saw fit to include a 10 by 10 multiplication table. The *Geometry* of Boethius consisted of only the statements, without proofs, of the simpler propositions in the first four books of Euclid's *Elements*.

From this beginning, the level of scholarship in Western Europe generally declined until the time of Gerbert, a Frenchman who became Pope Sylvester II (999–1003). Gerbert traveled to Spain to learn some of the mathematics and science of the Arab world, and apparently rekindled an emphasis on rudimentary mathematical instruction in the church schools that were, until the emergence of universities in the late twelfth century, the main centers of learning in Western Europe. Even so, geometry in this period consisted only of the enumeration of the first few theorems of Euclid, without any logical sequence or semblance of proofs. The Pythagorean theorem had apparently been long since forgotten.

The most famous eleventh century mathematician (in Western Europe) seems to have been Franco of Liege, who wrote a much-quoted book on the quadrature of the circle. Thinking that the approximation $11/14 = (1/4)(22/7)$ was an exact value for the ratio of the area of a circle to that of the circumscribed square, he thought he could solve the classical problem of "squaring the circle" by somehow constructing a square with area equal to that of a 11 by 14 rectangle, and hence to that of a circle of radius 7. It is to this latter construction of a square that his work is largely devoted! Clearly an infusion of ancient Greek knowledge was necessary to enable Western Europe to rise above the intellectual morass of the dark ages.

The Arab Connection

The Greek cultural heritage was to some extent preserved throughout the middle ages by the shrinking Byzantine Empire centered at Constantinople. However, it was the Arab hegemony in the Mediterranean area that effectively nutured this heritage during the centuries of the dark ages, and finally transmitted it to Western Europe.

During the seventh and eighth centuries, the Muslim empire rapidly extended its domination along the southern crescent of the Mediterranean, from Persia and Syria in the east to Spain and Morocco in the west. A century of conquest was followed during the second half of the eighth century by the rapid development of an Islamic culture that avidly absorbed the knowledge of the newly conquered lands. The eastern capital of Baghdad became a new cosmopolitan center, a new Alexandria, where the ancient sciences of Greece, India, and Mesopotamia were studied.

The mathematician and astronomer al-Khowarizmi worked at Baghdad during the early ninth century, and wrote historically important textbooks on arithmetic and algebra. The first of these was an exposition of the Hindu art of reckoning. The Hindus in India were fascinated by numbers and computations, but had little interest in geometry and deductive proofs. This division of interests worked to their practical advantage, for it permitted them to calculate freely with rational numbers and irrational roots alike, oblivious to the fine distinctions of commensurability and logical subtleties that had impeded the Greeks. They combined in a single system of numeration three separate features of various previous systems—decimal (base 10) numerals, positional notation, and a zero symbol. This "Hindu-Arabic numeration," essentially our present system, came into widespread use through the influence of al-Khowarizmi's arithmetic.

Al-Khowarizmi's second book, entitled *Al-jabr wa'l muqabalah*, has provided us with the modern word "algebra." Apparently "al-jabr" referred to the transposition of subtracted terms to the other side of an equation, and "muqabalah" to the cancellation of equal terms on the two sides of an equation. The first six chapters of the *Al-jabr* list routine procedures, in terms of illustrative examples, for the solution of those linear and quadratic equations with positive coefficients that have positive roots—negative roots are not considered. For a translation of these brief chapters, see Edward Grant's medieval science source book ([10], pp. 106–111).

Al-Khowarizmi's treatment was entirely rhetorical or verbal, with numbers even written out in word form. In this respect it was a retrogression from the *Arithmetica* of Diophantus (third century A.D.). Diophantus, whose problems dealt mainly with number theory rather than algebra in the present sense, had introduced abbreviations for powers of the unknown, such as Δ^γ for the square, K^γ for the cube, $\Delta^\gamma\Delta$ for the square-square or fourth power, ΔK^γ for the square-cube or fifth power, etc. The chief merit of al-Khowarizmi's exposition was its return to the Babylonian and Hindu tradition of working routinely with quantities as "mere" numbers rather than geometric magnitudes, and of reducing the solution of equations to operational procedures or algorithms. Indeed, the word "algorithm" is derived from the author's name: al-Khowarizmi→algorismi→algorithm.

Al-Khowarizmi's equations involved three kinds of quantities: roots, squares, and numbers (that is, x terms, x^2 terms, and constants). He stated the equation $x^2 + 10x = 39$ as "a square and ten of its roots are equal to thirty-nine." His solution reads essentially as follows: Take half the number of roots, that is, five, and multiply this by itself to obtain twenty-five.

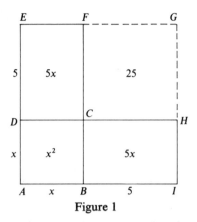

Figure 1

Add this to the thirty-nine, giving sixty-four. Take the square root, or eight, and subtract from it half the number of roots, namely five. The result, three, is the desired root. In terms of modern notation, he is saying that the (positive) root of $x^2 + 2bx = c$ is $-b + \sqrt{b^2 + c}$.

Al-Khowarizmi verifies his algorithms by means of geometric constructions that indicate Greek influence. To explain the solution of the equation $x^2 + 10x = 39$, he starts with a square $ABCD$ of area x^2 (Fig. 1). The equation calls for the addition of $10x$, so he adds the rectangles $CDEF$ and $BCHI$, each of area $5x$. He "completes the square" by adding the square $CFGH$ of area 25 (the square of "half the number of roots"). Since $x^2 + 10x$ equals 39, it follows that the larger square $AEGI$ has area $39 + 25 = 64$. Hence its edge is 8, so $x = 8 - 5 = 3$.

EXERCISE 1. Solve the equation $x^2 + 8x = 65$ (for its positive root) by means of a construction similar to Figure 1.

During the ninth and tenth centuries Euclid's *Elements* and many of the works of Archimedes, Apollonius, and Ptolemy were translated from Greek into Arabic. Some of them are extant today only because of these Arabic translations. The Greek masterpieces were studied carefully, and alternative proofs and generalizations (e.g., of the Pythagorean theorem) were produced, indicating a reasonable level of understanding of Greek mathematics in Islam.

Arab mathematical science reached its apex in the eleventh century. Al-Haitham (ca. 965–1039), known in the West as Alhazen, wrote an influential treatise on geometrical optics and extended some of Archimedes' volume results. For example, he showed that, if a segment of a parabola is revolved about its base (rather than about its axis, as in Archimedes' *On Conoids*), then the volume of the solid obtained is 8/15 that of the circumscribed cylinder. This computation required formulas for the sums of the first n cubes and fourth powers whereas Archimedes had

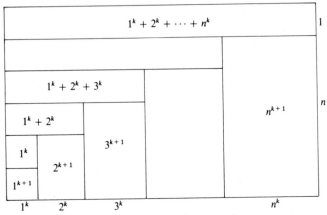

Figure 2

used only the formulas for the sums of the first n integers and of their squares. The following exercise outlines Alhazen's ingenious geometric derivation of these formulas, which were to play a continuing role in the development of the calculus. It is based on Figure 2, from which we read off the formula

$$(n+1) \sum_{i=1}^{n} i^k = \sum_{i=1}^{n} i^{k+1} + \sum_{p=1}^{n} \left(\sum_{i=1}^{p} i^k \right). \tag{1}$$

EXERCISE 2. (a) Substitute $k=1$ and the formula

$$1 + 2 + \cdots + n = \tfrac{1}{2}n^2 + \tfrac{1}{2}n \tag{2}$$

into (1) to derive the formula

$$1^2 + 2^2 + \cdots + n^2 = \tfrac{1}{3}n^3 + \tfrac{1}{2}n^2 + \tfrac{1}{6}n. \tag{3}$$

(b) Substitute $k=2$ and equation (3) into (1) to derive the formula

$$1^3 + 2^3 + \cdots + n^3 = \tfrac{1}{4}n^4 + \tfrac{1}{2}n^3 + \tfrac{1}{4}n^2. \tag{4}$$

(c) Substitute $k=3$ and equation (4) into (1) to derive the formula

$$1^4 + 2^4 + \cdots + n^4 = \tfrac{1}{5}n^5 + \tfrac{1}{2}n^4 + \tfrac{1}{3}n^3 - \tfrac{1}{30}n. \tag{5}$$

(d) Apply the above formulas to show that

$$\sum_{i=1}^{n} (n^2 - i^2)^2 = n^5 - 2n^2 \sum_{i=1}^{n} i^2 - \sum_{i=1}^{n} i^4 = \tfrac{8}{15}n^5 - \tfrac{1}{2}n^4 - \tfrac{1}{30}n. \tag{6}$$

Add n^4 to both sides of (6) to obtain

$$\sum_{i=0}^{n-1} (n^2 - i^2)^2 = \tfrac{8}{15}n^5 + \tfrac{1}{2}n^4 - \tfrac{1}{30}n. \tag{7}$$

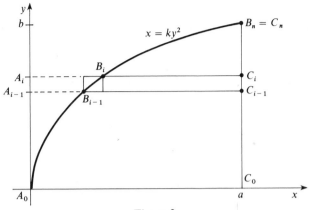

Figure 3

EXERCISE 3. Let the segment $A_0 C_0 C_n$ (Fig. 3) of the parabola $x = ky^2$, over the interval $[0, a]$ on the x-axis, be revolved around the ordinate $C_0 C_n$ to obtain the solid of revolution S. Denote by Z the circumscribed cylinder with radius a and height b, where $a = kb^2$. Let the points $A_1, \ldots, A_{n-1}$ divide the interval $[0, b]$ on the y-axis into n equal subintervals of length $h = b/n$. Let $B_1, \ldots, B_{n-1}$ and $C_1, \ldots, C_{n-1}$ be the corresponding points on the parabola and the ordinate $C_0 C_n$, respectively. Finally, let P be the union of the cylinders with radius $B_i C_i$ and height $C_{i-1} C_i$, and Q the union of the cylinders with radius $B_{i-1} C_{i-1}$ and height $C_{i-1} C_i$, $i = 1, \ldots, n$. Then $P \subset S \subset Q$.

   (a) Show that

$$v(P) = \sum_{i=1}^{n} \pi k^2 h^5 (n^2 - i^2)^2$$

and

$$v(Q) = \sum_{i=0}^{n-1} \pi k^2 h^5 (n^2 - i^2)^2.$$

   (b) Apply Equations (6) and (7) to show that

$$v(P) < \tfrac{8}{15} v(Z) < v(Q).$$

Hint: Calculate $v(P)/v(Z)$ and $v(Q)/v(Z)$.

   (c) Conclude that $v(S) = 8v(Z)/15$ as desired.

For approximately four centuries the Muslim world preserved the Greek mathematical tradition and enriched it with the addition of Eastern elements of arithmetic and algebra. By the twelfth century Arabic science had begun to decline but, fortunately, Western Europe had emerged from its dark ages with an appetite for new knowledge. The *Elements* of Euclid was translated from Arabic into Latin in 1142 by Adelard of Bath, and Robert of Chester produced a Latin translation of al-Khowarizmi's *Algebra* in 1145. The most prolific of a small army of Latin translators in Spain, after the reclamation of Toledo from the Moors, was Gerard of Cremona

(1114–1187) who produced an improved translation of the *Elements* as well as some of Archimedes' works, including *Measurement of the Circle* and parts of *On the Sphere and Cylinder*. In 1269 William of Moerbeke published a Latin translation of the extant Greek corpus of Archimedean treatises.

These translations served to reestablish quantitative science in Western Europe, at least at the level of elementary algebra and geometry. However, the works of Archimedes were too sophisticated for immediate or widespread assimilation, and did not bear significant fruit until the sixteenth and seventeenth centuries. The principal medieval contributions to later progress in mathematics stemmed from the Scholastic speculations on continuity and variability.

Medieval Speculations on Motion and Variability

The age of Latin translation brought the ancient wisdom to a now dynamic Europe with flourishing new universities. Whereas the gap between Boethius and Archimedes would take several centuries to bridge, the vast intellectual system encompassed by the scientific and philosophical treatises of Aristotle was more accessible. Indeed, the absorption and assimilation of Aristotelian thought that took place in the thirteenth century makes it the major turning point in Western intellectual history between the fourth century B.C.—the heroic age of Plato, Aristotle, and Euclid—and the scientific revolution of the seventeenth century.

Aristotle's treatise on *Physics* had explored the nature of the infinite, the existence of indivisibles or infinitesimals, and the divisibility of continuous quantities—time, motion, and geometric magnitudes. Having pointed out that "motion is supposed to belong to the class of things which are *continuous*; and the *infinite* presents itself first in the continuous—that is how it comes about that 'infinite' is often used in definitions of the continuous ('what is infinitely divisible is continuous')" [Book III, Chapt. 1], Aristotle charged scholars "to discuss the infinite and to inquire whether there *is* such a thing or not, and, if there is, *what* it is" [Book III, Chapt. 4].

The medieval Scholastic philosophers responded to this challenge with evident relish. Their detailed (and often interminably prolix) speculations and disputations on the infinite, the nature of the continuum, and the existence of indivisibles were more philosophical than mathematical in character, and were generally inconclusive from a scientific viewpoint. However, they frequently showed a keen appreciation of logical difficulties that were not finally resolved until the late nineteenth century. This "sub-mathematical" activity of the late middle ages no doubt enhanced the acceptability of infinitesimal techniques that were officially taboo in Greek mathematics but were freely used to great advantage in the seventeenth century.

Of more immediate consequence were the quantitative studies of change and motion that began in the early fourteenth century. The concept of continuous variation of quantities had played no role in the mathematics of the Greeks—their quantities were either numerical and discrete or geometric and static. Their algebra dealt with constants rather than variables, and their geometry treated fixed and unchanging geometric figures. They studied only *uniform* (linear or circular) motion, so such concepts as acceleration and instantaneous velocity had no meaning to the Greeks. In short, Greek science did not discuss phenomena of change or variability in quantitative terms.

The problem of quantifying change was attacked during the second quarter of the fourteenth century by a group of logicians and natural philosophers at Merton College in Oxford, including Thomas Bradwardine (the "Doctor profundus" of his day and later Archbishop of Canterbury) and Richard Swineshead (known to medievals as *the* Calculator, as Aristotle was *the* Philosopher and Paul *the* Apostle). They were concerned specifically with what was then referred to as the *latitude of forms*, and might today be described as the intensity of qualities. In Aristotelian philosophy, *qualities* were attributes that admit of intensity (at a point of a body or at an instant in time), such as hotness and density. Intensive (or local) qualities were distinguished from the corresponding extensive (or global) quantities, such as heat and weight. Analogously, (instantaneous) velocity was regarded as a quality, the intensity of motion; the corresponding quantity was the total motion, i.e. the distance covered. The Merton scholars sought to study variations in the intensity of a quality, from point to point of a body, or from point to point in time. We will restrict our discussion here to their consideration of the case of motion and velocity.

They recognized that the heart of the matter is the framing of definitions of these terms that provide an adequate basis for quantitative analysis. They defined motion to be *uniform* (i.e. constant speed) if equal distances are described in equal times. *Uniform acceleration* was defined to be that for which equal increments of velocity are acquired in equal intervals of time. For even this simplest case of variable motion, a definition of *instantaneous velocity* was needed. Lacking the notion of limits of ratios, they could only define instantaneous velocity in terms of the distance that would be traversed by a point if it moved uniformly over a period of time with the same speed it possessed at the instant in question. Although circular in nature, this intuitive concept of instantaneous velocity sufficed for the derivation of correct results in the fourteenth century (as it did in the early days of calculus, and continues today to do in everyday scientific discourse).

The central result derived from these concepts was the Merton Rule of uniform acceleration, the mean speed theorem:

If a moving body is uniformly accelerated during a given time interval, then the total distance s traversed is that which it would move during the same time interval with a uniform velocity equal to the average of its initial velocity v_o and its final velocity v_f (namely, its instantaneous velocity at the midpoint of the time interval).

That is,

$$s = \tfrac{1}{2}(v_o + v_f)t, \tag{8}$$

where t is the length of the time interval. A number of lengthy and rhetorical but ingenious derivations of this theorem were given by the Merton scholars; accounts and discussions of them may be found in Chapter 5 of Clagett [5].

EXERCISE 4. Rewrite the Merton Rule (8) in the form

$$s = v_o t + \tfrac{1}{2}at^2. \tag{9}$$

What is the number a, and why is it constant (independent of t)?

EXERCISE 5. To identify (9) as a calculus result, obtain it by antidifferentiation, starting with $s''(t) = a$, $s'(0) = v_o$.

The Merton studies spread to France and Italy in the mid-fourteenth century. In his *Treatise on the Configurations of Qualities and Motions*, written in the 1350s, the Parisian scholastic Nicole Oresme introduced the important concept of graphical representations, or geometrical "configurations", of intensities of qualities. English translations of this work can be found in Grant's medieval science source book [10] or Clagett's comprehensive analysis of the medieval geometry of qualities and motions [7].

Oresme discusses mainly the case of a "linear" quality, one whose "extension" is measured by an interval (line segment) of either space (as in the case of a rod of variable density) or time (as in the case of a moving point). He proposes to measure the intensity of the quality at each point of the reference interval by a perpendicular line segment at that point, thereby constructing a graph with the reference interval as its base. As he says at the beginning of his treatise,

> Every measurable thing except numbers is imagined in the manner of continuous quantity. Therefore, for the mensuration of such a thing, it is necessary that points, lines, and surfaces, or their properties, be imagined. For in them (i.e., the geometrical entities), as the Philosopher has it, measure or ratio is initially found .... Therefore, every intensity which can be acquired successively ought to be imagined by a straight line perpendicularly erected on some point of the space or subject of the intensible thing, e.g., a quality .... And since the quantity or ratio of lines is better known and is more readily conceived by us—nay the line is in the first species of continua, therefore such intensity ought to be

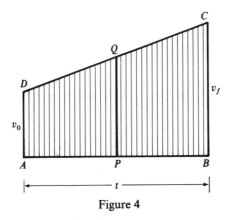

Figure 4

imagined by lines .... Therefore, equal intensities are designated by equal lines, a double intensity by a double line, and always in the same way if one proceeds proportionally.

He refers to the reference interval for a quality as its *longitude*, and its intensity at a point as its *latitude* or altitude there (perhaps adopting these terms from their geographical use). Finally, he specifies that the *quantity* of a linear quality is to be "imagined by" its *configuration* as described above.

For example, in the case of a uniformly accelerated motion during a time interval $[0, t]$ corresponding to the longitude AB in Figure 4, the latitude at each point P of AB is an ordinate PQ whose length is the velocity at the corresponding instant, so the upper edge CD of the configuration is simply a time-velocity graph. Oresme saw that the definition of uniform acceleration implies that CD is a straight line segment, so the configuration is a trapezoid with base $AB = t$ and heights $AD = v_o$ and $BC = v_f$. He assumed without explicit proof that the area s of this trapezoid equals the total distance traveled, perhaps on the basis of regarding this area as made up of very many vertical segments or indivisibles, each representing a velocity continued for a very short or infinitesimal time. At any rate, it follows immediately from the formula for the area of a trapezoid that

$$s = \tfrac{1}{2}(v_o + v_f)t, \tag{8}$$

so Oresme has provided the Merton Rule with a geometrical verification.

EXERCISE 6. Consider the case of uniformly accelerated motion with $v_o = 0$, so the trapezoid reduces to a triangle (Fig. 5). Subdivide the base AB into n equal time subintervals, and denote by $s_1, s_2, s_3, \ldots, s_n$ the distances traveled during these successive subintervals. Show that these distances are proportional to the odd numbers $1, 3, 5, \ldots, 2n - 1$, that is,

$$\frac{s_1}{1} = \frac{s_2}{3} = \frac{s_3}{5} = \cdots = \frac{s_n}{2n - 1}.$$

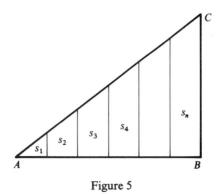

Figure 5

This "law of odd numbers," for the distances traveled in successive equal intervals of time under constant or uniform acceleration, was important for Galileo's later empirical verification that freely falling bodies (near the surface of the earth) experience constant acceleration.

In his *Treatise on Configurations* Oresme introduced, at least implicitly, four innovative ideas:

(1) the measurement of diverse types of physical variables (such as temperature, density, velocity) by means of line segments (in lieu of real numbers, following Greek precepts);

(2) some notion of a functional relationship between variables (e.g., velocity as a function of time);

(3) a diagrammatic or graphical representation of such a functional relationship. This may be regarded as a partial step towards the introduction of a coordinate system;

(4) a conceptual process of "integration" or continuous summation to calculate distance as the area under a velocity-time graph, albeit Oresme only had the technical machinery to perform this calculation in the case of uniformly accelerated motion.

Mature versions of these incipient ideas played key roles in the seventeenth century development of the calculus. The work of Oresme and the Merton scholars on motion was widely disseminated in Europe for the next two centuries, and undoubtedly led to the work of Galileo (once thought to have been original with him), who assembled the medieval components into a new science of mechanics. For example, the *Third Day* of Galileo's *Discourses on Two New Sciences* (1638) begins with the mean speed theorem, with a proof and accompanying geometric diagram that are strikingly similar to those of Oresme, and proceeds to the distance formula $s = at^2/2$ for uniformly accelerated motion from rest (see Equation (9)), from which the law of odd numbers is then derived (as in Exercise 6). Chapter II of Clagett [7], entitled "The Configuration Doctrine in Historical Perspective," gives a detailed account of the origins and subsequent influence of the medieval geometry of motions.

Medieval Infinite Series Summations

The subject of infinite series fascinated medieval philosophers and mathematicians, appealing to both their interest in the infinite and their disputatious delight with apparent paradoxes. The work of the Merton scholars on the latitude of forms led naturally to various infinite series problems. For example, Swineshead solved a problem that, when stated in terms of motion, reads as follows.

> If a point moves throughout the first half of a certain time interval with a constant velocity, throughout the next quarter of the interval at double the initial velocity, throughout the following eighth at triple the initial velocity, and so on ad infinitum; then the average velocity during the whole time interval will be double the initial velocity.

Taking both the time interval and the initial velocity as unity, this is equivalent to the summation

$$\frac{1}{2} + \frac{2}{4} + \frac{3}{8} + \cdots + \frac{n}{2^n} + \cdots = 2. \tag{10}$$

Swineshead gave a long and tedious verbal proof of (10). It is equivalent to arguing that the effect of doubling the velocity during the last half of the interval is equivalent to that of doubling it during the first half of the interval; the additional effect (over doubling) of tripling the velocity during the last quarter of the interval is equivalent to that of doubling it during the second subinterval (of length one-fourth); the additional effect (over tripling) of quadrupling it during the last eighth of the interval is equivalent to that of doubling it during the third subinterval (of length one-eight); and so on ad infinitum. Hence the total cumulative effect is the same as that of doubling the initial velocity during all of the subintervals.

This appears to be the first infinite series summation, other than geometric series such as

$$1 + \frac{1}{4} + \cdots + \frac{1}{4^n} + \cdots = \frac{4}{3}, \tag{11}$$

which Archimedes effectively used in the *Quadrature of the Parabola* (albeit without actually extending the sum to infinity). In a tract written around 1350, Oresme gave the more general geometric series

$$\frac{a}{k} + \frac{a}{k}\left(1 - \frac{1}{k}\right) + \cdots + \frac{a}{k}\left(1 - \frac{1}{k}\right)^n + \cdots = a \tag{12}$$

(where k is an integer greater than one) which includes (11) as a special case (why?). He stated (12) verbally as follows:

> If an aliquot part [one kth] should be taken from some quantity [a], and from the first remainder such a part is taken, and from the second

remainder such a part is taken, and so on into infinity, such a quantity
would be consumed exactly—no more, no less—by such a mode of
subtraction ([10], p. 133).

The proof is that, after the first part is subtracted, the remainder is
$a(1-1/k)$; after a kth part of this is subtracted, the (second) remainder is
$a(1-1/k)^2$; etc. Each subtraction multiplies the previous remainder by
$(1-1/k)$, so the nth remainder is $a(1-1/k)^n$. This means that

$$\frac{a}{k}\left[1+\left(1-\frac{1}{k}\right)+ \cdots +\left(1-\frac{1}{k}\right)^{n-1}\right] + a\left(1-\frac{1}{k}\right)^n = a.$$

Since $a(1-1/k)^n$ obviously approaches zero as n goes to infinity, (12)
follows, as desired. Oresme finds interesting the corollary that "if one-
thousandth part of a foot were taken away [or removed], then [if] one-
thousandth part of the remainder of this foot [were removed], and so on
into infinity, exactly one foot would be subtracted from this [original
foot]."

In the same tract Oresme proves that the harmonic series

$$1+\frac{1}{2}+\frac{1}{3}+ \cdots +\frac{1}{n}+ \cdots$$

diverges, meaning that if the successive terms were added one-by-one then,
as he puts it, "the whole would become infinite." In proof of this he points
out that the sum of $\frac{1}{3}$ and $\frac{1}{4}$ is greater than $\frac{1}{2}$, as is the sum of the next four
terms $\frac{1}{5}$ through $\frac{1}{8}$, as is the sum of the next eight terms $\frac{1}{9}$ through $\frac{1}{16}$, etc.
This was the first example of a divergent series which "has a chance" of
converging because its terms approach zero.

In his *Treatise on Configurations*, Oresme gave a geometric method for
summing series (10). Figure 6 shows two dissections of the configuration or
graph of Swineshead's motion, with velocity 1 during the first half of the
unit time interval, velocity 2 during the next quarter, velocity 3 during the
next eight, etc. Since it is clear that $a(A_n) = n/2^n$ and $a(B_n) = 1/2^n$ for each

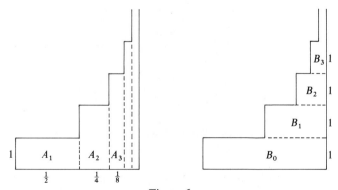

Figure 6

n, we see that

$$\frac{1}{2} + \frac{2}{4} + \frac{3}{8} + \cdots + \frac{n}{2^n} + \cdots = \sum_{n=1}^{\infty} a(A_n)$$

$$= \sum_{n=0}^{\infty} a(B_n)$$

$$= 1 + \frac{1}{2} + \frac{1}{4} + \cdots + \frac{1}{2^n} + \cdots$$

$$= 2,$$

since the latter series is the geometric series (12) with $a = 1$, $k = 2$.

EXERCISE 7. Apply (12) to show that

$$\frac{3}{4} + \frac{3}{16} + \frac{3}{64} + \cdots + \frac{3}{4^n} + \cdots = 1.$$

EXERCISE 8. Apply Oresme's geometric method to show that

$$\frac{3}{4} + 2 \cdot \frac{3}{16} + 3 \cdot \frac{3}{64} + \cdots + n \cdot \frac{3}{4^n} + \cdots = \frac{4}{3}.$$

Think of a motion with velocity 1 during the first $\frac{3}{4}$ of an unit time interval, with velocity 2 during the next $\frac{3}{16}$ of the interval, velocity 3 during the next $\frac{3}{64}$, etc.

The study of infinite series continued during the fifteenth and sixteenth centuries in the mode of Swineshead and Oresme, without significant advance over their exclusively verbal and geometrical techniques. The principal contribution of these early infinite series investigations lay not in the particular results obtained, but in the encouragement of a new point of view—the free acceptance of infinite processes in mathematics. Medieval currents of thought thus prepared the way for the more significant work on infinite series and processes of the seventeenth century, when a more potent arsenal of arithmetic and algebraic techniques was available.

The Analytic Art of Viète

The scientific and cultural Renaissance of the fifteenth and sixteenth centuries is often associated with the increased availability of the ancient Greek classics resulting from the invention of the printing press. In mathematics, however, the Renaissance consisted largely of rapid progress in the area of algebra, and this progress stemmed less from restored classical traditions than from the practical arithmetic and algebra of late medieval commercial circles that was based on problem-solving methods dating back to al-Khowarizmi. The publication in 1545 of Cardan's *Ars Magna* served to publicize the solutions by del Ferro and Tartaglia of cubic equations and by Ferrari of quartic equations. These exciting discoveries stimulated an accelerated development of algebraic techniques.

The algebra of this time was still largely verbal rather than symbolic, although the use of abbreviations (such as the Italian *p* and *m* for plus and minus) and symbols (such as the German + and −) was gradually emerging. In terms of its problems, the algebra of the early sixteenth century concentrated on finding the unknown in a given equation with specific numerical coefficients. As a consequence, algebra was still essentially a "bag of tricks" rather than a general method, because every special case required a different trick. The idea of studying a general equation representing a whole class of equations had not yet made its appearance.

In order to study the general cubic equation, for example, it is necessary to distinguish between the roles of the unknown variable whose value is sought and of the coefficients which are parameters in the problem—their values are unspecified even though they are assumed known in advance. This crucial idea, of a clear-cut distinction between parameters (known) and variables (unknown), was contributed by the Frenchman François Viète (1540–1603). In his *Introduction to the Analytic Art* of 1591 (see the book of Klein [11] for an English translation) he wrote:

> In order that [the setting up of equations] be aided by some art, it is necessary that the given [parameters] be distinguished from the unknown [variables] being sought by a constant, perpetual, and highly conspicuous convention (*symbolo*), such as by designating the [variables] being sought by the letter *A* or some other vowel *E, I, O, U, V*, and the given [parameters] by the letters *B, G, D*, or other consonants.

Thus Viète systematically used vowels for variables and consonants for parameters. In the designation of algebraic operations his notation was "syncopated," involving a combination of verbal abbreviations (such as *A quadratus* and *A cubus* for A^2 and A^3) and symbols (such as + and −). For example he would write

$$A \ cub + B \ plano \ in \ A \ aequatur \ C \ in \ A \ quad + D \ solido$$

for

$$A^3 + BA = CA^2 + D.$$

The terms *plano* and *solido* are included to preserve homogeneity of degree in the equation; the symbol = for equality (*aequalis*) was not yet in common use, although it had been introduced in 1557 by the Englishman Robert Recorde. The transition to a fully symbolic algebraic notation took place during the interval between Viète and Descartes.

Although his operational notation was still somewhat primitive by modern standards, Viète's clarification of the distinct roles of variables and parameters was an important step towards the seventeenth century reorientation of mathematics, from the study of particular problems to the search for general methods. The ability to deal with parameters as such focused attention on solution procedures rather than specific solutions themselves, and on questions concerning relationships between different problems. This shift in emphasis, from the particular to the general, was a necessary ingredient for the algorithmic approach that characterizes the calculus.

The Analytic Geometry of Descartes and Fermat

The final step in preparation for the new infinitesimal mathematics, and the most far-reaching one, was the origination of analytic geometry by Rene Descartes (1596–1650) and Pierre de Fermat (1601–1665). Descartes' *Geometry* was published in 1637 as one of three appendices to his *Discourse on the Method* (of Reasoning Well and Seeking Truth in the Sciences). In the same year Fermat sent to his correspondents in Paris his *Introduction to Plane and Solid Loci*. These two essays established the foundations for analytic geometry. Although Fermat's work was more systematic in some respects, it was not actually published until 1679 after his death, and for this reason we speak today of Cartesian geometry rather than Fermatian geometry.

The central idea of analytic geometry is the correspondence between an equation $f(x, y) = 0$ and the locus (generally a curve) consisting of all those points whose coordinates (x, y) relative to two fixed perpendicular axes satisfy the equation. Actually, neither Descartes nor Fermat systematically used two coordinate axes in the way now standard. The closest either came is indicated by Fermat's guiding principle:

> Whenever in a final equation two unknown quantities are found, we have a locus, the extremity of one of these describing a line, straight or curved.

For Fermat (as well as Descartes) the two unknown quantities in an equation were line segments rather than numbers. One of these was measured to the right from a reference point on a horizontal axis, and the second was placed as a vertical ordinate at the endpoint of the first (Fig. 7). Fermat's principle then says that the endpoint of the ordinate describes the curve corresponding to the given equation. Descartes' general practice was similar, so both, in fact, dealt with "ordinate geometry" rather than coordinate geometry.

Fermat adhered to the algebraic notation of Viète, and designated his variables as A and E instead of x and y. However, Descartes used the fully symbolic algebraic notation that is standard today (or, more accurately, we use Descartes' notation), with the single exception that he wrote ∞ instead of $=$ for equality. He standardized the exponential notation for powers,

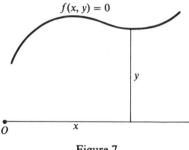

Figure 7

and initiated the common practice of using letters near the beginning of the alphabet for parameters and those near the end for variables.

The aim of both Descartes and Fermat was to apply the methods of Renaissance algebra to the solution of problems in geometry. Descartes stated the plan as follows ([13], pp. 6–8):

> If, then, we wish to solve any problem, we first suppose the problem already affected, and give names [symbols] to all the lines that seem needful for its construction—to those that are unknown as well as to those that are known. Then, making no distinction between known and unknown lines, we must unravel the difficulty in any way that shows most naturally the relations between these lines, until we find it possible to express a single quantity in two ways. This will constitute an equation, since the terms of one of these two expressions are together equal to the terms of the other.

Thus Descartes started with a geometrical problem, ordinarily involving a given curve, defined either as a static locus in the usual Greek fashion or in terms of uniform continuous motion (as the Archimedean spiral). His procedure was to translate the geometrical problem into the language of an algebraic equation, then simplify and finally solve this equation.

Whereas Descartes ordinarily began with a curve and derived its algebraic equation, Fermat ordinarily began with an algebraic equation and derived from it the geometric properties of the corresponding curve. For example, he started with the general second degree equation in two variables,

$$ax^2 + bxy + cy^2 + dx + ey + f = 0, \qquad (13)$$

showed by translation and rotation techniques that its locus is a conic section (except for degenerate cases), and classified the various cases of (13) as to whether this conic section is an ellipse, hyperbola, or parabola. A discussion of this work can be found in Chapter III of Mahoney's mathematical biography of Fermat [12].

Thus the works of Descartes and Fermat, taken together, encompass the two complementary aspects of analytic geometry—studying equations *by means of* curves, and studying curves *defined by* equations. An important common feature of their work in analytic geometry was their concentration on *indeterminate* equations involving *continuous* variables. Viète, for example, had studied only *determinate* equations, in which the "variable", although unknown, is actually a fixed constant to be found.

The notion of a variable, as first emphasized explicitly by Descartes and Fermat, was indispensable to the development of the calculus—the subject can hardly be discussed except in terms of continuous variables. Moreover, analytic geometry opened up a vast virgin territory of new curves to be studied, and called for the invention of algorithmic techniques for their systematic investigation. Whereas the Greek geometers had suffered from a paucity of known curves, a new curve could now be introduced by the simple act of writing down a new equation. In this way, analytic geometry provided both a much broadened field of play for the infinitesimal techniques of the seventeenth century, and the technical machinery needed for their elucidation.

References

[1] M. E. Baron, *The Origins of the Infinitesimal Calculus*. London: Pergamon, 1969, Chapter 2.
[2] C. B. Boyer, *History of the Calculus*. New York: Dover, 1959, Chapter III.
[3] C. B. Boyer, *History of Analytic Geometry*. New York: Scripta Mathematica, 1956, Chapters III–V.
[4] C. B. Boyer, *A History of Mathematics*. New York: Wiley, 1968, Chapters XI–XVII.
[5] M. Clagett, *The Science of Mechanics in the Middle Ages*. University of Wisconsin Press, 1959, Chapters 4–6.
[6] M. Clagett, *Archimedes in the Middle Ages*, Vol. 1. University of Wisconsin Press, 1964.
[7] M. Clagett, *Nicole Oresme and the Medieval Geometry of Qualities and Motions*. University of Wisconsin Press, 1968, Chapters I and II.
[8] E. J. Dijksterhuis, *The Mechanization of the World Picture*. Oxford University Press, 1961, Part II.
[9] E. Grant, *Physical Science in the Middle Ages*. New York: Wiley, 1971, Chapters I–IV.
[10] E. Grant, *A Source Book in Medieval Science*. Cambridge, MA: Harvard University Press, 1974.
[11] J. Klein, *Greek Mathematical Thought and the Origins of Algebra*. MIT Press, 1968.
[12] M. S. Mahoney, *The Mathematical Career of Pierre de Fermat*. Princeton, NJ: Princeton University Press, 1973. Chapters II and III.
[13] D. E. Smith and M. L. Latham, *The Geometry of Rene Descartes*. Chicago: Open Court, 1925 (Dover reprint).
[14] B. L. van der Waerden, *Science Awakening*. Oxford University Press, 1961, Chapter VIII.

4

Early Indivisibles and Infinitesimal Techniques

Introduction

During the late middle ages Euclid's *Elements* and the works of Archimedes had been extant, but not always generally accessible and never fully mastered. The sixteenth century saw, finally, the wide dissemination and serious study of these Greek mathematical masterworks. By the latter part of the century, the understanding of Archimedes' work had reached the point that further progress along the lines of classical Greek mathematics was possible. During the century preceding Newton and Leibniz the method of exhaustion was refined and applied by numerous mathematicians to a wide variety of new quadrature, cubature, and rectification problems (see the reviews of this work by Baron [2], Chapter 3, and Whiteside [12], pp. 331–348).

Although Archimedes' accomplishments provided the chief inspiration for the resumption of mathematical progress, the time was ripe for the development of simpler new methods, ones that could be applied to the investigation of area and volume problems with greater ease than could the method of exhaustion with its tedious double *reductio ad absurdum* proofs. While continuing to regard Archimedean proofs as the ultimate models of rigor and precision, the Renaissance mathematical mind was more interested in quick new results and methods of rapid discovery than in the stringent requirements of rigorous proof. The common view of the period was expressed in 1657 by Huygens as follows:

> In order to achieve the confidence of the experts it is not of great interest
> whether we give an absolute demonstration or such a foundation of it that
> after having seen it they do not doubt that a perfect demonstration can be

given. I am willing to concede that it should appear in a clear, elegant, and ingenious form, as in all works of Archimedes. But the first and most important thing is the mode of discovery itself, which men of learning delight in knowing. Hence it seems that we must above all follow that method by which this can be understood and presented most concisely and clearly. We then save ourselves the labor of writing, and others that of reading—those others who have no time to take notice of the enormous quantity of geometrical inventions which increase from day to day and in this learned century seem to grow beyond bounds if they must use the prolix and perfect method of the Ancients. (Struik [11], p. 189).

The Greek "horror of the infinite" had prevented the development of a usable theory of limits to replace the ubiquitous double *reductio ad absurdum* proofs. But, as a result of the medieval scholastic speculations on infinity and the continuum, seventeenth century mathematicians were no longer reluctant to introduce infinitesimal techniques. Whereas the Greek insistence on absolute rigor had banished irrational magnitudes from the field of number (and hence number from geometry), irrational numbers now came to be freely employed, even though they still had no logical basis as numbers, and could only be interpreted rigorously as geometrical magnitudes. In addition, the symbolic algebra of Viète and Descartes facilitated the development of formal techniques that emphasized computational method more than logical proof. Finally, the algebraic representation of curves (analytic geometry) permitted the rapid and easy formulation of new and diverse area and volume problems for investigation.

This rich amalgam of mathematical ingredients—Archimedean problems, algebraic computational techniques, and the free use of intuitive concepts of the infinite—produced a profusion of powerful (if loosely based) infinitesimal methods for the solution of area and volume problems during the "century of anticipation" preceding the time of Newton and Leibniz. As we will see in this chapter, these developments constituted a gradual *arithmetization* of problems whose treatment in antiquity had been wholly geometric in character and approach.

Johann Kepler (1571–1630)

Kepler is most famous for his discovery of the laws of planetary motion, to the effect that (I) a planet moves along an elliptical orbit with the sun at one focus of the ellipse, in such a way that (II) the radius vector from the sun to the planet sweeps out area at a constant rate, with (III) the squares of the periods of revolution of any two planets being proportional to the cubes of the major semi-axes of their orbits. Newton later showed that

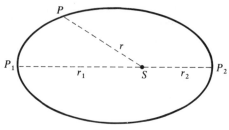

Figure 1

these three laws, which Kepler deduced from observational data, follow from the inverse-square law of gravitation.

The second law (of areas), published in 1609, was derived by means of a curious combination of compensating errors. Kepler first deduced from astronomical observations that, when a planet is at either of its apsides—the nearest and farthest points on its orbit from the sun, its velocity v is inversely proportional to its distance r from the sun. That is, if v_1 and v_2 are its velocities at the aphelion P_1 and perihelion P_2 (Fig. 1), then there is a constant k such that

$$v_1 = \frac{k}{r_1} \quad \text{and} \quad v_2 = \frac{k}{r_2}. \tag{1}$$

He next purported to prove that this is true at every point P of the orbit,

$$v = \frac{k}{r}. \tag{2}$$

However, this "theorem" is false—actually the velocity at P is inversely proportional (as it turns out) to the perpendicular distance from S to the tangent line to the ellipse at P (this distance being equal to r only at the apsides where the tangent line and radius vector happen to be perpendicular).

Nevertheless, Kepler proceeded on the basis of the incorrect relation (2). In order to calculate the time t required to traverse an arc $\overset{\frown}{PQ}$ of its orbit

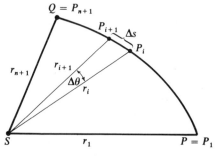

Figure 2

(Fig. 2), he divided the arc into a large number of subarcs of equal lengths Δs. If r_i is the distance SP_i from the sun to the initial point P_i of the ith subarc $\overset{\frown}{P_iP_{i+1}}$, v_i the velocity at P_i, and t_i the time required for the planet to traverse this subarc, then (2) gives

$$t = \sum_{i=1}^{n} t_i \cong \sum_{i=1}^{n} \frac{\Delta s}{v_i} = \frac{1}{k} \sum_{i=1}^{n} r_i \Delta s. \tag{3}$$

Thus the (incorrect) relation $v = k/r$ implies that the time t is proportional to the sum $\sum_1^n r_i$ of the radii.

At this point Kepler mistakenly assumes that the same sum is proportional to the area SPQ, saying

> Since I was aware that there exists an infinite number of points on the orbit and accordingly an infinite number of distances [from the sun] the idea occurred to me that the sum of these distances is contained in the *area* of the orbit. For I remember that in the same manner Archimedes too divided the circle into an infinite number of triangles (quoted by Koestler [8], p. 327).

His thought here is reminiscent of the heuristic derivation of the circle area formula by the fifth century B.C. Greeks (rather than Archimedes). If the ith piece or "indivisible" of the area were a triangle with base r_i and height Δs, then it would follow that

$$A \cong \frac{1}{2} \sum_{i=1}^{n} r_i \Delta s, \tag{4}$$

and this together with (3) would imply that

$$A = ht \qquad (h = k/2). \tag{5}$$

This is how Kepler actually obtained his (correct) second law to the effect that the area swept out is proportional to the time elapsed. For additional discussion of this comedy of errors, see the article by Aiton [1] or the books of Dreyer ([7], pp. 387–388) and Koestler ([8], pp. 327–328).

EXERCISE 1. Explain why (4) is false in general. Note that (4) gives

$$A = \frac{1}{2} \int r\, ds$$

in integral notation, whereas the (correct) area formula in polar coordinates is

$$A = \frac{1}{2} \int r^2 d\theta.$$

Does Δs equal $r\Delta\theta$ for a small segment of an arbitrary curve?

Kepler's more systematic work on the calculation of areas and volumes by infinitesimal techniques was undertaken for more prosaic reasons than the study of the harmony of the celestial spheres—the original motivation

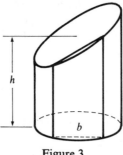

Figure 3

for his astronomical investigations. His treatise *Nova stereometria doliorum vinariorum* (New solid geometry of wine barrels), published in 1615, was intended to enable wine merchants to accurately gauge the volumes of their barrels. This work concentrates on solids of revolution, and includes determinations of the (exact or approximate) volumes of over ninety such solids.

Kepler's approach in the *stereometria* is to dissect a given solid into an (apparently) infinite number of infinitesimal pieces, or solid "indivisibles", of a size and shape convenient to the solution of the particular problem. For example, he regards the sphere as composed of an infinite number of pyramids, each having its vertex at the center and its base on the surface of the sphere, and height equal to the radius r of the sphere. Adding up the volumes of these pyramids, the formula for the volume of a pyramid immediately gives $V = Ar/3 = 4\pi r^3/3$, where $A = 4\pi r^2$ is the surface area of the sphere.

EXERCISE 2. Derive similarly the formula for the volume of a circular cone by considering a dissection of its base into infinitely many infinitesimal triangles.

EXERCISE 3. Consider a cylindrical segment with nonparallel bases as in Figure 3. Derive the volume formula $V = \pi r^2 h$ by considering it to be the sum of infinitely many thin vertical slices as indicated. By the formula for the area of a trapezoid the volume of such a slice is bht, where t is its thickness.

Kepler showed that the volume of an anchor ring or torus, generated by revolving a circle of radius a around a vertical axis at a distance b from its center, is equal to the product of the area of the circle and the distance traveled by its center (the theorem of Pappus). That is,

$$V = (\pi a^2)(2\pi b) = 2\pi^2 a^2 b. \qquad (6)$$

He derived (6) by dissecting the torus into infinitely many thin vertical circular slices by means of planes through the axis of revolution. Each such slice is thinner on the inside (nearest the axis) and thicker on the outside. (Fig. 4). Kepler assumes the volume of such a slice is $\pi a^2 t$, where $t = (t_1 +$

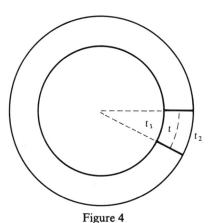

Figure 4

$t_2)/2$, the average of its minimum and maximum thickness. Then t is the thickness of the slice at its center, so the volume of the torus is $V = (\pi a^2)(\Sigma t) = (\pi a^2)(2\pi b)$.

EXERCISE 4. Derive the formula $\pi a^2 t$ for the volume of a vertical circular slice of the torus, by considering it to be composed of narrow horizontal slices with trapezoidal cross-sections.

EXERCISE 5. Give an alternative derivation of (6) as follows. Dissect the torus into infinitely many thin vertical cylindrical shells, corresponding to a dissection of the generating circle into narrow vertical rectangular strips (Fig. 5). Calculate pairwise the volumes of cylindrical shells corresponding to pairs of strips that are symmetrically located relative to the vertical diameter of the circle.

An English translation of part of the *stereometria* may be found in Struik's source book ([11], pp. 192–197). Confident that his results could be established rigorously if necessary, Kepler indulged in free play with infinitesimals to calculate the volumes of a wide variety of solids of revolution, saying "We could obtain absolute and in all respects perfect demonstrations from these books of Archimedes themselves, were we not repelled by the thorny reading thereof."

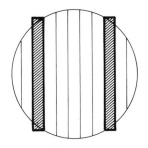

Figure 5

Cavalieri's Indivisibles

The systematic use of infinitesimal techniques for area and volume computation was popularized by two influential books written by Bonaventura Cavalieri (1598–1647)—his *Geometria indivisibilibus* (Geometry of indivisibles) of 1635 and his *Exercitationes geometricae sex* (Six geometrical exercises) of 1647. English translations of brief but illustrative sections of these two lengthy works may be found in Struik's source book ([11], pp. 209–219).

Cavalieri's methods differed in two significant ways from those of Kepler. Firstly, Kepler imagined a given geometrical figure to be decomposed into infinitesimal figures, whose areas or volumes he then added up in some *ad hoc* way to obtain the area or volume of the given figure. However, Cavalieri proceeded by setting up a one-to-one correspondence between the indivisible elements of *two* given geometrical figures. If corresponding indivisibles of the two given figures had a certain (constant) ratio, he concluded that the areas or volumes of the given figures had the same ratio. Typically the area or volume of one of the figures was known in advance, so this gave the other.

Secondly, Kepler thought of a geometrical figure as being composed of indivisibles of the same dimension (i.e., infinitesimal areas or volumes), as might be conceived to result from some process of successive subdivision leading eventually to ultimate indivisible units. However, Cavalieri generally considered a geometrical figure to be composed of an indefinitely large number of indivisibles of lower dimension. Thus he regarded an area as consisting of parallel and equidistant line segments, and a volume as consisting of parallel and equidistant plane sections, without making it entirely clear whether these indivisible units have thickness or not. Usually they appeared not to, but on at least one occasion he suggested that they might, mentioning the analogy of the parallel threads in a piece of cloth, or the parallel pages filling up the thickness of a book. In contrast to medieval speculators, he was less interested in questions as to the precise nature or existence of indivisibles, than in their pragmatic use as a device for obtaining computational results. Rigor, he wrote in the *Exercitationes*, is the affair of philosophy rather than mathematics.

Cavalieri's method of comparing two geometrical figures by comparing the indivisibles of one with the indivisibles of another is based on a principle that is still known as Cavalieri's Theorem:

> If two solids have equal altitudes, and if sections made by planes parallel
> to the bases and at equal distances from them are always in a given ratio,
> then the volumes of the solids are also in this ratio.

He attempted to prove this theorem by a superposition argument that involved moving one of the solids piece by piece so as to superimpose it on the other one.

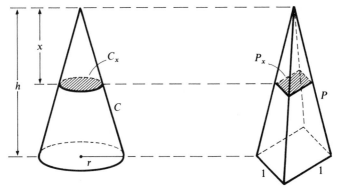

Figure 6

The practical effect of Cavalieri's Theorem is to gloss over or "hide" the role of limit processes in volume computations. For example, if two triangular pyramids have the same altitude h and the same base area A, then an easy similarity argument shows that their triangular cross-sections at equal heights have equal areas, so Cavalieri's Theorem implies that the two pyramids have equal volumes. Recall (from Chapter 1) that this result is the principal step in the proof of the pyramid volume formula $V = Ah/3$.

To derive the formula for the volume of a circular cone C with base radius r and height h, we compare it with a pyramid P with height h and unit square base. If C_x and P_x are the sections indicated in Figure 6, then a similarity computation gives

$$a(C_x) = \frac{\pi r^2 x^2}{h^2} \quad \text{and} \quad a(P_x) = \frac{x^2}{h^2}.$$

Thus $a(C_x) = \pi r^2 a(P_x)$, so Cavalieri's Theorem implies that $v(C) = \pi r^2 v(P) = \pi r^2 h/3$, since $v(P) = h/3$.

EXERCISE 6. Derive the formula for the volume of a sphere by comparing a hemisphere of radius r with the solid that is obtained from a cylinder of radius and height r by removing an inverted cone whose base is the top of the cylinder and whose vertex is the center of the base of the cylinder.

EXERCISE 7. A spherical ring is obtained from a solid sphere by boring out a cylindrical hole whose axis is the vertical diameter of the sphere. Find the volume of the spherical ring by comparing it with a sphere whose diameter is equal to the height of the ring.

EXERCISE 8. Consider the solid intersection of the unit cylinders $x^2 + z^2 = 1$ and $y^2 + z^2 = 1$ along the x- and y-axes. Find its volume by comparing it with the solid that is obtained from a rectangular parallelepiped with unit square base and height 2, by removing two square pyramids having the top and bottom of the parallelepiped as their bases, and having a common vertex at the center of the parallelepiped. Compare your answer with the result Archimedes obtained using his mechanical method (Chapter 2).

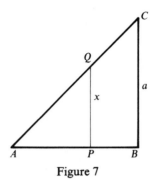

Figure 7

In addition to his technique of comparing two solids by comparing their cross-sections, Cavalieri devised a method of calculating the volume of a single solid in terms of its cross-sections. This latter method was based on a formal procedure for computing what may be referred to as "sums of powers of lines" in a triangle parallel to its base. This procedure, though far from rigorous, led Cavalieri to a correct result equivalent to the basic integral

$$\int_0^a x^n \, dx = \frac{a^{n+1}}{n+1}. \tag{7}$$

For example, consider the triangle ABC with base and height a, and typical vertical section PQ of length x (Fig. 7). Then, in Cavalieri's sense of indivisibles, the triangle is the sum of all such segments, so we might write

$$a(\triangle ABC) = \sum_A^B x,$$

a concise notation for what Cavalieri said in verbose geometrical terminology. If P is a pyramid with vertex A and its base being a square on BC, then its cross-section at a distance x from the vertex has area x^2, so we similarly write

$$v(P) = \sum_A^B x^2,$$

thinking of the pyramid as the sum of its cross-sections. The same sum, of the squares of the lines in the triangle, represents also the area under the parabola $y = x^2$ (Fig. 8), since its typical vertical section has length x^2. If Q is the solid obtained by revolving the parabola around its base AB, then its cross-section at a distance x from the "vertex" A has area πx^4, so we write

$$v(Q) = \sum_A^B \pi x^4 = \pi \sum_A^B x^4.$$

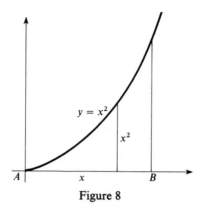

Figure 8

These examples indicate the way in which a wide variety of area and volume problems can be solved in terms of formal sums of powers of lines in a triangle.

To outline Cavalieri's method for computing these formal sums, we start with a square $ABCD$ with edgelength a, divided into two triangles by its diagonal AC (Fig. 9). If x and y denote the lengths of typical sections PQ and QR of these congruent triangles, then $x+y=a$, so

$$\sum_{A}^{B} a = \sum_{A}^{B} (x+y) = \sum_{A}^{B} x + \sum_{A}^{B} y = 2\sum_{A}^{B} x,$$

because $\sum x = \sum y$ by symmetry. Hence

$$\sum_{A}^{B} x = \frac{1}{2} \sum_{A}^{B} a = \frac{1}{2} a^2 \tag{8}$$

because $\sum a$ represents the area of the square.

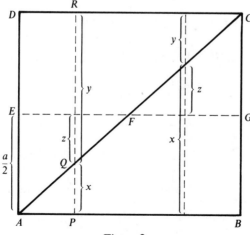

Figure 9

To compute Σx^2 we start with

$$\sum_A^B a^2 = \sum_A^B (x+y)^2 = \sum x^2 + 2\sum xy + \sum y^2$$

$$= 2\sum x^2 + 2\sum xy \quad \text{(symmetry)}$$

$$= 2\sum x^2 + 2\sum \left(\frac{a^2}{4} - z^2\right),$$

where $x = (a/2) - z$, $y = (a/2) + z$ (see Fig. 9). Hence

$$\sum a^2 = 4\sum x^2 - 4\sum z^2. \tag{9}$$

Here Σz^2 is a sum of squares of lines in the *two* triangles AEF and CFG. But the sum of z^2 over *one* of these triangles represents the volume of a pyramid with dimensions one-half those of the pyramid whose volume is $\Sigma_A^B x^2$. Therefore

$$\sum z^2 = 2 \cdot \frac{1}{8} \sum_A^B x^2 = \frac{1}{4} \sum_A^B x^2.$$

Substitution of this result into (9) gives

$$\sum_A^B x^2 = \frac{1}{3} \sum_A^B a^2 = \frac{1}{3} a^3, \tag{10}$$

because Σa^2 represents the volume of a cube with edge a.

Proceeding to sums of cubes of lines, we have

$$\sum a^3 = \sum (x+y)^3 = \sum x^3 + 3\sum x^2y + 3\sum xy^2 + \sum y^3$$

$$\sum a^3 = 2\sum x^3 + 6\sum x^2y \quad \text{(symmetry)}. \tag{11}$$

To evaluate Σx^2y we resort to the following trick.

$$\sum a^3 = a\sum a^2 = a\left(2\sum x^2 + 2\sum xy\right)$$

$$= a\left(\frac{2}{3}\sum a^2 + 2\sum xy\right)$$

$$= \frac{2}{3}\sum a^3 + 2\sum (x+y)xy$$

$$\sum a^3 = \frac{2}{3}\sum a^3 + 4\sum x^2y \quad \text{(symmetry)}.$$

Therefore $\Sigma x^2y = \Sigma a^3/12$, and substitution of this result into (11) gives

$$\sum_A^B x^3 = \frac{1}{4} \sum_A^B a^3 = \frac{1}{4} a^4. \tag{12}$$

Formulas (8), (10) and (12) are the first three instances of the general formula

$$\sum_A^B x^n = \frac{a^{n+1}}{n+1}, \tag{13}$$

equivalent to (7), which Cavalieri inferred after verifying it case by case up to $n = 9$. On the basis of this result, he could immediately write down the area under the curve $y = x^n$ (n a positive integer) over the unit interval,

$$A = \sum_0^1 x^n = \frac{1}{n+1},$$

and the volume of the solid obtained by revolving this area around the x-axis,

$$V = \pi \sum_0^1 x^{2n} = \frac{\pi}{2n+1}.$$

This unification and generalization of previous results constituted a giant step towards the development of the algorithmic procedures of the calculus.

EXERCISE 9. Expand $\Sigma a^4 = \Sigma(x+y)^4$ to derive the result

$$\sum x^4 = \frac{1}{5} \sum a^4 = \frac{1}{5} a^5.$$

EXERCISE 10. Consider the area in the xy-plane bounded by the line $y = 1$ and the parabola $y = x^2$. Let P be the solid obtained by revolving this area around the line $y = 1$. Considering P as the sum of its circular cross-sections, apply Cavalieri's results to obtain $v(P) = 16\pi/15$.

Arithmetical Quadratures

We saw in Chapter 2 that Archimedes used the formulas for sums of integers and their squares,

$$1 + 2 + \cdots + n = \frac{n}{2}(n+1) \tag{14}$$

and

$$1^2 + 2^2 + \cdots + n^2 = \frac{n}{6}(n+1)(2n+1), \tag{15}$$

to establish quadrature results equivalent to the integrals

$$\int_0^a x\,dx = \frac{a^2}{2} \quad \text{and} \quad \int_0^a x^2\,dx = \frac{a^3}{3}.$$

Actually, all that is needed for these two quadratures are the immediate consequences

$$\lim_{n\to\infty} \frac{1+2+\cdots+n}{n^2} = \frac{1}{2}$$

and

$$\lim_{n\to\infty} \frac{1^2+2^2+\cdots+n^2}{n^3} = \frac{1}{3}$$

of formulas (14) and (15).

During the two decades following the publication of Cavalieri's first book in 1635, the French mathematicians Fermat, Pascal, and Roberval gave more or less rigorous proofs of Cavalieri's (conjectured) general formula

$$\int_0^a x^k \, dx = \frac{a^{k+1}}{k+1} \tag{16}$$

for the area under the generalized parabola $y = x^k$ (k a positive integer). Each of their proofs made use of the limit

$$\lim_{n\to\infty} \frac{1^k+2^k+\cdots+n^k}{n^{k+1}} = \frac{1}{k+1}, \tag{17}$$

involving the sum of the kth powers of the first n positive integers, to replace Cavalieri's intuitive arguments in terms of geometrical indivisibles with explicit arithmetical computations.

To see how it is that the arithmetical limit (17) implies the area formula (16), we subdivide the interval $[0, a]$ into n equal subintervals of length a/n, and construct the usual inscribed and circumscribed polygons P_n and Q_n (Fig. 10). P_n consists of rectangles with base a/n and heights $(a/n)^k, (2a/n)^k, \ldots, ((n-1)a/n)^k$, and Q_n consists of rectangles with

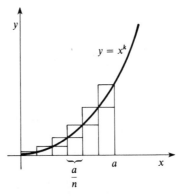

Figure 10

base a/n and heights $(a/n)^k$, $(2a/n)^k$, ..., $(na/n)^k$. Adding up the areas of these rectangles, we find that

$$a(P_n) = \frac{a^{k+1}}{n^{k+1}}\left(1^2 + 2^k + \cdots + (n-1)^k\right)$$

and

$$a(Q_n) = \frac{a^{k+1}}{n^{k+1}}\left(1^k + 2^k + \cdots + n^k\right).$$

Denoting by S the region under the curve $y = x^k$ over the interval $[0, a]$, we see that

$$a^{k+1}\frac{1^k + \cdots + n^k}{n^{k+1}} - \frac{a^{k+1}}{n} < a(S) < a^{k+1}\frac{1^k + \cdots + n^k}{n^{k+1}}.$$

Taking limits as $n \to \infty$, it now follows from (17) that $a(S) = a^{k+1}/(k+1)$ as desired.

EXERCISE 11. Show that (16) also follows from the inequality

$$1^k + 2^k + \cdots + (n-1)^k < \frac{n^{k+1}}{k+1} < 1^k + 2^k + \cdots + n^k.$$

Fermat derived formulas for $\Sigma_{i=1}^n i^k$ from a theorem concerning figurate numbers. The nth triangular number (the first type of figurate numbers) is

$$1 + 2 + \cdots + n = \frac{n}{2}(n+1).$$

The nth pyramidal number is the sum of the first n triangular numbers. In general, the nth figurate number of type k is the sum of the first n figurate numbers of type $k-1$. In a letter written in 1636 (see Mahoney [9], pp. 229–232), Fermat stated without proof that the nth figurate number of type k is given by

$$\sum_{i=1}^n \frac{i(i+1)\cdots(i+k-1)}{k!} = \frac{n(n+1)\cdots(n+k)}{(k+1)!}. \qquad (18)$$

EXERCISE 12. Regarding k as a fixed positive integer, prove formula (18) by induction on n.

Let us write

$$i(i+1)\cdots(i+k-1) = i^k + a_1 i^{k-1} + \cdots + a_{k-1}i, \qquad (19)$$

where the coefficients $a_1, \ldots, a_{k-1}$ are constants (depending, however, upon k). Then (18) becomes

$$\frac{1}{k!}\left[\sum_1^n i^k + a_1 \sum_1^n i^{k-1} + \cdots + a_{k-1}\sum_1^n i\right] = \frac{n(n+1)\cdots(n+k)}{(k+1)!},$$

from which we can solve for the recursion formula

$$\sum_{i=1}^{n} i^k = \frac{n(n+1)\cdots(n+k)}{k+1} - \left[a_1 \sum_{1}^{n} i^{k-1} + \cdots + a_{k-1} \sum_{1}^{n} i \right] \quad (20)$$

which gives the sum of the kth powers of the first n integers in terms of the sums of lower powers.

EXERCISE 13. Starting with $\sum_{1}^{n} i = n(n+1)/2$, apply the recursion formula (20) to compute $\sum_{1}^{n} i^2$, $\sum_{1}^{n} i^3$, and $\sum_{1}^{n} i^4$. You will have to compute the a_j's from (19), separately for each value of k.

EXERCISE 14. Apply (20) to prove by induction on k that

$$\sum_{i=1}^{n} i^k = \frac{n^{k+1}}{k+1} + \text{lower powers of } n.$$

Note that this fact suffices to establish the limit in (17).

In 1654 Blaise Pascal discovered the following more explicit recursion formula for sums of kth powers:

$$\binom{k+1}{k} \sum_{i=1}^{n} i^k + \binom{k+1}{k-1} \sum_{1}^{n} i^{k-1} + \cdots + \binom{k+1}{1} \sum_{1}^{n} i$$
$$= (n+1)^{k+1} - n - 1 \quad (21)$$

where $\binom{p}{q} = p!/q!(p-q)!$ is the usual binomial coefficient. See the article of Boyer ([5], p. 239) for Pascal's rhetorical statement of this formula, which he deduced by incomplete induction from number relationships in the "Pascal triangle".

EXERCISE 15. Apply the binomial formula to establish formula (21), starting with the following trick.

$$(n+1)^{k+1} - 1 = \sum_{i=1}^{n} \left[(i+1)^{k+1} - i^{k+1} \right]$$
$$= \sum_{i=1}^{n} \left[\sum_{p=0}^{k} \binom{k+1}{p} i^p \right].$$

EXERCISE 16. Starting with $\sum_{1}^{n} i = n(n+1)/2$, apply the recursion formula (21) to compute $\sum_{1}^{n} i^2$, $\sum_{1}^{n} i^3$, and $\sum_{1}^{n} i^4$.

EXERCISE 17. Use the binomial formula to expand $(n+1)^{k+1}$ on the right-hand side of (21), and then prove by induction on k that

$$\sum_{i=1}^{n} i^k = \frac{n^{k+1}}{k+1} + \frac{n^k}{2} + \text{lower powers of } n.$$

EXERCISE 18. Deduce from the previous exercise that, if n is sufficiently large, then

$$1^k + 2^k + \cdots + (n-1)^k < \frac{n^{k+1}}{k+1} < 1^k + 2^k + \cdots + n^k.$$

After stating formula (21) in verbal form, Pascal went on to remark that

> Any person at all familiar with the doctrine of indivisibles will perceive the results that one can draw from the above for the determination of curvilinear areas. Nothing is easier, in fact, than to obtain immediately the quadratures of all the types of parabolas and the measures of numberless other magnitudes.
>
> If then we extend to continuous quantities the results found for numbers, we will be able to state the following rules: ... The sum of a certain number of lines is to the square of the largest as 1 is to 2. The sum of the squares is to the cube of the largest as 1 is to 3. ... The sum of like powers of a certain number of lines is to the power of the next higher degree of the greatest of these as unity is to the exponent of this latter power (quoted from Boyer [5], p. 240).

Pascal's idea here appears to be that, when n is very large, the lower powers of n are negligible in comparison with the first term $n^{k+1}/(k+1)$ in the formula for $\sum_1^n i^k$. Therefore, when the area under the curve $y = x^k$ over $[0, a]$ is subdivided into a very large number n of (almost) rectangular strips of width $w = a/n$, it seems apparent that the area under the curve is

$$\left[w^k + (2w)^k + \cdots + (nw)^k \right] w = w^{k+1} \sum_{i=1}^n i^k$$

$$\cong \frac{(nw)^{k+1}}{k+1} = \frac{a^{k+1}}{k+1}.$$

This is essentially an abbreviation of our earlier derivation of this result by the method of exhaustion.

The Integration of Fractional Powers

The quadrature of curves of the form $y = x^k$, with k not necessarily a positive integer, was first attacked systematically by John Wallis (1616–1703) who was the Savilian professor of geometry at Oxford. In fact, rational and negative exponents were introduced by Wallis in his *Arithmetica Infinitorum* (The Arithmetic of Infinites) of 1655, which (as we will see in Chapter 7) had a decisive influence on Newton's early mathematical development.

On the basis of computations with arithmetical indivisibles similar to those described in the previous section, Wallis knew that the area under the curve $y = x^k$ (k a positive integer) over the unit interval is given by

$$\int_0^1 x^k \, dx = \lim_{n \to \infty} \frac{0^k + 1^k + \cdots + n^k}{n^k + n^k + \cdots + n^k}.$$

His approach to the determination of this limit was empirical. For example, in the case $k = 3$ he noted that

$$\frac{0^3 + 1^3}{1^3 + 1^3} = \frac{2}{4} = \frac{1}{4} + \frac{1}{4};$$

$$\frac{0^3 + 1^3 + 2^3}{2^3 + 2^3 + 2^3} = \frac{9}{24} = \frac{1}{4} + \frac{1}{8};$$

$$\frac{0^3 + 1^3 + 2^3 + 3^3}{3^3 + 3^3 + 3^3 + 3^3} = \frac{36}{108} = \frac{1}{4} + \frac{1}{12};$$

$$\frac{0^3 + 1^3 + 2^3 + 3^3 + 4^3}{4^3 + 4^3 + 4^3 + 4^3 + 4^3} = \frac{100}{320} = \frac{1}{4} + \frac{1}{16};$$

$$\frac{0^3 + 1^3 + \cdots + 5^3}{5^3 + 5^3 + \cdots + 5^3} = \frac{225}{750} = \frac{1}{4} + \frac{1}{20};$$

$$\frac{0^3 + 1^3 + \cdots + 6^3}{6^3 + 6^3 + \cdots + 6^3} = \frac{441}{1512} = \frac{1}{4} + \frac{1}{24}.$$

On the basis of this numerical evidence he concluded that

$$\frac{0^3 + 1^3 + \cdots + n^3}{n^3 + n^3 + \cdots + n^3} = \frac{1}{4} + \frac{1}{4n},$$

so the limit as $n \to \infty$ is $\frac{1}{4}$. After carrying out such computations for several small values of k he inferred (without further proof) that

$$\lim_{n \to \infty} \frac{0^k + 1^k + \cdots + n^k}{n^k + n^k + \cdots + n^k} = \frac{1}{k+1} \tag{22}$$

for all non-negative integral values of k.

In order to describe his next step, let us define the *index* $I\{\phi\}$ of a function ϕ by the equation

$$\lim_{n \to \infty} \frac{\phi(0) + \phi(1) + \cdots + \phi(n)}{\phi(n) + \phi(n) + \cdots + \phi(n)} = \frac{1}{I\{\phi\} + 1}, \tag{23}$$

assuming the limit exists. Then equation (22) simply says that the index of $\phi(x) = x^k$ is $I\{x^k\} = k$. Wallis then noted that, given a *geometric* progression of positive integral powers of x (such as $1, x^2, x^4, x^6$), the corresponding sequence of indexes is an *arithmetic* progression (0, 2, 4, 6). From this trivial observation he leapt boldly to the assumption that the same conclusion would follow for a geometric progression such as

$$1, \quad \sqrt[q]{x}, \quad (\sqrt[q]{x})^2, \quad \cdots, \quad (\sqrt[q]{x})^{q-1}, \quad x.$$

That is, the sequence of indexes

$$0 = I\{1\}, \quad I\{\sqrt[q]{x}\,\}, \quad \cdots, \quad I\{(\sqrt[q]{x}\,)^p\}, \quad \cdots, \quad I\{x\} = 1$$

should be an arithmetic progression, so it would follow that

$$I\{(\sqrt[q]{x}\,)^p\} = \frac{p}{q}, \quad (p \text{ and } q \text{ integers})$$

and hence from (23) that

$$\lim_{n \to \infty} \frac{(\sqrt[q]{0}\,)^p + (\sqrt[q]{1}\,)^p + \cdots + (\sqrt[q]{n}\,)^p}{(\sqrt[q]{n}\,)^p + (\sqrt[q]{n}\,)^p + \cdots + (\sqrt[q]{n}\,)^p} = \frac{1}{(p/q)+1} = \frac{q}{p+q}. \quad (24)$$

It was on these highly speculative grounds that Wallis associated the index or exponent p/q with the power $(\sqrt[q]{x}\,)^p$, leading to the now standard notation $(\sqrt[q]{x}\,)^p = x^{p/q}$. He also introduced irrational exponents, asserting that "If we suppose the index irrational, say $\sqrt{3}$, then the ratio is as 1 to $1 + \sqrt{3}$, etc."

EXERCISE 19. Show by exhaustion or indivisibles that the area under the curve $y = x^{p/q}$ over the unit interval is equal to the limit in Equation (24).

Wallis was able to verify (24) only for the special case $p = 1$. In this case it follows from Figure 11 and Exercise 19 that

$$\int_0^1 x^{1/q}dx + \int_0^1 x^q dx = 1$$

so

$$\int_0^1 x^{1/q}dx = 1 - \frac{1}{q+1} = \frac{1}{(1/q)+1}$$

as desired.

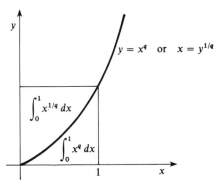

Figure 11

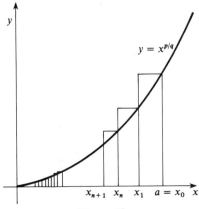

Figure 12

See the article by Nunn [10] for an English paraphrase of the *Arithmetica Infinitorum*. Struik's source book also contains a translation of part of this work ([11], pp. 244–247). Wallis' conjecture that

$$\int_0^a x^{p/q} dx = \frac{a^{(p/q)+1}}{(p/q)+1} = \frac{q}{p+q} a^{(p+q)/q} \tag{25}$$

if p/q is a positive rational number was established by Fermat and by Evangelista Torricelli (1608–1647), who was a disciple of Galileo and Cavalieri. Although the investigations of Fermat and Torricelli predated that of Wallis—for example, see Mahoney's account of Fermat's work ([9], pp. 243–252)—they were not published until somewhat later.

Fermat began by subdividing the interval $[0, a]$ into an infinite sequence of subintervals with endpoints $\{x_n\}_0^\infty$, where $x_n = ar^n$ and $0 < r < 1$. If a rectangle with height $x_n^{p/q}$ is erected on the nth subinterval $[x_{n+1}, x_n]$ (see Fig. 12), then the sum of the areas of this sequence of rectangles is

$$A(r) = \sum_{n=0}^\infty x_n^{p/q}(x_n - x_{n+1})$$

$$= \sum_{n=0}^\infty (ar^n)^{p/q}(ar^n - ar^{n+1})$$

$$= a^{(p+q)/q}(1-r) \sum_{n=0}^\infty r^{n(p+q)/q}$$

$$= a^{(p+q)/q}(1-r) \sum_{n=0}^\infty s^n \qquad (s = r^{(p+q)/q})$$

$$= a^{(p+q)/q} \frac{1-r}{1-s}$$

$$= a^{(p+q)/q} \frac{1-t^q}{1-t^{p+q}} \qquad (t = r^{1/q})$$

$$A(r) = a^{(p+q)/q} \frac{1+t+\cdots+t^{q-1}}{1+t+\cdots+t^{p+q-1}}, \tag{26}$$

since $(1-t)(1+t+\cdots+t^{k-1})=1-t^k$. Now the area under the curve is evidently the limit of $A(r)$ as r approaches 1 (and hence $t \to 1$), so equation (25) follows by taking the limit in (26).

EXERCISE 20. Show similarly that the area under the generalized hyperbola $y = x^{-p/q}$ ($p/q > 1$) over the semi-infinite interval $[a, \infty)$ is given by

$$\int_a^\infty x^{-p/q}\, dx = \frac{q}{p-q}\, a^{-(p-q)/q}.$$

Subdivide the base $[a, \infty)$ into an infinite sequence of subintervals with endpoints $\{x_n\}_0^\infty$, where $x = ar^n$ and now $r > 1$.

EXERCISE 21. (a) Consider the generalized hyperbola $y = x^{-p/q}$, where p/q is a positive rational number not equal to one, and the areas A_1, A_2, A_3, A_4, A_5 indicated in Figure 13. Torricelli showed by an exhaustion proof that

$$\frac{A_3 + A_4}{A_2 + A_4} = \frac{p}{q}. \qquad (*)$$

Use calculus to verify (*) by computing the integrals

$$A_2 + A_4 = \int_a^b x^{-p/q}\, dx$$

and

$$A_3 + A_4 = \int_{b^{-p/q}}^{a^{-p/q}} y^{-q/p}\, dy.$$

(b) Derive the integral

$$\int_a^b x^{-p/q}\, dx = A_2 + A_4 = \frac{b^{-(p-q)/q} - a^{-(p-q)/q}}{(q-p)/q}$$

from (*) and the obvious fact that

$$A_1 + A_2 + A_3 + A_4 + A_5 = ba^{-p/q}.$$

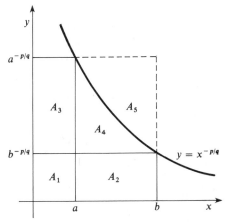

Figure 13

The First Rectification of a Curve

We have seen that the quadrature of certain curvilinear figures such as segments of parabolas dates back to ancient times. However, it was long thought that a segment of an algebraic curve could never have the same length as a constructible *straight* line segment. That is, the *rectification problem*—of constructing a straight line segment equal in length to a given curve—was thought to be impossible for algebraic curves. But in the late 1650s infinitesimal techniques were applied to show that this pessimism had been unjustified.

The first rectification of a curve was that of the "semi-cubical parabola" $y^2 = x^3$ in 1657 by the Englishman William Neil (who was then twenty years old, and apparently was never heard from again). To describe his procedure for computing the length of the segment of this curve (Fig. 14) that lies over the interval $0 \leqslant x \leqslant a$, we subdivide this interval into an indefinitely large number n of infinitesimal subintervals, the ith one being $[x_{i-1}, x_i]$. If s_i denotes the length of the (almost straight) piece of the curve $y = x^{3/2}$ joining the corresponding points (x_{i-1}, y_{i-1}) and (x_i, y_i) then

$$s_i \cong \left[(x_i - x_{i-1})^2 + (y_i - y_{i-1})^2 \right]^{1/2}, \qquad (27)$$

so the length of the curve is given by

$$s \cong \sum_{i=1}^{n} \left[(x_i - x_{i-1})^2 + (y_i - y_{i-1})^2 \right]^{1/2}. \qquad (28)$$

In order to compute the sum in (28), Neil introduced as an auxiliary curve the parabola $z = x^{1/2}$ (Fig. 15). If A_i denotes the area under this parabola over the interval $[0, x_i]$, then we know from the general quadrature result (25) that $A_i = 2x_i^{3/2}/3$. Therefore we obtain

$$
\begin{aligned}
y_i - y_{i-1} &= x_i^{3/2} - x_{i-1}^{3/2} \\
&= \tfrac{3}{2}(A_i - A_{i-1}) \\
y_i - y_{i-1} &\cong \tfrac{3}{2} z_i (x_i - x_{i-1})
\end{aligned}
\qquad (29)
$$

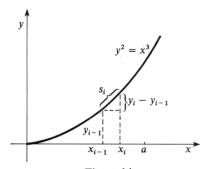

Figure 14

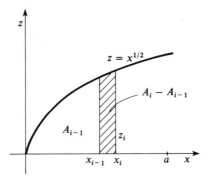

Figure 15

by approximating the strip of area over $[x_{i-1}, x_i]$ with a rectangle of height $z_i = x_i^{1/2}$. Substitution of (29) into (28) then gives

$$s \cong \sum_{i=1}^{n} \left[1 + \left(\frac{y_i - y_{i-1}}{x_i - x_{i-1}} \right)^2 \right]^{1/2} (x_i - x_{i-1})$$

$$\cong \sum_{i=1}^{n} \left(1 + \frac{9}{4} z_i^2 \right)^{1/2} (x_i - x_{i-1})$$

$$s \cong \sum_{i=1}^{n} \frac{3}{2} \left(x_i + \frac{4}{9} \right)^{1/2} (x_i - x_{i-1}). \tag{30}$$

At this point we recognize the sum in (30) as that which gives (in the limit) the area of the segment of the parabola $y = (3/2)(x + 4/9)^{1/2}$ lying over the interval $[0, a]$. By translation this is the same as the area of the segment of the parabola $y = 3x^{1/2}/2$ lying over the interval $[4/9, a + 4/9]$ (see Fig. 16). From the general quadrature result (25) we therefore obtain

$$s = \tfrac{3}{2} \left[\tfrac{2}{3} (a + 4/9)^{3/2} - \tfrac{2}{3} (4/9)^{3/2} \right]$$

$$= \frac{(9a + 4)^{3/2} - 8}{27}.$$

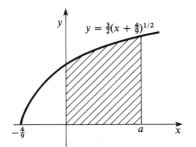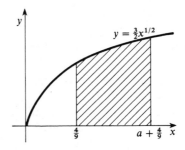

Figure 16

It is instructive to phrase Neil's procedure in more general terms. In order to calculate the length of the curve $y = f(x)$ over $[0, a]$, we first need an auxiliary curve $z = g(x)$ for which the area A_i over $[0, x_i]$ is

$$A_i = \int_0^{x_i} g(x)\,dx = f(x_i) = y_i. \tag{31}$$

Then it follows that

$$y_i - y_{i-1} = A_i - A_{i-1} \cong g(x_i)(x_i - x_{i-1}),$$

so "characteristic triangles" give

$$s \cong \sum_{i=1}^n \left[1 + (g(x))^2\right]^{1/2}(x_i - x_{i-1}),$$

$$s = \int_0^a \sqrt{1 + [g(x)]^2}\,dx.$$

Thinking of the fundamental theorem of calculus, we see that the proper choice of the auxiliary curve (so as to give (31)) is $g(x) = f'(x)$. Thus a combination of quadratures and tangents via the characteristic triangle is implicit in Neil's construction for the particular case $f(x) = x^{3/2}$.

EXERCISE 22. Apply this method to show that the rectification of a parabola is equivalent to the quadrature of a hyperbola. In particular, the length s of $y = x^2$ over $[0, a]$ is given by

$$s = \int_0^a \sqrt{1 + 4x^2}\,dx.$$

Summary

During the middle decades of the seventeenth century infinitesimal techniques or indivisibles, motivated by attempts to relax the rigor of the classical method of exhaustion, were applied to establish the basic quadrature result

$$\int_0^a x^k\,dx = \frac{a^{k+1}}{k+1}.$$

It was this result itself, rather than the particular methods used to derive it, that was of lasting importance. For by 1660 the early direct methods of quadrature were rapidly approaching obsolescence, soon to be superseded by indirect methods based on the interplay between quadrature and tangent methods. Neil's rectification was, at least implicitly, an early (and perhaps the first) example of this interplay between the two distinct aspects of the emerging calculus.

References

[1] E. J. Aiton, Kepler's second law of planetary motion. *Isis* **60**, 75–90, 1969.
[2] M. E. Baron, *The Origins of the Infinitesimal Calculus*. Oxford: Pergamon, 1969, Chapters 3–6.
[3] C. B. Boyer, *The History of the Calculus*. New York: Dover, 1959, Chapter 4.
[4] C. B. Boyer, Cavalieri, limits and discarded infinitesimals. *Scr Math* **8**, 79–91, 1941.
[5] C. B. Boyer, Pascal's formula for the sums of the powers of the integers. *Scr Math* **9**, 237–244, 1943.
[6] S. A. Christensen, The first determination of the length of a curve. *Bibl Math* N.S. **I**, 76–80, 1887.
[7] J. L. E. Dreyer, *A History of Astronomy from Thales to Kepler*. New York: Dover, 1953.
[8] A. Koestler, *The Sleepwalkers*. New York: Macmillan, 1968.
[9] M. S. Mahoney, *The Mathematical Career of Pierre de Fermat*. Princeton, NJ: Princeton University Press, 1973, Chapter 5.
[10] T. P. Nunn, The arithmetic of infinites. *Math Gaz* **5**, 345–356, 1909–1911.
[11] D. J. Struik, *A Source Book in Mathematics, 1200–1800*. Cambridge, MA: Harvard University Press, 1969.
[12] D. T. Whiteside, Patterns of mathematical thought in the later 17th century. *Arch Hist Exact Sci* **1**, 179–388, 1960–1962.

5 Early Tangent Constructions

Introduction

In modern calculus courses the treatment of differentiation and the construction of tangent lines to curves usually precede the treatment of integration and the calculation of areas under curves. This is a reversal of the historical sequence of discovery; as we have seen in the preceding chapters, the calculation of curvilinear areas dates back to ancient times. However, apart from simple constructions of tangent lines to conic sections (with the static Greek view of a tangent line as a line touching the curve in only one point), and the isolated example of Archimedes' construction of the tangent to his spiral, tangent lines were not studied until the middle decades of the seventeenth century.

Then, beginning about 1635, a number of different methods for the construction of tangent lines to general curves were rapidly discovered and investigated. It was the combination of these new tangent methods with area problems and techniques, during the last third of the seventeenth century, that produced the calculus as a new unified method of mathematical analysis.

Fermat's Pseudo-equality Methods

Fermat was the first to solve maximum-minimum problems by somehow taking into account the characteristic behavior of a function near its extreme values. For example, in order to determine how to subdivide a segment of length b into two segments x and $b - x$ whose product

122

$x(b - x) = bx - x^2$ is maximal (that is, to find the rectangle with perimeter $2b$ that has maximal area), he proceeded as follows. First he substituted $x + e$ (he used A, E instead of x, e) for the unknown x, and then wrote down the following "pseudo-equality" to compare the resulting expression with the original one:

$$b(x + e) - (x + e)^2 = bx + be - x^2 - 2xe - e^2 \sim bx - x^2.$$

After cancelling equal terms, he divided through by e to obtain

$$2x + e \sim b.$$

Finally he discarded the remaining term containing e, transforming the pseudo-equality into the true equality

$$x = \frac{b}{2}$$

that gives the value of x which makes $bx - x^2$ maximal.

Unfortunately, Fermat never explained the logical basis for this method with sufficient clarity or completeness to prevent disagreements between historical scholars as to precisely what he meant or intended. Two recent and contrasting views may be found in the book by Mahoney ([5], Chapter 4) and the article by Strømholm [6].

An explanation, one that perhaps is closer to modern perceptions than those of Fermat, might be given as follows. If $f(x)$ is a maximum (or minimum) value of the function f, then it seems on intuitive or pictorial grounds that the value of f changes very slowly near x (see Figure 1). Hence, if e is quite small, then $f(x)$ and $f(x + e)$ are approximately equal,

$$f(x + e) \sim f(x),$$
$$f(x + e) - f(x) \sim 0.$$

If $f(x)$ is a polynomial, then $f(x + e) - f(x)$ will be divisible by e, so we carry out this division, obtaining

$$\frac{f(x + e) - f(x)}{e} \sim 0.$$

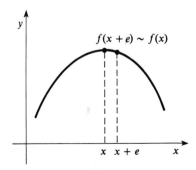

Figure 1

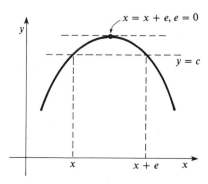

Figure 2

But the limit of this quotient as $e \to 0$ is the modern definition of the derivative. Consequently Fermat's suppression of the remaining terms that involve e amounts to writing $f'(x) = 0$.

However, it must be emphasized that Fermat did not explicitly require that e be "small", and said nothing at all about taking the *limit* as e approaches 0. On at least one occasion, he treated x and $x + e$ in a purely algebraic manner as distinct roots of the equation $f(x) = c$ (see Figure 2). Writing $f(x + e) = f(x)$, he cancelled equal terms, divided through by e, and finally discarded the remaining terms involving e, on account of the fact that the two roots are equal (so $e = 0$) when $c = f(x)$ is the maximum value of f.

EXERCISE 1. If $f(x) = \sum a_i x^i$ is a polynomial, verify that $f(x + e) - f(x)$ is divisible by e.

EXERCISE 2. Apply Fermat's method formally to find the maximum value of $f(x) = bx^2 - x^3$ for $0 \leqslant x \leqslant b$.

Fermat used a similar "pseudo-equality" technique to construct tangent lines. From the similar triangles in Figure 3, we read off the proportion

$$\frac{s + e}{s} = \frac{k}{f(x)}.$$

Upon substituting $k \sim f(x + e)$, we solve for the *sub-tangent s*,

$$s \sim \frac{ef(x)}{f(x + e) - f(x)}. \tag{1}$$

If we cancel the e in the numerator into the $f(x + e) - f(x)$ in the denominator (assuming f is a polynomial so Exercise 1 applies), and discard the remaining terms in the denominator that involve e, we then obtain an expression for the sub-tangent. In modern terms, this corre-

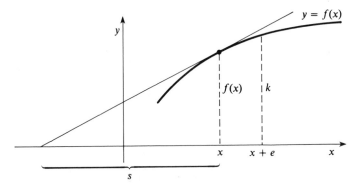

Figure 3

sponds to writing

$$s \sim \frac{f(x)}{[f(x+e) - f(x)]/e},$$

and then taking the limit as $e \to 0$ to obtain

$$s = \frac{f(x)}{f'(x)}. \tag{2}$$

Since the slope of the tangent line is $f(x)/s$, Equation (2) identifies the slope of the tangent line to the curve $y = f(x)$ with the derivative $f'(x)$.
  For example, with $f(x) = x^2$, Equation (1) gives

$$s \sim \frac{ex^2}{(x+e)^2 - x^2} = \frac{x^2}{2x+e}.$$

Suppression of the remaining e yields $s = x/2$, so the slope of the tangent line to the parabola $y = x^2$ is

$$f'(x) = \frac{f(x)}{s} = \frac{x^2}{x/2} = 2x.$$

EXERCISE 3. Apply the above method of Fermat to show that the sub-tangent to $y = x^n$ is $s = x/n$, so the slope of the tangent line is nx^{n-1}.

Descartes' Circle Method

Descartes devised a method of constructing tangent lines that was algebraic rather than infinitesimal in character. Although Fermat's approach struck closer to the infinitesimal heart of the matter, Descartes' algebraic approach probably exerted a greater influence on the immediate development of the calculus. Descartes' appreciation of the importance of the

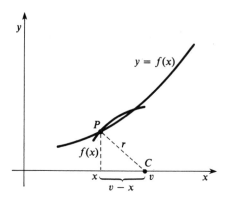

Figure 4

problem of constructing tangent lines was expressed in his statement that it is not only "the most useful and general problem that I know but even that I have ever desired to know in geometry."

His method of finding the tangent line to the curve $y = f(x)$ at the point $P(x, f(x))$ involved first locating the point $C(v, 0)$ of intersection with the x-axis of the normal line to the curve at P. The tangent line can then be taken as the perpendicular through P to the normal line.

In general, a circle with center $C(v, 0)$ and radius $r = CP$ will intersect the curve $y = f(x)$ in a second point near P (see Fig. 4). If, however, CP is the normal to the curve at P, then the point P should be a "double point" of intersection of the curve $y = f(x)$ and the circle $y^2 + (x - v)^2 = r^2$. Assuming that $[f(x)]^2$ is a polynomial, this means that the equation

$$[f(x)]^2 + (v - x)^2 = r^2 \tag{3}$$

(with v and r fixed) will have the coordinate x of P as a double root.

Now a polynomial which has a double root, say $x = e$, must be of the form $(x - e)^2 \Sigma c_i x^i$. Descartes imposed the condition that Equation (3) have a double root by writing

$$[f(x)]^2 + (v - x)^2 - r^2 = (x - e)^2 \Sigma c_i x^i. \tag{4}$$

By equating like powers of x, he then solved for v in terms of the root $e = x$. The slope of the tangent line at P is then $(v - x)/f(x)$ (the negative reciprocal of the slope $-f(x)/(v - x)$ of the normal CP in Figure 4).

For example, consider the parabola $y^2 = kx$, or $y = f(x) = \sqrt{kx}$. Then Equation (3) is

$$kx + (v - x)^2 - r^2 = 0.$$

This is a 2nd degree equation, so the right-hand side of (4) should be a polynomial of degree 2, hence

$$kx + (v - x)^2 - r^2 = (x - e)^2.$$

Equating coefficients of x gives $k - 2v = -2e$, or $v = e + \frac{1}{2}k$. Substituting $e = x$, the subnormal $v - x$ is $\frac{1}{2}k$, and the slope of the tangent line to the parabola at $(x, \sqrt{kx})$ is

$$\frac{v - x}{f(x)} = \frac{k/2}{\sqrt{kx}} = \frac{1}{2}\sqrt{\frac{k}{x}} \, .$$

For the parabola $y = x^2$, we write Equation (4) in the form

$$x^4 + (v - x)^2 - r^2 = (x - e)^2(x^2 + ax + b),$$

since the left-hand side is a 4th degree polynomial with leading coefficient 1. Expansion gives

$$x^4 + x^2 - 2vx + (v^2 - r^2) = x^4 + (a - 2e)x^3 + (b - 2ae + e^2)x^2$$
$$+ (ae^2 - 2be)x + be^2.$$

Equating coefficients gives the equations

$$a - 2e = 0$$
$$b - 2ae + e^2 = 1$$
$$ae^2 - 2be = -2v$$

which we solve for $v = 2e^3 + e$. Substituting $e = x$, the subnormal is $v - x = 2x^3$, and the slope of the tangent line to the parabola at the point (x, x^2) is

$$\frac{v - x}{f(x)} = \frac{2x^3}{x^2} = 2x.$$

EXERCISE 4. If $y = x^{3/2}$, apply Descartes' method to show that the subnormal is $v - x = 3x^2/2$ and the slope of the tangent line is $3x^{1/2}/2$.

The Rules of Hudde and Sluse

As may be guessed from the examples in the preceding section, the direct application of Descartes' circle method to any but the very simplest curves leads to prohibitively tedious algebraic computations. However, formal algorithms for the constructions of tangents were discovered in the 1650s by the Dutch mathematicians Johann Hudde and René François de Sluse Their mechanical rules made possible the routine computation of the slopes of tangent lines to arbitrary algebraic curves.

Hudde's rule provides a convenient means of determining the double roots that Descartes' circle method calls for. Given a polynomial

$$F(x) = \sum_{i=0}^{n} a_i x^i,$$

a second polynomial $F^*(x)$ is constructed as follows. The terms of $F(x)$, arranged in order of increasing degree, are multiplied in turn by the terms of an arbitrary arithmetic progression

$$a, a + b, a + 2b, \ldots, a + nb. \tag{5}$$

The resulting polynomial is

$$F^*(x) = \sum_{i=0}^{n} a_i(a + ib)x^i. \tag{6}$$

Note that if $a = 0$, $b = 1$, so the term in x^i is multiplied by i, then $F^*(x) = \sum ia_i x^i = xF'(x)$ where $F'(x) = \sum ia_i x^{i-1}$, the now-familiar derivative of the polynomial $F(x)$. In general,

$$F^*(x) = aF(x) + bxF'(x). \tag{7}$$

Hudde's rule states that any *double* root of $F(x) = 0$ must be a root of $F^*(x) = 0$. This algebraic fact can be established easily as follows. If e is a double root of $F(x)$, then we can write

$$F(x) = (x - e)^2 \sum c_i x^i = \sum c_i(x^{i+2} - 2ex^{i+1} + e^2 x^i).$$

If $A_i = a + bi$, then

$$\begin{aligned}
F^*(x) &= \sum c_i(A_{i+2}x^{i+2} - 2eA_{i+1}x^{i+1} + e^2 A_i x^i) \\
&= \sum c_i[(A_i + 2b)x^2 - 2e(A_i + b)x + e^2 A_i]x^i \\
&= \sum c_i[A_i(x - e)^2 + 2bx(x - e)]x^i,
\end{aligned}$$

whence it is clear that e is a root of $F^*(x)$.

In particular, any double root of the polynomial $F(x)$ must be a root of its derivative $F'(x)$. It was, in fact, this appearance of $F'(x)$, in the wholly algebraic (*not* infinitesimal) context of Hudde's rule, that first brought out the computational importance of what we now call the derivative of a polynomial.

For example, to apply Hudde's rule to the parabola $y^2 = kx$, we start with the Cartesian circle condition

$$F(x) = kx + (v - x)^2 - r^2 = 0$$

as before. Taking $a = 0$, $b = 1$ so $F^*(x) = xF'(x)$, Hudde's rule states that x is a double root of $F(x)$ only if it is a root of

$$F^*(x) = (1)kx + (0)v^2 - (1)2vx + 2(x^2) - (0)r^2 = 0,$$

$$rx - 2vx + 2x^2 = 0,$$

from which we solve for the subnormal $v - x = \frac{1}{2}k$. Hence the slope of the tangent line is $(v - x)/\sqrt{kx} = \frac{1}{2}\sqrt{k/x}$ as before.

The computation of the slope of the tangent line to $y = x^n$ by direct application of Descartes' circle method would be extremely tedious if $n > 2$. The circle condition is

$$F(x) = x^{2n} + (v - x)^2 - r^2 = 0,$$

so

$$F^*(x) = (2n)x^{2n} + (0)v^2 - (1)2vx + (2)x^2 - (0)r^2 = 0,$$

or

$$2nx^{2n} - 2vx + 2x^2 = 0.$$

We immediately solve this equation for the subnormal $v - x = nx^{2n-1}$, so the slope of the tangent line is

$$\frac{v - x}{x^n} = \frac{nx^{2n-1}}{x^n} = nx^{n-1},$$

the familiar derivative of x^n. Note that this computation remains valid if n is half of a positive integer.

EXERCISE 5. Apply the Cartesian circle method using Hudde's rule to show that the slope of the tangent line to $y = (x^2 + 1)^{3/2}$ is $3x(x^2 + 1)^{1/2}$.

Note that Hudde's rule for applying the Cartesian circle condition amounts (in the language of derivatives) to the following. Write

$$F(x) = [f(x)]^2 + (v - x)^2 - r^2 = 0,$$

and then solve the equation $F'(x) = 0$ for x in terms of v. Thus, thinking of a fixed point $C(v, 0)$ on the x-axis, we are finding x such that the distance from C to the point $P(x, f(x))$ of the curve $y = f(x)$ is extremal.

Hudde's rule can also be applied directly to maximum-minimum problems. Recall that Fermat started with the observation that the maximum (or minimum) value M of $f(x)$ occurs at a double root of the equation $f(x) = M$ or

$$F(x) = f(x) - M = 0,$$

and hence at a root of $F^*(x) = 0$. From Equation (5) we therefore see that a (local) maximum or minimum of $f(x)$ occurs at a root of the equation $f'(x) = 0$ (just as we learn in elementary calculus).

The combination of Hudde's rule and the Cartesian circle method applied only to algebraic curves that could be described in *explicit* form, $y = f(x)$. However, Sluse stated an even more mechanical rule that applied equally well (and even more easily) to algebraic curves described in *implicit* form, $f(x, y) = 0$, where

$$f(x, y) = \sum c_{ij}x^i y^j$$

is a polynomial in x and y. Given the arithmetic progression (5), define

$$f_x^*(x, y) = \sum (a + bi)c_{ij}x^i y^j \quad \text{and} \quad f_y^* = \sum (a + bj)c_{ij}x^i y^j. \quad (8)$$

Then *Sluse's rule* states that the slope m of the tangent line at a point (x, y) on the curve $f(x, y) = 0$ is given by

$$m = -\frac{y}{x} \cdot \frac{f_x^*}{f_y^*}. \quad (9)$$

EXERCISE 6. Apply Sluse's rule to $y = \sum c_i x^i$ to obtain $m = \sum ic_i x^{i-1}$. Thus Sluse's rule provides a completely algorithmic approach to derivatives of polynomials.

EXERCISE 7. (a) Show that

$$f_x^*(x, y) = af(x, y) + bxf_x(x, y) \quad \text{and} \quad f_y^*(x, y) = af(x, y) + byf_y(x, y)$$

where f_x and f_y denote the partial derivatives

$$\frac{\partial f}{\partial x} = \sum ic_{ij}x^{i-1}y^j \quad \text{and} \quad \frac{\partial f}{\partial y} = \sum jc_{ij}x^i y^{j-1}.$$

(b) Conclude that Sluse's rule (9) is equivalent to the familiar result $dy/dx = -(\partial f/\partial x)/(\partial f/\partial y)$ that is obtained by using the chain rule to differentiate $f(x, y) = 0$ with respect to x,

$$\frac{\partial f}{\partial x} + \frac{\partial f}{\partial y}\frac{dy}{dx} = 0.$$

As an application of Sluse's rule, consider the folium of Descartes,

$$f(x, y) = x^3 + y^3 - 3xy = 0$$

(which, as a matter of fact, was originally proposed by Descartes as a challenge for Fermat to find its tangent line). Taking $a = 0$, $b = 1$ in (8) and (9), we immediately obtain

$$m = -\frac{y}{x} \cdot \frac{3x^3 - 3xy}{3y^3 - 3xy} = \frac{y - x^2}{y^2 - x}.$$

EXERCISE 8. Write down the slope m given by Sluse's rule for the folium of Descartes, but taking $a = 2$, $b = 1$ in (8). Reconcile this answer and the one obtained above.

Sluse's rule was published in the 1673 *Philosophical Transactions* without explanation as to how he had discovered it some years earlier. If it was not deduced merely by inference from particular examples, a plausible possibility is that it was derived from Hudde's rule. An obvious connection between the two rules is suggested by their verbal statements in terms of arithmetic progressions.

From an algebraic point of view, a tangent line to the curve $f(x, y) = \sum c_{ij} x^i y^j = 0$ is a straight line $y = mx + k$ that intersects the curve in a *double* point. That is, the polynomial

$$F(x) = f(x, mx + k) = 0$$

should have a *double* root. But

$$F^*(x) = f_x^* + \frac{mx}{y} f_y^*, \qquad (10)$$

so Hudde's double root condition $F^*(x) = 0$ immediately implies Sluse's rule (9).

By the linearity of the operations that produce F^*, f_x^*, f_y^* from F, f, it suffices to establish (10) in the special case $f(x, y) = x^i y^j$. Then

$$F(x) = x^i (mx + k)^j = \sum_{p=0}^{j} \binom{j}{p} m^p x^{i+p} k^{j-p}$$

by the binomial formula, where $\binom{j}{p} = j!/p!(j-p)!$. Therefore, with $a = 0$, $b = 1$, we obtain

$$F^*(x) = \sum_{p=0}^{j} \binom{j}{p} m^p (i+p) x^{i+p} k^{j-p}$$

$$= i x^i \sum_{p=0}^{j} \binom{j}{p} (mx)^p k^{j-p} + mx^{i+1} \sum_{p=1}^{j} p \binom{j}{p} m^{p-1} x^{p-1} k^{j-p}$$

$$= i x^i (mx + k)^j + mx^{i+1} \sum_{p=1}^{j} j \binom{j-1}{p-1} m^{p-1} x^{p-1} k^{j-p}$$

$$= i x^i y^j + mj x^{i+1} \sum_{p=0}^{j-1} (mx)^p k^{j-1-p}$$

$$= i x^i y^j + mj x^{i+1} (mx + k)^{j-1}$$

$$= i x^i y^j + mj x^{i+1} y^{j-1}$$

$$F^*(x) = f_x^* + \frac{mx}{y} f_y^*,$$

as desired. We have used the fact that

$$p \binom{j}{p} = \frac{p(j!)}{p!(j-p)!} = \frac{j(j-1)!}{(p-1)!(j-p)!} = j \binom{j-1}{p-1}.$$

Whatever may have been the means by which Sluse's rule was first discovered, the principal significance of the rules of Sluse and Hudde lay in the fact that they provided general algorithms by which tangents to algebraic curves could be constructed in a routine manner. It was no longer necessary to resort to special devices adapted to particular curves, nor to give in every case a complete demonstration of the process. For these reasons, the rules of Sluse and Hudde were perhaps the first methods to exhibit fully the algorithmic approach that is a distinctive feature of the calculus.

Infinitesimal Tangent Methods

The introduction in the 1650s of the algebraic rules of Hudde and Sluse was soon followed by infinitesimal derivations of these and similar methods. These newer derivations and methods owed more to the ideas of Fermat than those of Descartes, and involved the concept of a tangent line at the point P of a curve as the limiting position of a secant line PQ as Q approaches P along the curve.

One such method was described by Isaac Barrow (1630–1677) in his *Geometrical Lectures* that were published in 1670 but delivered at Cambridge in the mid 1660s (and probably attended in 1664–65 by one Isaac Newton). Barrow was appointed in 1663 as the first Lucasian Professor of Mathematics at Cambridge, and resigned this chair in 1669 in favor of Newton (and perhaps also to qualify for administrative advancement).

The bulk of Barrow's published lectures treat tangent and quadrature problems from a somewhat classical and geometrical rather than analytical point of view. For example, he generally adopts the Greek definition of a tangent line to a curve as a straight line that touches the curve at a single point. However, at the close of Lecture X, he writes,

> We have now finished in some fashion the first part, as we declared, of our subject. Supplementary to this we add, in the form of appendices, a method for finding tangents by calculation frequently used by us. Although I hardly know, after so many well-known and well-worn methods of the kind above, whether there is any advantage in doing so. Yet I do so on the advice of a friend [who turned out to be Newton]: and all the more willingly, because it seems to be more profitable and general than those which I have discussed ([2], p. 119).

He proceeds to describe what is apparently his own modification of a method that Fermat had devised (but not published) to construct tangent lines to a curve defined implicitly by $f(x, y) = 0$ (for discussions of Fermat's method, see the articles by Coolidge [3], pp. 452–453 and Jensen [4]). Considering an "indefinitely small arc" MN of the curve (Fig. 5), he writes $M(x, y)$ and $N(x + e, y + a)$ for their coordinates, and sets

$$f(x + e, y + a) = f(x, y) = 0, \qquad (11)$$

since M and N are both points of the curve. He then deletes "all terms containing a power of a or e, or products of these (for these terms have no value)." Finally, ignoring the distinction between the "indefinitely small arc" MN and the straight line segment $\overline{MN}$, he notes the similarity of the

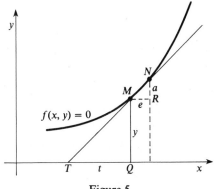

Figure 5

triangle TQM and the "characteristic triangle" MNR, and solves (11) (with the higher degree terms in a and e deleted) for the slope $y/t = a/e$ of the tangent line at M. Thus Barrow employs the concept of the "characteristic triangle"—essentially the idea of the tangent line as the limiting position of the secant line as a and e approach 0—and takes the limit by the expedient of neglecting "higher order infinitesimals."

For example, for the folium of Descartes, $f(x, y) = x^3 + y^3 - 3xy = 0$, we would write

$$(x+e)^3 + (y+a)^3 - 3(x+e)(y+a) = x^3 + y^3 - 3xy,$$
$$3x^2e + 3xe^2 + e^3 + 3y^2a + 3ya^2 + a^3 - 3xa - 3ye - 3ae = 0,$$

delete all higher degree terms in a and e to obtain

$$3x^2e + 3y^2a - 3xa - 3ye = 0,$$

and finally solve for the slope

$$m = \frac{a}{e} = \frac{y - x^2}{y^2 - x}.$$

EXERCISE 9. Apply Barrow's method to the curve $y = x^n$ to obtain the slope $a/e = nx^{n-1}$ of its tangent line.

In general, given the curve

$$f(x, y) = \sum c_{ij}x^iy^j = 0,$$

we write

$$\sum c_{ij}(x+e)^i(y+a)^j = \sum c_{ij}x^iy^j.$$

Expansion of $(x+e)^i$ and $(y+a)^j$ by the binomial formula gives

$$\sum c_{ij}(x^i + ix^{i-1}e + \cdots)(y^j + jy^{j-1}a + \cdots) = \sum c_{ij}x^iy^j.$$

Neglecting higher order terms in a and e, we obtain

$$\sum c_{ij}(ix^{i-1}y^je + jx^iy^{j-1}a) = 0,$$

so the slope of the tangent line is

$$m = \frac{a}{e} = -\frac{\sum ic_{ij}x^{i-1}y^j}{\sum jc_{ij}x^iy^{j-1}} = -\frac{\partial f/\partial x}{\partial f/\partial y}.$$

Thus Barrow's approach yields an analytical derivation of Sluse's rule.

Composition of Instantaneous Motions

During the 1630s and 1640s an approach to tangent lines that stemmed from the intuitive concept of instantaneous motion was developed by Torricelli and especially by Gilles Persone de Roberval, who was a professor at the College Royal (France) from 1634 until his death in 1675. Their idea (not itself a new one) was to consider a curve as the path of a moving point, and the tangent line as the line of instantaneous motion of the moving point. If the motion of the point generating the curve is the resultant or combination of two sufficiently simple motions, then the instantaneous line of motion can be determined by composition of the constituent motions.

The parallelogram law for the addition of *constant* velocity vectors was well-known. That is, if the points P and Q move along two intersecting straight lines with constant velocity vectors $\bar{u}$ and $\bar{v}$, respectively, and these two lines are taken as x- and y-axes, then the motion of the point R, whose x- and y-coordinates are given by P and Q, has velocity vector $\bar{w} = \bar{u} + \bar{v}$ (Fig. 6).

Roberval took the further step of applying the parallelogram law to *instantaneous* velocity vectors. That is, if the motion of a point is compounded of two simpler motions, he assumed that its instantaneous velocity vector is the parallelogram sum of the instantaneous velocity vectors corresponding to the two simpler motions.

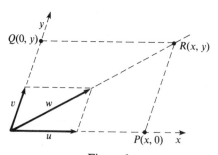

Figure 6

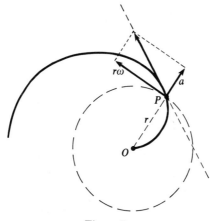

Figure 7

For example, consider the Archimedean spiral given in polar coordinates by $r = at$, $\theta = \omega t$. The motion of the point $P(at, \omega t)$ along the spiral may be regarded as the resultant of a radial motion (away from the origin) and an angular motion. To find the tangent to the spiral at P, we therefore construct a radial vector of length a (the radial speed) and a vector of length $r\omega$ (the angular speed) tangential to the circle of radius r through P. The diagonal of the parallelogram determined by these two vectors is the velocity vector at P, and therefore determines the tangent line to the spiral at P (Fig. 7).

An outstanding success of the instantaneous motion approach was the determination of the tangent to the cycloid. Consider a circle of radius a that is initially tangent to the x-axis at the origin, and thereafter rolls along the x-axis to the right with unit angular speed (one radian/sec). Then the cycloid is the trajectory of the point P on the circle that was initially at the origin, and is given in rectangular coordinates by $x = a(t - \sin t)$, $y = a(1 - \cos t)$. See Figure 8.

Roberval regarded the motion of the point P along the cycloid as compounded of (1) uniform translation to the right with speed a, and (2) clockwise rotation with unit angular speed, centered at time t at the point

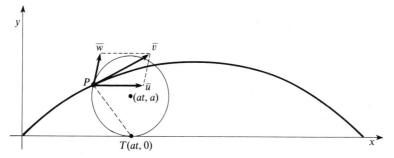

Figure 8

(*at*, *a*). The corresponding instantaneous velocity vectors are given in rectangular coordinates by

$$\bar{u} = (a, 0) \quad \text{(translation)}$$

and

$$\bar{w} = (-a \cos t, a \sin t) \quad \text{(rotation)}.$$

Their parallelogram sum (given in rectangular coordinates by coordinate-wise addition) is the velocity vector

$$\bar{v} = (a(1 - \cos t), a \sin t),$$

which determines the tangent line to the cycloid at *P*. Note that this result is the same as that obtained by coordinate-wise differentiation of the position vector $(a(t - \sin t), a(1 - \cos t))$. From the modern viewpoint, this latter observation is what verifies (in this example, at least) the validity of the process of combining instantaneous velocity vectors by parallelogram addition.

EXERCISE 10. Prove that the tangent vector to the cycloid, calculated above, is perpendicular to the line through the point *P* on the cycloid and the point *T* of contact between the rolling circle and the *x*-axis.

According to the focus-directrix definition, the parabola $y^2 = 4px$ is the locus of a point that is equidistant from the directrix $x = -p$ and the focus $F(p, 0)$. A point *P* moving along the parabola subject to this condition has equal components of velocity directed away from the directrix and away from the focus. It therefore appears that the tangent line to the parabola at

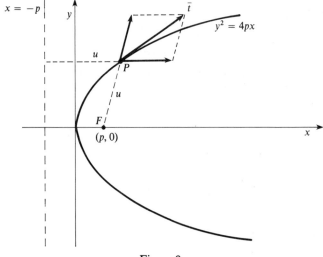

Figure 9

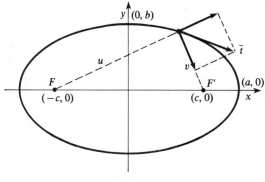

Figure 10

P bisects the angle between the line PF and the line through P perpendicular to the directrix (Fig. 9).

Similarly, an ellipse may be defined as the locus of a point, the sum of whose distances u and v from two foci F and F', respectively, is constant, $u + v = 2a$. If a point P moves along the ellipse subject to this condition, then the rate of increase of u and the rate of decrease of v must be equal. Thus the point undergoes motions of equal magnitude away from F and towards F'. It therefore appears that the tangent line to the ellipse at P bisects the angle between two unit vectors at P, one directed away from F and the other directed towards F' (Fig. 10).

The following two exercises indicate that it was something of a stroke of good fortune that Roberval obtained the correct tangent lines to the parabola and ellipse by this method, for it gives in each case only the correct direction and *not* the correct magnitude of the velocity vector.

EXERCISE 11. Let u denote the distance of a moving point P on the parabola $y^2 = 4px$ from the directrix $x = -p$ and from the focus $(p, 0)$. If the point moves in such a way that $u' = x' = 1$ (unit horizontal speed), show that the tangent vector $\bar{t}$ shown in Figure 9 is $\bar{t} = (2x/(x + p), y/(x + p))$, while the actual velocity vector of P is $\bar{v} = (1, \sqrt{p/x})$. Then show that $\bar{t}$ and $\bar{v}$ point in the same direction, but

$$|\bar{t}| = 2\sqrt{\frac{x}{x + p}} \quad \text{while} \quad |\bar{v}| = \sqrt{\frac{x + p}{x}} .$$

EXERCISE 12. Let u and v denote the distances of the point P on the ellipse $x^2/a^2 + y^2/b^2 = 1$ from the foci $F(-c, 0)$ and $F'(c, 0)$, respectively $(c = \sqrt{a^2 - b^2})$. Then $u = \sqrt{(x + c)^2 + y^2}$ and $v = \sqrt{(x - c)^2 + y^2}$. Consider the point P as moving clockwise around the ellipse subject to the condition $u + v = 2a$, with $u' = 1$, $v' = -1$. When P is at the point $(0, b)$, show that the vector $\bar{t}$ shown in Figure 10 is $\bar{t} = (2c/a, 0)$, while the actual velocity vector $\bar{v} = (x', y')$ is $\bar{v} = (a/c, 0)$. Hint: Subtraction of the equations $(x + c)^2 + y^2 = u^2$ and $(x - c)^2 + y^2 = v^2$ gives $4xc = u^2 - v^2$.

The Relationship Between Quadratures and Tangents

The application of time and motion concepts to the study of curves led both Torricelli and Barrow to at least an intuitive understanding of the inverse relationship between tangent and quadrature problems, that is, between the operations of differentiation and integration.

On the one hand, medieval investigations and the subsequent work of Galileo suggested that the motion of a point, along a straight line with varying velocity, be represented by means of a graph of its velocity versus time. Indivisibles considerations then indicated that the total distance traveled by the point would equal the area under the velocity-time curve, because the distance traveled during an infinitesimal element of time would equal the product of this time element and the instantaneous velocity (Fig. 11).

For example, if the point began its motion at time $t = 0$ and moved with velocity $v = t^n$ at time t, the distance y traveled would equal the area under the curve $v = t^n$, so

$$y = \frac{t^{n+1}}{n+1}. \tag{12}$$

On the other hand, the same motion could be represented by a graph of position versus time. If a point moves along the curve $y = y(t)$ with horizontal speed 1 and vertical speed v (the velocity of the point whose motion is represented in Fig. 11), the velocity vector of *this* point will be the resultant of a horizontal vector of length 1 and a vertical vector of length v (Fig. 12). Consequently the slope of the tangent line to the position curve $y = y(t)$ will be the velocity v.

For example, if the distance traveled in time t is given by Equation (12), then the velocity must be

$$v = t^n, \tag{13}$$

because t^n is the slope of the tangent line to the curve $y = t^{n+1}/(n+1)$.

Thus Equations (12) and (13) imply each other. Whereas the two facts that

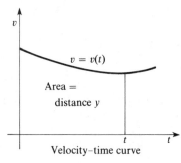

Figure 11

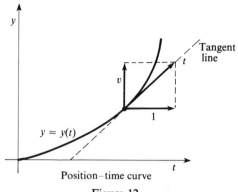

Figure 12

(a) the area under the curve $y = x^n$ is $x^{n+1}/(n+1)$, and
(b) the tangent line to the curve $y = x^{n+1}/(n+1)$ has slope x^n,

had originally been deduced from entirely separate considerations, the relationship between Figures 11 and 12 showed that each of these facts followed from the other.

Specifically, the relationship between these two figures is that the slope of the tangent line to the area curve $y = y(t)$ (Fig. 12) is equal to the ordinate of the original curve $v = v(t)$ (Fig. 11). This is an embryonic formulation of the fundamental theorem of calculus—the rate of change of the area under a curve is equal to its ordinate. As we will see in Chapter 8, this idea was Newton's starting point for the development of an algorithmic calculus. Chapters 6 and 7 will be devoted to the historical introduction of two additional analytical tools that played important roles in the computational machinery of the calculus—logarithms and infinite series.

Neither Torricelli nor Barrow exploited for computational purposes even an intuitive form of the fundamental theorem of calculus. Although Barrow began his published *Geometrical Lectures* with a treatment of curves that was based on motion concepts, he ended with formally stated results having a rigidly geometric and static character. His statement in Lecture X of the fundamental theorem may be described as follows (see Struik's source book [8], pp. 253–263 for an English translation of the pertinent passage).

For convenience let the y- and z-axes be oppositely oriented as shown in Figure 13. Given an increasing positive function $y = f(x)$, denote by $z = A(x)$ the area between the curve $y = f(x)$ and the segment $[0, x]$ along the x-axis. Given a point $D(x_0, 0)$ on the x-axis, let T be the point on the x-axis such that $DT = DF/DE = A(x_0)/f(x_0)$. Then Barrow asserts that the line TF touches the curve $z = A(x)$ only at the point $F(x_0, A(x_0))$.

Note that the slope of TF is

$$\frac{DF}{DT} = \frac{A(x_0)}{A(x_0)/f(x_0)} = f(x_0).$$

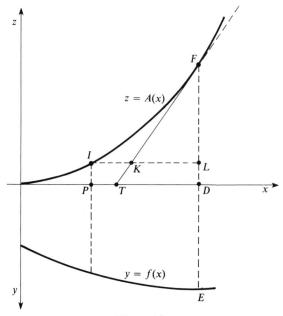

Figure 13

If Barrow were asserting that TF is the tangent line to the curve $z = A(x)$ in an analytical sense, with appropriately defined slope $A'(x_0)$, this result would therefore amount to the conclusion that $A'(x_0) = f(x_0)$, the fundamental theorem of calculus. However, he only asserts (and proves) that TF is tangent to $z = A(x)$ in the ancient Greek sense of a straight line that touches the curve at only one point.

To prove this, he considers a point $I(x_1, A(x_1))$ on the curve with $x_1 < x_0$, and proceeds to show that the point K, of intersection of the horizontal line IL with TF, lies to the right of I as shown (Fig. 13). To see this, note that $LF/LK = DF/DT = DE$ (by definition of point T), so $LF = LK \times DE$. But

$$LF = DF - PI = A(x_0) - A(x_1) < DP \times DE$$

because $f(x)$ is an increasing function. Therefore $LK \times DE < DP \times DE$, so $LK < DP = LI$, as desired. The case $x_1 > x_0$ is similar.

Thus we see that Barrow's result, which can and has been interpreted as an early statement of the fundamental theorem of calculus, was in reality formulated and established by him in a spirit more akin to classical Euclidean geometry than the emerging calculus of computational algorithms and processes.

It may be added that a similar result with similar proof was published slightly earlier in 1668 by the great young Scottish mathematician James

Gregory (1638–1675), who apparently duplicated (in his unpublished work) some of the key discoveries of Newton and Leibniz, but died prematurely before winning proper recognition for his work.

References

[1] M. E. Baron, *The Origins of the Infinitesimal Calculus*. London: Pergamon, 1969, Chapters 5 and 6.
[2] Isaac Barrow, *Geometrical Lectures*, edited by J. M. Child. Chicago: Open Court, 1916.
[3] J. L. Coolidge, The story of tangents. *Am Math Mon* **58**, 449–462, 1951.
[4] C. Jensen, Pierre Fermat's method of determining tangents of curves and its application to the conchoid and the quadratrix. *Centaurus* **14**, 72–85, 1969.
[5] M. S. Mahoney, *The Mathematical Career of Pierre de Fermat*. Princeton, NJ: Princeton University Press, 1973, Chapter 4.
[6] P. Strømholm, Fermat's methods of maxima and minima and of tangents. A reconstruction. *Arch His Exact Sci* **5**, 47–69, 1968.
[7] D. J. Struik, *A Source Book in Mathematics, 1200–1800*. Cambridge, MA: Harvard University Press, 1969.
[8] D. T. Whiteside, Patterns of mathematical thought in the later 17th century. *Arch Hist Exact Sci* **1**, 179–388, 1960–62, Chapters X and XI.

6 Napier's Wonderful Logarithms

John Napier (1550–1617)

The late sixteenth century was an age of numerical computation, as developments in astronomy and navigation called for increasingly accurate and lengthy trigonometric computations. Georg Joachim Rheticus (1514–1576) began the computation of a great collection of 15-place trigonometric tables which were completed and published by Otho in 1596 and by Pitiscus in 1613. The urgent need, for some device to shorten the labor of tedious multiplications and divisions with many decimal places, was met through the invention of logarithms by Napier and others around the turn of the seventeenth century.

John Napier was the eighth baron (or laird) of Merchiston. He is said to have regarded his book *A Plaine Discovery of the Whole Revelation of Saint John* (1593) as his most important contribution. This polemical tract contained proofs in Euclidean fashion that the Pope was the Antichrist and that the world was due to end in the year 1786. With this theological work behind him, he began in 1594 the work that was to revolutionize the practical art of numerical computation. This labor occupied a twenty-year period spent in the isolation of Merchiston castle near Edinburgh in the south of Scotland.

Napier's logarithmic tables first appeared in 1614 in a small book entitled *Mirifici Logarithmorum Canonis Descriptio* (Description of the Wonderful Canon of Logarithms), which contained only an introduction and guide to the computational use of the tables. The method of computation of the tables themselves, and to a lesser extent the reasoning upon which they were based, were summarized in the *Mirifici Logarithmorum Canonis Constructio* (Construction of the Wonderful Canon of Logarithms), the first written of the two books, but published posthumously in 1619. Extracts from an 1889 English translation of the *Constructio* by W. R. Macdonald may be found in the Napier tercentennary memorial volume ([NT], pp. 25–32) or in D. J. Struik's mathematics source book ([11], pp. 11–21).

The practical advantages, of using logarithms to convert tedious multiplications and divisions to comparatively simple additions and subtractions, were immediately obvious. For example, when Kepler received Napier's tables of 1614, he enthusiastically employed them in the enormous computations that led to the discovery of his third law of planetary motion.

Today we think of the logarithm $\log_a x$ of the number x (with base a) as the power to which a must be raised to obtain x. However, in order to properly gauge the magnitude of Napier's accomplishment, it is important to realize that fractional powers and exponential notation had in Napier's time not yet been developed. Neither was the decimal point system of numeration generally accepted. Indeed, it was Napier's systematic use of decimal points that was largely responsible for the general adoption of decimal point notation during the seventeenth century.

In particular, we think of the logarithm as a *function*, or even as the inverse of an exponential function. However, Napier's computations were based on a clear understanding of a particular functional relationship at a time when the general concept of a function was still unknown. Indeed, the logarithm function played a prototype role in the development of this general concept. Also, as we shall see in this chapter, the study of logarithms led to the calculation of hyperbolic areas (such as the area under the rectangular hyperbola $xy = 1$). In these ways the logarithm function, in addition to its computational importance, played a significant role in the historical development of the calculus.

The Original Motivation

The object of Napier's "wonderful canon of logarithms" was to reduce the tedious operation of multiplication to the much simpler operation of addition by means of the correspondence between an arithmetic series and a geometric series. In his *Arithmetica Integra* of 1544, Michael Stifel (with whose work Napier is likely to have been familiar) set down side-by-side the arithmetic and geometric series

0	1	2	3	4	5	6	7	8	$\cdots$
1	2	4	8	16	32	64	128	256	$\cdots$

and pointed out that addition in the upper (arithmetic) series corresponds to multiplication in the lower (geometric) series (see pp. 85–86 of D. E.

Smith's article "The law of exponents in the works of the sixteenth century" in [NT]). For example $3+5=8$ corresponds to $8 \times 32 = 256$. Although the lack of exponential notation prevented Stifel from writing $2^3 \cdot 2^5 = 2^{3+5}$, he referred to the upper numbers as "exponents" of the lower numbers.

In order that a correspondence between arithmetic and geometric series be useful for practical computations, it was obviously necessary that the common ratio between successive terms in the geometric series be close to unity, in order that the gaps between successive terms would remain small. Napier began with this common ratio as 0.9999999 $(=1-10^{-7}$ in modern exponential notation), and the First Table of the *Constructio* consists of the first 100 terms of the geometric series with first term $10,000,000$ $(=10^7)$, that is, the numbers

$$10^7(1-10^{-7})^n, \qquad n = 0, 1, 2, \ldots, 100.$$

He obtained each term from the previous one by an easy subtraction, as follows.

$$
\begin{array}{r}
10000000.0000000 \\
-1.0000000 \\
\hline
9999999.0000000 \\
-0.9999999 \\
\hline
9999998.0000001 \\
\text{continued up to} \\
\hline
9999900.0004950
\end{array}
$$

He called the numbers $0, 1, \ldots, 100$ the *logarithms* $(=$ ratio numbers$)$ of the numbers thereby obtained, e.g. 100 is the logarithm of 9999900.0004950. Thus his original idea of the logarithm of a number x (less than 10^7) was the number of times that 10^7 must be multiplied by $(1-10^{-7})$ to yield x. Hence let us write $y = \text{Nlog } x$ (the Naperian logarithm of x) if

$$x = 10^7(1-10^{-7})^y.$$

Note first that $\text{Nlog } 10^7 = 0$, and that $\text{Nlog } x$ increases as x decreases, in contrast with modern natural logarithms. Thus the frequent designation of natural logarithms as "Naperian logarithms" is inaccurate.

Next note that, if

$$x' = 10^7(1-10^{-7})^{y'},$$

then

$$\frac{x}{x'} = (1-10^{-7})^{y-y'},$$

so the *difference* of the logarithms of x and x' depends only on the *ratio* of x and x' (hence the name logarithm = ratio number, of Greek origin). As Napier says in Art. 36 of the *Constructio*, "the logarithms of similarly proportioned sines are equidifferent."

It follows that if $x_1, x_2, \ldots, x_n$ is a *geometric* progression, then the sequence of logarithms

$$\text{Nlog } x_1, \quad \text{Nlog } x_2, \quad \ldots, \quad \text{Nlog } x_n$$

is an *arithmetic* progression. This fact was the basis for Napier's computation of his table of logarithms.

Obviously Napier could not simply continue the First Table in the above fashion, because it would require over 6,900,000 steps to reduce the first term 10,000,000 by a factor of two to 5,000,000.

EXERCISE 1. Use modern logarithms to compute the exact number of steps that would be required to reach 5,000,000 in this manner. That is, for what n is $(1 - 10^{-7})^n = \frac{1}{2}$?

In the Second Table of the *Constructio* Napier computes the first 50 terms of the geometric series with common ratio $(1 - 10^{-5})$, that is, the numbers

$$10^7(1 - 10^{-5})^r, \quad r = 0, 1, 2, \ldots, 50.$$

Again, the successive terms are computed by easy successive subtractions.

$$
\begin{array}{r}
10000000.000000 \\
-100.000000 \\
\hline
9999900.000000 \\
-99.999000 \\
\hline
9999800.001000 \\
\text{continued up to} \\
\hline
9995001.224804
\end{array}
$$

(Napier erroneously has 9995001.222927 for the last term here.)

With a common ratio of $(1 - 10^{-5})$, it would still require over 69,000 steps to reach 5,000,000. Of course Napier does not intend to continue in this way, either. These first two tables are to be used only to interpolate between the entries in his Third Table, which has 21 rows and 69 columns, the element in the pth row and qth column being

$$10^7\left(1 - \frac{1}{2000}\right)^{p-1}\left(1 - \frac{1}{100}\right)^{q-1}.$$

Thus each row is a geometric progression with 69 terms and common factor $(1 - 1/100)$, while each column is a geometric progression with 21 terms and common factor $(1 - 1/2000)$. Finally, the ratio of the first number in each column, to the last number in that same column, is $(1 - 1/2000)^{20}$, which is approximately equal to $1 - 1/100 = 99/100$, and the last number in each column is approximately equal to the first number in the next column. The simplified version of this table printed below is based on Hobson's exposition [8].

First column	2nd column		69th column
10000000.0000	9900000.0000	$\cdots$	5048858.8900
9995000.0000	9895050.0000	$\cdots$	5046334.4605
9990002.5000	9890102.4750	$\cdots$	5043811.2932
$\vdots$	$\vdots$		$\vdots$
9900473.5780	9801468.8423	$\cdots$	4998609.4034

As Napier says, "in the Third Table [between 10000000 and approximately 5000000] you have sixty-eight numbers interpolated, in the proportion of 100 to 99 [between successive terms], and between each two of these you have twenty numbers interpolated in the proportion of 10000 to 9995." These $21 \times 69 = 1449$ numbers, fairly evenly interspersed in the interval [5000000, 10000000], constitute the basic reference points in the sophisticated and ingenious interpolation scheme that follows in the *Constructio*.

The logarithms of the numbers in the Third Table above could be approximated by linear interpolation. Since the numbers in each row and those in each column form a geometric progression, their logarithms form an arithmetic progression. Therefore the logarithm of the element in the pth row and qth column is

$$(p - 1) \text{ Nlog } 9995000 + (q - 1) \text{ Nlog } 9900000, \tag{1}$$

so it suffices to compute the logarithms of 9995000 and 9900000. This is the purpose of the First and Second Tables.

Extrapolating linearly from the last element of the First Table, we obtain

$$\text{Nlog } 9999900 \cong 100 \times (100/99.999505)$$
$$= 100.000495.$$

Hence if the ratio of two numbers (such as two successive terms in the Second Table) is 100000/99999, then the difference of their logarithms is 100.000495. From the last term of the Second Table we therefore obtain

$$\text{Nlog } 9995001.22 \cong 50 \times 100.000495$$
$$= 5000.02475,$$

and linear extrapolation from this value gives

$$\text{Nlog } 9995000 \cong 5001.24506.$$

Hence if the ratio of two numbers (such as two successive terms of a column in the Third Table) is 10000/9995, then the difference of their logarithms is approximately 5001.245. From the last term of the first column of the Third Table we therefore obtain

$$\text{Nlog } 9900473.578 \cong 20(5001.245) = 100024.9.$$

Now

$$\begin{aligned} \text{Nlog } 9895523.34 &= \text{Nlog}(9900473.58)(.9995) \\ &\cong 100024.9 + 5001.25 \\ &= 105026.15. \end{aligned}$$

Finally linear interpolation between these last two logarithms gives

$$\text{Nlog } 9900000 = 100503.36.$$

We can now fill in the logarithms of the remaining terms of the Third Table using (1). For the last terms in the 69th column we obtain

$$\begin{aligned} \text{Nlog } 5001109.96 &= 19(5001.245) + 68(100503.36) \\ &= 6929252.14 \end{aligned}$$

and

$$\begin{aligned} \text{Nlog } 4998609.40 &= 20(5001.245) + 68(100503.36) \\ &= 6934253.38. \end{aligned}$$

Linear interpolation between the last two logarithms gives

$$\text{Nlog } 5000000 \cong 6931472.12.$$

Actually

$$\text{Nlog } 5000000 = 6931471.81,$$

so our computations based on linear interpolation are correct to seven significant figures. If the ratio of two numbers is 2, then the difference of their logarithms is 6931472.

EXERCISE 2. Once the logarithms of numbers between 10,000,000 and 5,000,000 have been computed, show how the value of Nlog 5,000,000 can be used to compute the values of logarithms of numbers less than 5,000,000.

The table of logarithms of the numbers in the Third Table was called by Napier his "radical table". By interpolation between values in the radical table he computed his principal table or "canon" of logarithms of sines of angles between 0° and 90° at intervals of one minute. It should be pointed out that his sine of an angle was the opposite side in a right triangle with hypotenuse 10,000,000, so his sines ranged from 0 to 10,000,000; the modern definition of trigonometric functions as ratios is due to Euler. Hence his sines of angles between 0° and 90° were numbers lying between 0 and 10^7.

For purpose of simple illustration we have outlined above a reconstruction of Napier's "radical table" using *linear* interpolation. However, Napier recognized (at least intuitively) the non-linearity of the logarithm function, and therefore employed a somewhat subtler method of interpolation that enabled him to assign upper and lower bounds to the value of each desired logarithm. His intent was to guarantee accuracy to 7 significant figures, although his numerical error at the end of the Second Table made his seventh place unreliable.

For the purpose of this non-linear interpolation, Napier required a continuous definition of the logarithm function, rather than a discrete definition based on geometric progressions. Our interest here is in his definition of the logarithm as a continuous function, rather than his precise manner of interpolation, so for further details concerning his interpolation scheme we refer the reader to the accounts of Coolidge [5] and Hobson [8].

Napier's Curious Definition

Napier's actual logarithmic definition was based on considerations of the continuous motion of points along straight lines, no doubt because intuitive conceptions of physical motion provided (at that time) the only usable basis for quantitative considerations of continuous variables. For a conjectured reconstruction of Napier's thought, from the original consideration of arithmetic and geometric progressions to the eventual definition in terms of continuous motion, see Lord Moulton's article "The Invention of Logarithms, Its Genesis and Growth" in [NT].

This definition involves two points moving along two different lines. The first point P starts at the initial point P_0 of a segment $P_0 O$ of length 10^7, with initial speed 10^7, and moves toward O, with its speed decreasing in such a way that it always equals the remaining distance PO. The second point L starts at the initial point L_0 of a half-line, and moves to the right with constant speed 10^7 (Fig. 1). Napier then defines the segment $y = L_0 L$ to be the *logarithm* of the segment $x = PO$. As he says in Art. 26 of the *Constructio*, "The logarithm of a given sine is that number which has increased arithmetically with the same velocity throughout as that with which radius began to decrease geometrically, and in the same time as radius has decreased to the given sine" ([11], p. 16).

It is informative to explore this somewhat obscure definition in terms of what we now call *natural* logarithms—log x being the power to which e

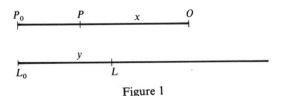

Figure 1

must be raised to give x. In calculus notation, the motion of the point P is described by the differential equation

$$\frac{dx}{dt} = -x$$

with the initial condition $x(0) = 10^7$, whose solution is

$$\log x = -t + \log 10^7$$

or

$$t = \log \frac{10^7}{x}.$$

The motion of the point L is therefore given by

$$y = 10^7 t = 10^7 \log \frac{10^7}{x}.$$

If we write $y = \text{Nog } x$ for Napier's logarithm of x, we therefore see that the relation between Nog x and the natural logarithm $\log x$ is given by

$$\text{Nog } x = 10^7 \log \frac{10^7}{x}. \tag{2}$$

It is clear from (2) that Napier's logarithms do not share the (now) usual properties of logarithms. For example, Nog $10^7 = 0$, while Nog x increases as x decreases in such a way that Nog $x \to \infty$ as $x \to 0$. Nevertheless Nog x has alternative properties that facilitate computation in a manner similar to the use of "ordinary" logarithms.

EXERCISE 3. Use (2) and the laws of logarithms ($\log xy = \log x + \log y$, $\log x^a = a \log x$) to show that

(i) Nog $xy = \text{Nog } x + \text{Nog } y - 10^7 \log 10^7$
(ii) Nog $x^a = a \text{ Nog } x + (1 - a)10^7 \log 10^7$
(iii) Nog $(x/y) = \text{Nog } x - \text{Nog } y + 10^7 \log 10^7$

Thus the computational use of a table of "Nogarithms" would involve continual addition or subtraction of multiples of $10^7 \log 10 = 23,025,851$.

On the basis of the above definition, Napier proceeded with his computations in essentially the following way. In successive time intervals of length 10^{-7}, starting at time $t = 0$, the point L with constant speed 10^7 moves a distance of 1 during each time interval, determining the points $L_1, L_2, L_3, \ldots$, with $L_0 L_n = n$ (Fig. 2). During the first of these very short time intervals the point P moves from P_0 to P_1 with a speed that is decreasing but still almost 10^7, so $P_0 P_1 \cong 1$ and $x_1 = P_1 O \cong 10^7 - 1 = 10^7(1 - 10^{-7})$. During the second time interval the speed of P is approximately $10^7(1 - 10^{-7})$, so

$$x_2 = OP_2 \cong 10^7(1 - 10^{-7}) - (1 - 10^{-7}) = 10^7(1 - 10^{-7})^2.$$

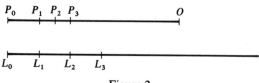

Figure 2

Continuing in this way, we find that $x_n = OP_n$ is approximately $10^7(1 - 10^{-7})^n$. Thus Napier's logarithm of $10^7(1 - 10^{-7})^n$ is approximately n,

$$\text{Nog } 10^7(1 - 10^{-7})^n \cong n.$$

That is, if $x = 10^7(1 - 10^{-7})^n$, then Nog $x \cong n$. If we now write $\tilde{\text{N}}\text{og } x = n$ (the tilde signifying approximation), then $\tilde{\text{N}}\text{og } x$ is the version of the logarithm that we denoted by Nlog x in the previous section, and this is the basis for the complicated interpolation scheme described there.

By means of ingenious approximations Napier guaranteed that his interpolations were accurate (except for the mistake mentioned previously) to 7 significant figures. Using (2) and the infinite series

$$\log(1 - x) = -x - \frac{x^2}{2} - \frac{x^3}{3} - \cdots$$

that will play an important role later, we see that, if $x = 10^7(1 - 10^{-7})^n$, then

$$\begin{aligned}
\text{Nog } x &= 10^7 \log \frac{10^7}{10^7(1 - 10^{-7})^n} \\
&= -10^7 n \log(1 - 10^{-7}) \\
&= n\left(1 + \frac{10^{-7}}{2} + \frac{10^{-14}}{3} + \cdots\right) \\
&= \left(1 + \frac{10^{-7}}{2} + \cdots\right) \tilde{\text{N}}\text{og } x \\
&\cong 1.00000005 \, \tilde{\text{N}}\text{og } x.
\end{aligned}$$

EXERCISE 4. If

$$x = 10^7(1 - 10^{-7})^m \quad \text{and} \quad y = 10^7(1 - 10^{-7})^n$$

so $\tilde{\text{N}}\text{og } x = m$ and $\tilde{\text{N}}\text{og } y = n$, show that

$$\tilde{\text{N}}\text{og } xy = \tilde{\text{N}}\text{og } x + \tilde{\text{N}}\text{og } y - Q$$

where

$$(1 - 10^{-7})^Q = 10^{-7}.$$

Then note that

$$Q = \log_{(1-10^{-7})}10^{-7} = \tilde{N}\text{og } 1$$
$$= -10^7 \log_{(1-10^{-7})^{10^7}}10^7 \quad \text{(why?)}$$
$$\cong -10^7 \log_{1/e}10^7$$
$$= 10^7 \log_e 10^7 = \text{Nog } 1.$$

Thus $\tilde{N}$og x obeys essentially the same additive law as Nog x.

EXERCISE 5. If $x = 10^7(1 - 10^{-7})^n$, show that

$$\tilde{N}\text{og } x = \log_{(1-10^{-7})}(10^{-7}x)$$
$$= 10^7 \log_{(1-10^{-7})^{10^7}}(10^{-7}x)$$
$$\cong 10^7 \log_{1/e}10^{-7}x$$
$$= 10^7 \log\frac{10^7}{x}.$$

Arithmetic and Geometric Progressions

As we have seen, the key idea employed by Napier in his logarithmic calculations was that of pairing the terms of an arithmetic progression with those of a geometric progression as in the following table. If x and y are two numbers whose product is desired, and they are terms of the geometric progression, $x = ar^m$ and $y = ar^n$, then their product divided by a, $xy/a = ar^{m+n}$ appears in the right-hand column opposite the term $(m + n)b$ in the left-hand column. If the number a is a power of 10 so that multiplication by a to obtain $xy = ax^m \cdot ax^n = a(ax^{m+n})$ involves simply a shift of the decimal point, the table therefore reduces the problem of multiplying x and y to the addition of the integers m and n.

Arithmetic Progression	Geometric Progression
b	ar
$2b$	ar^2
$3b$	ar^3
$\vdots$	$\vdots$
mb	ar^m
$\vdots$	$\vdots$
nb	ar^n
$\vdots$	$\vdots$
$(m+n)b$	ar^{m+n}
$\vdots$	$\vdots$

Napier's table involved such a pairing with $b = 1$, $a = 10^7$, $r = 1 - 10^{-7}$. This idea seems to have been "in the air" around the turn of the seventeenth century, for in work done simultaneously (but published later than Napier's), the Swiss instrument-maker Jost Bürgi constructed a similar table with $b = 10$, $a = 10^8$, $r = 1 + 10^{-4}$. Napier's value $r = 1 - 10^{-7}$ and Bürgi's value $r = 1 + 10^{-4}$ were both chosen very close to unity, so that the successive entries in the right-hand column (where x and y must be found if they are to be multiplied) would be very close together.

	Napier's Table		Burgi's Table
1	$10^7(1 - 10^{-7})$	$10 \cdot 1$	$10^8(1 + 10^{-4})$
2	$10^7(1 - 10^{-7})^2$	$10 \cdot 2$	$10^8(1 + 10^{-4})^2$
3	$10^7(1 - 10^{-7})^3$	$10 \cdot 3$	$10^8(1 + 10^{-4})^3$
$\vdots$	$\vdots$	$\vdots$	$\vdots$
n	$10^7(1 - 10^{-7})^n$	$10 \cdot n$	$10^8(1 + 10^{-4})^n$
$\vdots$	$\vdots$	$\vdots$	$\vdots$

Bürgi continued his table to 23,027 entries, because $(1 + 10^{-4})^{23,027} \cong 10$. Since n is the logarithm to the base $(1 + 10^{-4})$ of $(1 + 10^{-4})^n$, Bürgi's table was, except for the placement of decimal points, a table of antilogarithms to the base $(1 + 10^{-4})$.

By an appropriate shift of the decimal points in either Napier's or Bürgi's table, we can approximate natural (base e) logarithms. For example, write

$$\text{Bog } x = n \times 10^{-4} \quad \text{if } x = (1 + 10^{-4})^n,$$

and consider the following variant of Bürgi's table.

Bog x	x
1×10^{-4}	$(1 + 10^{-4})^1$
2×10^{-4}	$(1 + 10^{-4})^2$
$\vdots$	$\vdots$
$n \times 10^{-4}$	$(1 + 10^{-4})^n$
$\vdots$	$\vdots$

To see what "Bogs" really are, write $m = n \times 10^{-4}$. Then

$$x = (1 + 10^{-4})^n = \left[(1 + 10^{-4})^{10^4} \right]^m$$

so

$$\text{Bog } x = n \times 10^{-4}$$
$$= m$$
$$= \log_{[(1 + 10^{-4})^{10^4}]} x.$$

But $(1 + 10^{-4})^{10^4} = 2.718 \cong e$ to 4 significant figures, so Bog x is essentially the *natural* logarithm of x. This answers the question, properly posed by students, as to precisely what is "natural" about natural logarithms.

EXERCISE 6. Motivated by Napier's table, write nog $x = n \times 10^{-7}$ if $x = (1 - 10^{-7})^n$. Then show that nog x is essentially the logarithm of x with base $1/e$. *Hint*: $(1 - 10^{-7})^{10^7} \cong 1/e$.

EXERCISE 7. Show that Bogs satisfy the laws of logarithms,

$$\text{Bog } xy = \text{Bog } x + \text{Bog } y,$$
$$\text{Bog } x^a = a \text{ Bog } x.$$

The Introduction of Common Logarithms

In 1615 the English mathematics professor Henry Briggs visited Napier in Scotland, and their discussions led to Briggs' construction of a table of "improved" logarithms, ones for which the logarithm of one is zero and the logarithm of ten is one. These are now called "common" or base 10 logarithms.

Briggs immediately began the computation of these improved logarithms, which had more useful computational properties, and in 1624 published the *Arithmetica Logarithmica*, a table of 14-place common logarithms of the first 20,000 integers and of those from 90,000 to 100,000. The gap between 20,000 and 90,000 was filled by the Dutchman Adrian Vlacq, who published in 1628 the table of 10-place common logarithms from 1 to 100,000 that was to constitute the basis for nearly all logarithm tables for the next three centuries.

If Briggs' common logarithm Log x of x is defined in terms of Napier's logarithm by the transformation

$$\text{Log } x = \frac{\text{Nog } 1 - \text{Nog } x}{\text{Nog } 1 - \text{Nog } 10}, \tag{3}$$

then it is obvious that Log $1 = 0$ and Log $10 = 1$.

EXERCISE 8. Use the fact that Nog $xy = \text{Nog} x + \text{Nog } y - \text{Nog } 1$ and (3) to show that the common logarithm Log x satisfies the law of logarithms,

$$\text{Log } xy = \text{Log } x + \text{Log } y.$$

It follows from the law of logarithms that common logarithms enjoy the useful property that numbers differing only by location of the decimal point have logarithms differing by an integer,

$$\text{Log } 10^n x = n + \text{Log } x.$$

For example,

$$\text{Log } 2 = 0.30103,$$
$$\text{Log } 20 = 1.30103,$$
$$\text{Log } 200 = 2.30103.$$

Instead of using the above transformation (3) to convert Napier's table to a table of common logarithms, Briggs recomputed the whole table using a different method. He began by calculating successive square roots of 10. Starting with Log 10 = 1, he obtained the logarithm of each root by halving the logarithm of the previous root, as follows.

x	Log x
10	1.0000
$10^{1/2} = 3.16228$	0.5000
$10^{1/4} = 1.77828$	0.2500
$10^{1/8} = 1.33352$	0.1250
$10^{1/16} = 1.15478$	0.0625

After 54 such square root extractions (each carried out to 30 decimal places), he obtained a number $\alpha = (10)^{1/2^{54}}$ very slightly greater than one, for which

$$\text{Log } \alpha = \frac{1}{2^{54}}.$$

By repeated application of the law of logarithms, he then built up a table of logarithms of closely spaced numbers, the first table of "common logarithms."

Logarithms and Hyperbolic Areas

The tables of Napier and Briggs and their followers revolutionized the art of numerical computation. However, the importance of logarithms in the historical development of the calculus stems from a discovery published in 1647 by the Belgian Jesuit Gregory St. Vincent, that implies a surprising connection between the natural logarithm function and the rectangular hyperbola $xy = 1$.

If $[a, b]$ is a closed interval on the positive axis, denote by $A_{a,b}$ the area of the region that lies over this interval and under the hyperbola $xy = 1$ (Fig. 3). Then what Gregory discovered may be stated as follows. If $t > 0$, then

$$A_{ta, tb} = A_{a, b}. \tag{4}$$

To see why this is true, let

$$a = x_0 < x_1 < \cdots < x_{i-1} < x_i < \cdots < x_n = b$$

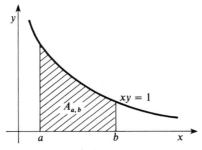

Figure 3

be equally-spaced points subdividing the interval $[a, b]$ into a large number n of sub-intervals, and above these sub-intervals construct inscribed and circumscribed rectangles as indicated in Fig. 4. Then the inscribed and circumscribed rectangles over the ith subinterval of $[a, b]$ have base $(b - a)/n$ and heights $1/x_i$ and $1/x_{i-1}$, respectively. Therefore

$$\sum_{i=1}^{n} \frac{b-a}{nx_i} \leqslant A_{a,b} \leqslant \sum_{i=1}^{n} \frac{b-a}{nx_{i-1}}. \tag{5}$$

Now the points

$$ta = tx_0 < \cdots < tx_{i-1} < tx_i < \cdots < tx_n = tb$$

similarly subdivide the interval $[ta, tb]$ into n equal subintervals. The inscribed and circumscribed rectangles over the ith subinterval $[tx_{i-1}, tx_i]$ of $[ta, tb]$ have base $(tb - ta)/n$ and heights $1/tx_i$ and $1/tx_{i-1}$, respectively. Hence their areas are equal to those of the inscribed and circumscribed rectangles over $[x_{i-1}, x_i]$. Therefore

$$\sum_{i=1}^{n} \frac{b-a}{nx_i} \leqslant A_{ta,tb} \leqslant \sum_{i=1}^{n} \frac{b-a}{nx_{i-1}}. \tag{6}$$

Comparison of (5) and (6) makes evident the truth of (4).

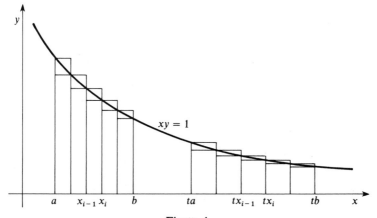

Figure 4

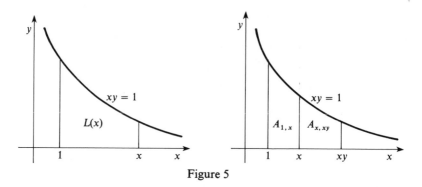

Figure 5

EXERCISE 9. Complete the above argument to rigorously prove by Archimedes' method of compression that $A_{a,\,b} = A_{ta,\,tb}$.

In reading through Gregory's *Opus Geometricum*, his friend A. A. de Sarasa noticed that Equation (4) implies that a certain area function associated with the hyperbola $xy = 1$ has the additive property that is characteristic of logarithms. Let

$$L(x) = \begin{cases} A_{1,\,x} & \text{if } x \geqslant 1, \\ -A_{x,\,1} & \text{if } 0 < x < 1. \end{cases}$$

Then $L(x)$ satisfies the "law of logarithms,"

$$L(xy) = L(x) + L(y). \tag{7}$$

For example, if x and y are both greater than 1, then

$$L(xy) = A_{1,\,xy}$$
$$= A_{1,\,x} + A_{x,\,xy} \quad \text{(see Fig. 5)}$$
$$= A_{1,\,x} + A_{1,\,y} \quad \text{(by Eq. (4))}$$
$$L(xy) = L(x) + L(y).$$

EXERCISE 10. Establish Equation (7) in the cases
$$0 < x \leqslant y \leqslant 1 \quad \text{and} \quad 0 < x \leqslant 1 \leqslant y.$$

EXERCISE 11. If $1 < a_1 < a_2 < \cdots < a_n < \cdots$ is a geometric progression, apply Equation (4) to show that
$$0 < L(a_1) < L(a_2) < \cdots < L(a_n) < \cdots$$
is an arithmetic progression.

Thus the hyperbolic area function $L(x)$ "looks like a logarithm," in that it provides a pairing between geometric and arithmetic progressions, so it is natural to inquire as to its relation to the natural logarithm function log x. It is, in fact, true that $L(x) = \log x$, although this relationship was not fully clarified until the time of Euler in the eighteenth century.

However, using a bit of calculus, we can "unmask" the function $L(x)$, by computing its derivative as follows.

$$
\begin{aligned}
L'(x) &= \lim_{h \to 0} \frac{L(x+h) - L(x)}{h} \\
&= \lim_{h \to 0} \frac{1}{h} L\left(1 + \frac{h}{x}\right) \quad \text{(using (7))} \\
&= \frac{1}{x} \lim_{h \to 0} \frac{x}{h} L\left(1 + \frac{h}{x}\right) \\
&= \frac{1}{x} \lim_{k \to 0} \frac{1}{k} L(1 + k) \quad\quad \left(k = \frac{h}{x}\right) \\
&= \frac{1}{x} \lim_{k \to 0} \frac{L(1+k) - L(1)}{k} \quad \text{(because } L(1) = 0) \\
&= \frac{L'(1)}{x}.
\end{aligned}
$$

It remains only to compute the single value

$$
L'(1) = \lim_{h \to 0} \frac{L(1+h)}{h} = \lim_{h \to 0} \frac{A_{1,\,1+h}}{h}
$$

of the derivative of L. Consulting Fig. 6, we see that

$$
\frac{h}{1+h} \leqslant A_{1,\,1+h} \leqslant h,
$$

so

$$
\frac{1}{1+h} \leqslant \frac{A_{1,\,1+h}}{h} \leqslant 1.
$$

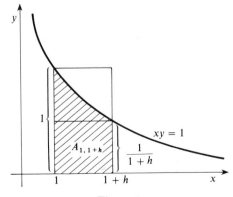

Figure 6

Taking the limit as $h \to 0$, it is clear that $L'(1) = 1$, so

$$L'(x) = \frac{1}{x}.$$

Since $L(x)$ and $\log x$ thus have the same derivative $1/x$, as well as the same value $L(1) = \log 1 = 0$ at $x = 1$, it follows by elementary calculus that $L(x) \equiv \log x$.

EXERCISE 12. Mimic the above computation of $L'(x)$ to show that the derivative of $\log x$ is

$$D \log x = \frac{\log e}{x},$$

where

$$e = \lim_{k \to 0} (1 + k)^{1/k}.$$

Newton's Logarithmic Computations

Although the precise relationship between logarithms and hyperbolic areas was not understood in the early seventeenth century (nor, for that matter, were natural logarithms recognized as logarithms to the base e), the general logarithmic character of the hyperbolic area function (as noticed by de Sarasa) served to stimulate the study of hyperbolic areas, and these investigations played a significant role in the introduction of infinite series and algorithmic calculus techniques, beginning in the 1650s and 1660s.

Apparently the first systematic computations of logarithms as hyperbolic areas were carried out by Newton in the mid 1660s. In a manuscript probably written in 1667 (see pp. 184–189 of Vol. II of Newton's *Mathematical Papers* cited in the references to Chapter 8), he starts with the hyperbola

$$y = \frac{1}{1 + x}, \qquad (x > -1)$$

and calculates the area $A(1 + x)$ lying under the hyperbola and over the interval $[0, x]$ (or the negative of this area if $-1 < x < 0$); see Figure 7. Writing

$$y = \frac{1}{1 + x} = 1 - x + x^2 - x^3 + \cdots,$$

this infinite series resulting from mechanical long division of $1 + x$ into 1 (as will be discussed in Chapter 7), he integrates term by term to obtain

$$A(1 + x) = x - \frac{x^2}{2} + \frac{x^3}{3} - \frac{x^4}{4} + \cdots. \tag{8}$$

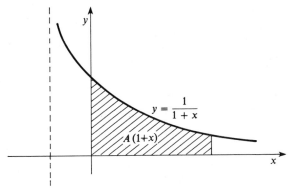

Figure 7

Of course $A(1 + x) = \log(1 + x)$, the natural logarithm of $1 + x$. Although Newton does not refer to $A(1 + x)$ as a logarithm, he recognizes its logarithmic character (perhaps from a direct or indirect acquaintance with the results of Gregory St. Vincent). For, referring to the points labeled in Figure 8, he says

> Now since the lines *ad*, *ae*, etc.: beare such respect to y^e [areas] *bcdf*, *bche*, etc: as numbers do to their logarithms; (viz: as ye lines *ad*, *ae*, etc.: increase in Geometrical Progression, so ye superfices *bcfd*, *bche*, etc.: increase in Arithmetical Progression): Therefore if any two or more of those lines multiplying or dividing one another doe produce some other line *ak*, their correspondent [areas], added or subtracted one to or from another shall produce y^e [area] *bcgk* correspondent to y^t line *ak*.

Thus Newton thinks of the hyperbolic area over $[0, x]$ as "correspondent" to the line of length $1 + x$ (hence *our* notation $A(1 + x)$), and he

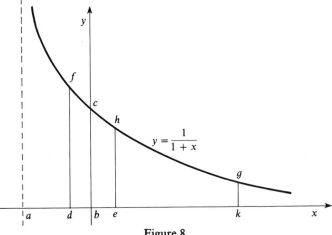

Figure 8

asserts that

$$A((1+x)(1+y)) = A(1+x) + A(1+y),$$
$$A\left(\frac{1+x}{1+y}\right) = A(1+x) - A(1+y),$$

the laws of logarithms. On the basis of these formulas he proceeds to calculate a small table of logarithms of integers.

First, taking $x = \pm 0.1$, ± 0.2 in (8), he calculates $A(0.8)$, $A(0.9)$, $A(1.1)$, $A(1.2)$ (to 57 decimal places!). Next he notes that

$$2 = \frac{1.2 \times 1.2}{0.8 \times 0.9}$$

$$3 = \frac{1.2 \times 2}{0.8}$$

$$5 = \frac{2 \times 2}{0.8}$$

$$11 = 10 \times 1.1$$
$$10 = 2 \times 5$$
$$100 = 10 \times 10$$

so that he can obtain $A(2)$, $A(3)$, $A(5)$, $A(11)$, $A(10)$, $A(100)$ merely by addition and subtraction, e.g.

$$A(2) = 2A(1.2) - A(0.8) - A(0.9).$$

Next he substitutes $x = \pm 0.02$, ± 0.001 into (8) to calculate $A(0.98)$, $A(1.02)$, $A(0.999)$, $A(1.001)$. This permits him to calculate the logarithms of 7, 13, 17, because

$$7 = \sqrt{\frac{100 \times 0.98}{2}} \; ,$$

(so $A(7) = \frac{1}{2}[A(100) + A(0.98) - A(2)]$)

$$13 = \frac{1000 \times 1.001}{7 \times 11} \; ,$$

$$17 = \frac{100 \times 1.02}{6} \; .$$

In order to check the accuracy of his computations, Newton calculates $A(0.9984)$ in two different ways: First, by substituting $x = -0.0016$ into (8), and then by noting the factorization

$$0.9984 = \frac{2^8 \times 3 \times 13}{10^5} \; ,$$

so

$$A(0.9984) = 8A(2) + A(3) + A(13) - 5A(10).$$

He finds (with evident pleasure) that the two results agree to more than 50 decimal places.

EXERCISE 13. In each of the following, start by finding the multiple of the given number that is closest to 1000.

(a) Express the logarithm of 37 in terms of those of 3, 10, and 0.999.
(b) Express the logarithm of 19 in terms of those of 2, 13, and 0.988. How would you compute the logarithm of 0.988?
(c) Express the logarithm of 31 in terms of those of 2, 10, and 0.992. How would you compute the logarithm of 0.992?

Mercator's Series for the Logarithm

The *Logarithmotechnia* of Nicolas Mercator (1620–1687) was published in 1668. The first two parts of this book were devoted entirely to the calculation of a table of common logarithms. Mercator's intuitive approach was to insert 10 million geometrical means (he called them *ratiunculae*) between 1 and 10; the logarithm of the number $x \in (1, 10)$ is then 10^{-7} times the number of *ratiunculae* between 1 and x.

EXERCISE 14. Consider the geometric sequence

$$1 = r^0, r^1, r^2, \ldots, r^n = 10.$$

If $x = r^k$, show that $\text{Log}_{10}x = k/n$.

To give an idea of Mercator's approach, he starts by calculating $\text{Log}_{10}1.005$ as follows. First he successively squares $g = 1.005$ and finds that $g^{256} < 10 < g^{512}$, and then he narrows this to

$$9.965774 = g^{461} < 10 < g^{462} = 10.015603.$$

Interpolation then gives $g^{461.6868} \cong 10$, so the number of *ratiunculae* between 1 and 1.005 is $10^7/461.6868 = 21{,}659.7$, and the common logarithm of 1.005 is 0.00216597 (his value; actually $\text{Log}_{10}1.005 = 0.00216606$). With this computation as a base, he proceeds to give directions for the practical computation of a complete table of common logarithms. See the article by Hofmann [9] for further details.

EXERCISE 15. By successively squaring on a pocket calculator, obtain the following powers of $g = 1.005$.

$g = 1.00500000$	$g^{16} = 1.08307115$
$g^2 = 1.01002500$	$g^{32} = 1.17304313$
$g^4 = 1.02015050$	$g^{64} = 1.37603017$
$g^8 = 1.04070705$	$g^{128} = 1.89345904$

Next calculate

$$g^{138} = g^{128}g^8g^2 = 1.99029078,$$
$$g^{139} = 1.005g^{138} = 2.00024224,$$

and then interpolate between these two values to obtain

$$g^{138.97565824} = 2.$$

Finally use the (corrected) value $\text{Log}_{10}1.005 = 0.00216606$ calculated by Mercator to obtain

$$\text{Log}_{10}2 = 138.97565824\ \text{Log}_{10}1.005 = 0.301030.$$

It is the very different third part of the *Logarithmotechnia* that is now of principal interest. Here Mercator finds his famous series (apparently used previously by Newton, as we have seen)

$$\log(1+x) = x - \frac{x^2}{2} + \frac{x^3}{3} - \frac{x^4}{4} + \cdots \tag{8}$$

for the area under the hyperbola $y = 1/(1+x)$ over the interval from 0 to x.

He starts by computing by long division the geometric series

$$y = \frac{1}{1+x} = 1 - x + \frac{x^2}{2} - \frac{x^3}{3} + \cdots . \tag{9}$$

It is sometimes incorrectly stated that Mercator obtained (8) from (9) by simple termwise integration, but he actually computed the area of the hyperbolic segment by a technique based on Cavalieri's indivisibles. Mercator only briefly alludes to the details, but a clearer exposition was presented by Wallis in his review of the *Logarithmotechnia* that was published in the *Philosophical Transactions* of 1668. In modern terms, the computation that Wallis outlines is roughly as follows (compare the discussion in Coolidge [4]).

Let us subdivide the interval $[0, x]$ into n equal subintervals each of length $h = x/n$, and construct the circumscribed rectangles based on these

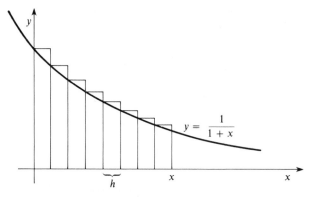

Figure 9

subintervals (Fig. 9), with heights

$$1, \quad \frac{1}{1+h}, \quad \frac{1}{1+2h}, \quad \cdots, \quad \frac{1}{1+(n-1)h}.$$

Expanding each of these heights in a geometric series, we find that the desired area

$$A \cong h + \sum_{j=1}^{n-1} \frac{h}{1+jh}$$

$$= h + h\left(\sum_{k=0}^{\infty} (-1)^k h^k\right) + h\left(\sum_{k=0}^{\infty} (-1)^k (2h)^k\right)$$

$$+ \cdots + h\left(\sum_{k=0}^{\infty} (-1)^k ((n-1)h)^k\right).$$

Collecting terms containing equal powers of h, we obtain

$$A \cong nh - h[h + 2h + \cdots + (n-1)h]$$
$$+ h[h^2 + (2h)^2 + \cdots + (n-1)^2 h^2]$$
$$\vdots$$
$$+ (-1)^k h[h^k + (2h)^k + \cdots + (n-1)^k h^k] + \cdots$$
$$= x - h^2[1 + 2 + \cdots + (n-1)]$$
$$+ h^3[1^2 + 2^2 + \cdots + (n-1)^2]$$
$$\vdots$$
$$+ (-1)^k h^{k+1}[1^k + 2^k + \cdots + (n-1)^k] + \cdots$$
$$= x - \frac{x^2}{n^2}\left(\sum_{i=1}^{n-1} i\right) + \frac{x^3}{n^3}\left(\sum_{i=1}^{n-1} i^2\right) + \cdots$$
$$+ (-1)^k \frac{x^{k+1}}{n^{k+1}}\left(\sum_{i=1}^{n-1} i^k\right) + \cdots,$$

substituting $h = x/n$.

Now in his *Arithmetica infinitorum* of 1656, Wallis had shown (by analogy with explicit computations for $k \le 10$) that

$$\lim_{n \to \infty} \frac{\sum i^k}{n^{k+1}} = \frac{1}{k+1} \quad (n \text{ terms in numerator}).$$

Taking the termwise limit as $n \to \infty$ of the last series above, we therefore obtain Mercator's series

$$A = x - \frac{x^2}{2} + \frac{x^3}{3} - \frac{x^4}{4} + \cdots . \qquad (8)$$

Wallis mentions that $x < 1$ is necessary for convergence.

As a consequence of the work of Gregory St. Vincent and de Sarasa, it seems to have been generally known in the 1660s that the area of a segment under the hyperbola $y = 1/x$ is proportional to the logarithm of the ratio of the ordinates at the ends of the segment. In a note by Mercator himself in the *Philosophical Transactions* of 1668, the logarithms determined by hyperbolic segments are referred to as *natural* logarithms, and he supplies the factor 0.43429 ($= 1/\log_e 10$) for transforming from natural to common logarithms (see Hofmann [9]).

EXERCISE 16. Show that the series (8) above is then the *natural* logarithm of $(1 + x)$.

EXERCISE 17. Let L_1 and L_2 be two "logarithm functions" having the property that $L_i(x) = \alpha L_i(y)$ if $x = y^\alpha$. Then show that the functions L_1 and L_2 are proportional, i.e.

$$\frac{L_1(x)}{L_2(x)} = \frac{L_1(y)}{L_2(y)}.$$

In regard to the inverse relation between the exponential and logarithmic concepts, Cajori ([2], p. 37) traces it back to Wallis' *Algebra* of 1685. Wallis considers the progressions

$$1, r, r^2, r^3, \ldots,$$
$$0, 1, 2, 3, \ldots,$$

and remarks that "These exponents they call logarithms, which are artificial numbers, so answering to the natural numbers, as that the addition and subduction (i.e. subtraction) of these answers to the multiplication and division of the natural numbers."

References

[NT]*Napier Tercentenary Memorial Volume* (edited by C. G. Knott). London: Longmans, 1915.
[1] C. B. Boyer, Fractional indices, exponents, and powers. *Natl Math Mag* **17**, 81–86, 1943.
[2] F. Cajori, History of the exponential and logarithmic concepts. *Am Math Mon* **20**, 35–47, 1913.
[3] H. S. Carslaw, The discovery of logarithms by Napier. *Math Gaz* **8**, 76–84, 115–119, 1915–16.
[4] J. L. Coolidge, The number *e*. *Am Math Mon* **57**, 591–602, 1950.
[5] J. L. Coolidge, *The Mathematics of Great Amateurs*. New York: Dover (reprint), 1963.
[6] J. W. L. Glaisher, On early tables of logarithms and early history of logarithms. *Q J Pure Appl Math* **48**, 151–192, 1920.
[7] N. T. Gridgeman, John Napier and the history of logarithms. *Scr Math* **29**, 49–65, 1973.

[8] E. W. Hobson, *John Napier and the Invention of Logarithms*. Cambridge University Press, 1914.

[9] J. E. Hofmann, On the discovery of the logarithmic series and its development in England up to Cotes. *Natl Math Mag* **14**, 37–45, 1939.

[10] J. F. Scott, *A History of Mathematics*. London: Taylor and Francis, 1960.

[11] D. J. Struik, *A Source Book in Mathematics, 1200–1800*. Cambridge, MA, Harvard University Press, 1969.

[12] O. Toeplitz, *The Calculus—A Genetic Approach*. Chicago: University of Chicago Press, 1963.

[13] D. T. Whiteside, Patterns of mathematical thought in the later seventeenth century. *Arch Hist Exact Sci* **1**, 179–388, 1960–62.

7 The Arithmetic of the Infinite

Introduction

Two main streams of discovery fueled the seventeenth century mathematical revolution and culminated in the synthesis of a powerful new infinitesimal analysis. One was the rich amalgam of specialized area and tangent methods from which the basic general algorithms of the calculus were distilled by Newton and Leibniz. The other centered on the development and application of infinite series techniques.

These two tools, the calculus and the analysis of infinite series, reinforced each other in their simultaneous development, because each served to broaden the range of application of the other. For example, in order to apply the early calculus methods to transcendental or "mechanical" functions, it was often necessary to express these functions as infinite series that could be differentiated or integrated termwise. Thus, if the function $f(x)$ could be "expanded" as an infinite (power) series,

$$f(x) = a_0 + a_1 x + a_2 x^2 + \cdots = \sum_{n=0}^{\infty} a_n x^n,$$

then presumably its derivative (or integral) could be calculated by differentiating (or integrating) each term of the series individually, just as though $f(x)$ were a (finite) polynomial in x. If the validity of this process is not critically questioned (and, in the seventeenth century, it was not), then the result is immediate,

$$f'(x) = a_1 + 2a_2 x + \cdots = \sum_{n=0}^{\infty} n a_n x^{n-1}.$$

In short, the elementary techniques of calculus, as they applied to simple

166

polynomials, could in this way be applied to any function for which an infinite power series expansion was available. At the same time, the termwise differentiation or integration of a known infinite series yielded a new one.

As an example, we saw in Chapter 6 that the quadrature of the hyperbola $y = 1/(1 + x)$, a problem that is not amenable to elementary exhaustion or indivisibles methods, was achieved by termwise integration of the geometric series

$$\frac{1}{1+x} = 1 - x + x^2 - x^3 + \cdots,$$

thereby yielding Mercator's series for the logarithm function,

$$\log(1 + x) = x - \frac{x^2}{2} + \frac{x^3}{3} - \frac{x^4}{4} + \cdots.$$

The infusion of infinite series into the "analytic art" of the seventeenth century raised immediate questions as to their behavior with respect to the ordinary algebraic processes of addition, subtraction, multiplication, division, and the extraction of roots. Could one properly manipulate infinite series in essentially the same ways that computations with ordinary algebraic expressions (i.e., polynomials) are carried out? We will see in this chapter that these questions were answered in the affirmative with the conclusion that (subject to convergence questions upon which the seventeenth century did not dwell) the algebra of infinite series obeys the same laws as the algebra of finite algebraic quantities.

The central event in this process of "legalizing" the use and enjoyment of infinite series was Newton's discovery of his famous binomial series. In modern notation, the *binomial series* takes the form

$$(1 + x)^\alpha = 1 + \binom{\alpha}{1}x + \binom{\alpha}{2}x^2 + \cdots = 1 + \sum_{n=1}^{\infty} \binom{\alpha}{n}x^n, \qquad (1)$$

where α is an arbitrary real number and the "binomial coefficients" are defined by

$$\binom{\alpha}{n} = \frac{\alpha(\alpha - 1) \cdot \cdots \cdot (\alpha - n + 1)}{n!}. \qquad (2)$$

The necessary condition $|x| < 1$ for the convergence of the binomial series was not stated by Newton.

In case the exponent α is a positive integer, (2) may be rewritten as

$$\binom{\alpha}{n} = \frac{\alpha!}{n!(\alpha - n)!} \qquad (3)$$

for $n \leqslant \alpha$, but is 0 for $n > \alpha$, so (1) reduces to a (finite) polynomial, such as the familiar cubic

$$(1 + x)^3 = 1 + 3x + 3x^2 + x^3.$$

However, if α is not a positive integer, then (1) gives an infinite series expansion of $(1+x)^\alpha$, such as

$$\frac{1}{1+x} = (1+x)^{-1} = 1 - x + x^2 - x^3 + \cdots$$

or

$$\sqrt{1+x} = (1+x)^{1/2} = 1 + \tfrac{1}{2}x - \tfrac{1}{8}x^2 + \tfrac{1}{16}x^3 + \cdots .$$

In order to appreciate the magnitude of Newton's accomplishment in formulating the binomial series in 1665, we must view it in the perspective of two pertinent historical facts. The first is that the use of non-integral exponents was then unknown. The modern exponential notation for positive integral powers (i.e., A^3 instead of Viète's "A cubus") had been introduced by Descartes in his *La Geometrie* of 1637, and was in general (if not universal) use by the 1660s. In his *Arithmetica infinitorum* of 1655 Wallis had mentioned negative and fractional "indices"—he spoke of the series $\sqrt{1}$, $\sqrt{8}$, $\sqrt{27}$, ... as having the "index 3/2," and the series $1/\sqrt{1}$, $1/\sqrt{2}$, $1/\sqrt{3}$, ... as having the "index $-1/2$." However, according to Cajori [3], the explicit use of fractional and negative exponents first appeared (publicly) in Newton's statement of his binomial series.

The second thing to remember is that the binomial formula for the case of positive integral powers was not then known in a form that suggested its generalization to negative or fractional powers. The binomial coefficients were not known in terms of a simple formula such as (3), but rather in terms of the entries in "Pascal's triangle" (which apparently dates back to medieval times), in which each entry is the sum of the two entries immediately above and to either side. The entries in the nth row (starting at 0 and counting down from the top) are then the coefficients of the powers of x in the binomial expansion of $(1+x)^n$.

```
              1
           1     1
        1     2     1
     1     3     3     1
  1     4     6     4     1
1     5    10    10     5     1
```

If the elements of Pascal's triangle are arranged as a matrix (Table 1) with ones in the 0th row and column, then the "law of formation" of this matrix gives the element $b_{p,q}$ in the pth row and qth column as

$$b_{p,q} = b_{p,q-1} + b_{p-1,q}, \tag{4}$$

the sum of the two elements immediately above and to the left of $b_{p,q}$. The binomial formula for positive integral exponents can then be written as

$$(1+x)^n = \sum_{p=0}^n b_{p,n-p} x^p. \tag{5}$$

Table 1

p	0	1	2	3	4
0	1	1	1	1	1
1	1	2	3	4	5
2	1	3	6	10	15
3	1	4	10	20	35
4	1	5	15	35	70

(column header: q)

EXERCISE 1. Apply (4) to add the row and column for $p=q=5$ to Table 1. Then use (5) to write down the binomial formula for $(1+x)^5$.

In essence, Newton attacked the problem, of generalizing (5) to the case where n is not a positive integer, as a problem of interpolating between the rows and column's of Pascal's triangle in the form of Table 1. That is, he sought some natural way, consistent with the law of formation (4), of inserting new rows and columns corresponding to non-integral values of p and q.

Newton modeled his interpolation procedure on the complex process by which Wallis a decade earlier had deduced his famous infinite product for π,

$$\frac{\pi}{2} = \frac{2}{1} \cdot \frac{2}{3} \cdot \frac{4}{3} \cdot \frac{4}{5} \cdot \frac{6}{5} \cdot \frac{6}{7} \cdot \ldots . \tag{6}$$

Nowadays Wallis' product is established as an easy consequence of the integrals

$$I_{2n} = \int_0^{\pi/2} \sin^{2n} x \, dx = \frac{\pi}{2} \cdot \frac{1}{2} \cdot \frac{3}{4} \cdot \ldots \cdot \frac{2n-1}{2n} \tag{7}$$

and

$$I_{2n+1} = \int_0^{\pi/2} \sin^{2n+1} x \, dx = \frac{2}{3} \cdot \frac{4}{5} \cdot \frac{6}{7} \cdot \ldots \cdot \frac{2n}{2n+1}, \tag{8}$$

and it is an important step in proving Stirling's asymptotic formula for the factorial,

$$n! \sim \sqrt{2\pi n} \left(\frac{n}{e}\right)^n \tag{9}$$

(meaning that the limit as $n \to \infty$ of the ratio of the two sides is one).

The interpolation procedure by which Wallis discovered (6) did not suffice to prove it, nor did Newton's interpolation procedure suffice to prove the binomial series—neither Newton nor anyone else proved it rigorously before the early nineteenth century. However, the mere discovery of the binomial series played an important role in establishing the use of infinite series as a working tool, and provided a cornucopia of new infinite series for use and application. In this chapter we describe the original and almost mystical investigations of Wallis and Newton, not merely as one of the more exotic byways in the history of mathematics, but also as a paradigm example of the frequently unexpected nature and sequence of mathematical invention, and of the crucial distinction between rigorous proof and the process of discovery that must precede it.

EXERCISE 2. Apply the reduction formula

$$\int \sin^n x \, dx = -\frac{1}{n}\sin^{n-1}x \cos x + \frac{n-1}{n}\int \sin^{n-2}x \, dx$$

(resulting from integration by parts) to obtain integrals (7) and (8).

EXERCISE 3. (a) Deduce from (7) and (8) that

$$\frac{\pi}{2} = \frac{2}{1}\cdot\frac{2}{3}\cdot\frac{4}{3}\cdot\frac{4}{5}\cdot\ \cdots\ \cdot\frac{2n}{2n-1}\cdot\frac{2n}{2n+1}\cdot\frac{I_{2n}}{I_{2n+1}}.$$

(b) Show that

$$1 < \frac{I_{2n}}{I_{2n+1}} \leqslant \frac{I_{2n-1}}{I_{2n+1}} = 1+\frac{1}{2n}.$$

(c) Derive Wallis' product from (a) and (b).

EXERCISE 4. Deduce from Wallis' product that

$$\lim_{n\to\infty}\frac{(n!)^2 2^{2n}}{(2n)!\sqrt{n}} = \sqrt{\pi} .$$

Hint: Multiply and divide the right-hand-side of

$$P_n = \frac{2}{1}\cdot\frac{2}{3}\cdot\frac{4}{3}\cdot\frac{4}{5}\cdot\ \cdots\ \cdot\frac{2n}{2n-1}\cdot\frac{2n}{2n+1}$$

to obtain

$$P_n = \frac{(n!)^4 2^{4n}}{[(2n)!]^2(2n+1)} .$$

EXERCISE 5. Write $n! = a_n\sqrt{n}\,(n/e)^n$, thereby defining a_n for each n. Assuming that $\lim_{n\to\infty}a_n = a\neq 0$, deduce from the previous exercise that $a=\sqrt{2\pi}$. This gives a weak form of Stirling's formula.

Wallis' Interpolation Scheme and Infinite Product

The last part of Wallis' *Arithmetica Infinitorum* (The Arithmetic of Infinites) of 1655 is an attempt to compute, using his arithmetical indivisibles, the area of a quadrant of the unit circle,

$$\frac{\pi}{4} = \int_0^1\sqrt{1-x^2}\ dx. \tag{10}$$

Of course neither of the symbols π and $\int$ were then in use. Wallis writes $\square$ for the reciprocal of the desired area, and in Proposition 121 sets up the limit sum that we would write as

$$\frac{1}{\square} = \lim_{n\to\infty} \frac{1}{n} \sum_{k=0}^{n} \sqrt{1 - \frac{k^2}{n^2}}$$

(a "Riemann sum" for the integral (10) corresponding to a subdivision of $[0, 1]$ into n equal subintervals).

Unable to directly compute this limit sum, he embarks on one of the more audacious investigations by analogy and intuition that has ever yielded a correct result (for anyone other than Euler), and winds up in the end with his infinite product (6) for $\pi/2$. As we have seen in Chapter 4, he knew from earlier work in the *Arithmetica infinitorum* that

$$\int_0^1 x^{p/q}\, dx = \frac{1}{(p/q)+1} = \frac{q}{p+q} \tag{11}$$

if p and q are positive integers. This formula suffices for the evaluation of any integral of the form

$$\int_0^1 (1 - x^{1/p})^q\, dx \tag{12}$$

if p and q are positive integers. For example,

$$\int_0^1 (1 - x^{1/3})^2\, dx = \int_0^1 (1 - 2x^{1/3} + x^{2/3})dx$$

$$= 1 - \frac{2}{\frac{1}{3}+1} + \frac{1}{\frac{2}{3}+1} = \frac{1}{10}.$$

Wallis' goal was to discover the "general law" or formula for the above integral in terms of p and q, and then substitute $p = q = \frac{1}{2}$ in this formula to obtain

$$\frac{1}{\square} = \int_0^1 (1 - x^2)^{1/2}\, dx.$$

For the purpose of recognizing the pattern, he found it more convenient to work with the reciprocal of the integral in (12),

$$f(p, q) = \frac{1}{\int_0^1 (1 - x^{1/p})^q\, dx}.$$

He began by computing the values of $f(p, q)$ for $p, q \leqslant 10$, and obtained the results shown in Table 2, where $a_{pq} = f(p, q)$ is tabulated in the pth row and qth column.

Table 2. $a_{pq} = f(p, q)$

p	0	1	2	3	4	...	10
				q			
0	1	1	1	1	1	...	1
1	1	2	3	4	5	...	11
2	1	3	6	10	15	...	66
3	1	4	10	20	35	...	286
4	1	5	15	35	70	...	1001
⋮	⋮	⋮	⋮	⋮	⋮		⋮
10	1	11	66	286	1001	...	184756

On the basis of these computed values for $p, q \leqslant 10$, Wallis took it as obvious that Table 2 is (for all p and q) simply a table of binomial coefficients. That is, each entry in the table is the sum of the one above it and the one to its left (compare Table 2 and Table 1).

EXERCISE 6. (For those familiar with the gamma and beta functions). Substitute $x = y^p$ to obtain

$$\frac{1}{a_{pq}} = \int_0^1 (1 - x^{1/p})^q \, dx = p \int_0^1 (1 - y)^q y^{p-1} \, dy$$

$$= pB(p, q+1) = \frac{\Gamma(p+1)\Gamma(q+1)}{\Gamma(p+q+1)}$$

$$= \frac{p!q!}{(p+q)!} = \frac{1}{\binom{p+q}{p}}.$$

In addition to explaining the evident diagonal symmetry of Table 2, Exercise 6 provides formulas that can be used to interpolate between the elements of a given row or column of the table. Wallis actually writes down these formulas on the basis of regarding the rows as sequences of "figurate" numbers.

For example, the second row ($p = 2$) consists of the "triangular" numbers

$$1, 3, 6, 10, 15, \ldots,$$

for which

$$a_{2, q} = \tfrac{1}{2}(q+1)(q+2).$$

Similarly, the third row ($p = 3$) consists of the "pyramidal" numbers

$$1, 4, 10, 20, 35, \ldots,$$

for which

$$a_{3, q} = \tfrac{1}{6}(q+1)(q+2)(q+3).$$

In general,

$$a_{p,q} = \frac{1}{p!}(q+1)(q+2)\cdots(q+p). \tag{13}$$

EXERCISE 7. Conclude from (13) that

$$a_{p,q} = \frac{p+q}{p}\, a_{p,q-1}. \tag{14}$$

Now Wallis wants to expand Table 2 by interpolation, to insert rows and columns corresponding to half-integral values of p and q (including in particular $p=q=\frac{1}{2}$ for which $a_{1/2,1/2}=\Box$). To begin with, he inserts half-integral values for q into (13) to interpolate between the elements of the pth row (p integral) of Table 2. For example,

$$a_{2,1/2} = \tfrac{1}{2}\left(\tfrac{1}{2}+1\right)\left(\tfrac{1}{2}+2\right) = \tfrac{15}{8},$$
$$a_{3,5/2} = \tfrac{1}{6}\left(\tfrac{5}{2}+1\right)\left(\tfrac{5}{2}+2\right)\left(\tfrac{5}{2}+3\right) = \tfrac{693}{48}.$$

By diagonal symmetry, this at the same time inserts values $a_{p,q}$ (p half-integral) between the elements of the qth column (q integral) of Table 2. The result of this interpolation of values of $a_{p,q}$ for either p or q (but not both) half-integral is the expanded Table 3 below, in which the interpolated values are printed in boldface. We have also inserted the unknown value $a_{1/2,1/2}=\Box$.

What remained for Wallis at this point was the crucial step of "filling in the blanks" in Table 3. To simplify the description of this final interpolation let us write

$$m = 2p, \qquad n = 2q, \qquad b_{m,n} = a_{p,q} = a_{m/2,n/2}.$$

Table 3. $a_{p,q}=f(p,q)=b_{2p,2q}$

n		0	1	2	3	4	5	6	$\cdots$
						q			
m	p	0	$\frac{1}{2}$	1	$\frac{3}{2}$	2	$\frac{5}{2}$	3	$\cdots$
0	0	1	**1**	1	**1**	1	**1**	1	$\cdots$
1	$\frac{1}{2}$	**1**	$\Box$	$\frac{3}{2}$		$\frac{15}{8}$		$\frac{105}{48}$	$\cdots$
2	1	1	$\frac{3}{2}$	2	$\frac{5}{2}$	3	$\frac{7}{2}$	4	$\cdots$
3	$\frac{3}{2}$	**1**		$\frac{5}{2}$		$\frac{35}{8}$		$\frac{315}{48}$	$\cdots$
4	2	1	$\frac{15}{8}$	3	$\frac{35}{8}$	6	$\frac{63}{8}$	10	$\cdots$
5	$\frac{5}{2}$	**1**		$\frac{7}{2}$		$\frac{63}{8}$		$\frac{693}{48}$	$\cdots$
6	3	1	$\frac{105}{48}$	4	$\frac{315}{48}$	10	$\frac{693}{48}$	20	$\cdots$

If m and n are even integers, then it follows from (14) that

$$b_{mn} = a_{m/2,\,n/2}$$
$$= \frac{m/2 + n/2}{m/2} a_{m/2,\,(n/2)-1}$$
$$b_{mn} = \frac{m+n}{n} b_{m,\,n-2}. \tag{15}$$

However, Wallis noted that Equation (15) is also satisfied by those elements a_{mn} for m or n odd that were inserted in Table 3 in the previous step. For example, from $b_{12} = \frac{3}{2}$ we obtain

$$b_{14} = \tfrac{5}{4} \times \tfrac{3}{2} = \tfrac{15}{8},$$

and from $b_{41} = \frac{15}{8}$ we obtain

$$b_{43} = \tfrac{7}{3} \times \tfrac{15}{8} = \tfrac{35}{8}.$$

EXERCISE 8. Use elementary properties of the gamma function and the fact that

$$a_{pq} = \frac{\Gamma(p+q+1)}{\Gamma(p+1)\Gamma(q+1)}$$

to establish (15) for all integers $m \geqslant 0$, $n \geqslant 2$.

Wallis then used (15) to fill in the remaining elements in the row $m = 1$ (and by symmetry the column $n = 1$) in terms of $\square$. For example,

$$b_{13} = \tfrac{4}{3} b_{11} = \tfrac{4}{3}\square,$$

and

$$b_{15} = \tfrac{6}{5} b_{13} = \tfrac{8}{5}\square.$$

Finally, he filled in the remaining blanks in Table 3 by using the "fact" that

$$b_{m,\,n} = b_{m,\,n-2} + b_{m-2,\,n}. \tag{16}$$

For m and n even, Equation (16) is just the familiar law of formation, $a_{p,\,q} = a_{p,\,q-1} + a_{p-1,\,q}$, of Table 2 as Pascal's triangle; Wallis simply assumed "by analogy" that (16) holds also when m or n or both are odd. For example, having already computed $b_{13} = b_{31} = \frac{4}{3}\square$, (16) gives

$$b_{33} = \tfrac{4}{3}\square + \tfrac{4}{3}\square = \tfrac{8}{3}\square.$$

Again,

$$b_{35} = b_{33} + b_{15}$$
$$= \tfrac{8}{3}\square + \tfrac{8}{5}\square = \tfrac{64}{15}\square, \text{ etc.}$$

EXERCISE 9. Use elementary properties of the gamma function and the fact that

$$a_{pq} = \frac{\Gamma(p+q+1)}{\Gamma(p+1)\Gamma(q+1)}$$

to establish (16) for all integers $m, n \geqslant 2$.

Now the final computation of $\square = b_{1,1}$ comes from the completed row for $m = 1$:

n

m	0	1	2	3	4	5	6	$\cdots$
1	1	$\square$	$\frac{3}{2}$	$\frac{4}{3}\square$	$\frac{15}{8}$	$\frac{8}{5}\square$	$\frac{105}{48}$	$\cdots$

From formula (15) with $m = 1$,

$$b_{1,n} = \frac{n+1}{n} b_{1,n-2},$$

it follows easily by induction that

$$b_{1,n} = 1 \times \frac{3}{2} \times \frac{5}{4} \times \cdots \times \frac{n+1}{n} \tag{17}$$

if n is *even*, while

$$b_{1,n} = \frac{\square}{2} \times \frac{2}{1} \times \frac{4}{3} \times \cdots \times \frac{n+1}{n} \tag{18}$$

if n is *odd*.

In addition it is clear from the definition

$$b_{1,n} = \frac{1}{\int_0^1 (1-x^2)^{n/2}\, dx}$$

that the sequence is monotone increasing,

$$b_{1,1} < b_{1,2} < b_{1,3} < \cdots < b_{1,n} < b_{1,n+1} < \cdots.$$

If we substitute (17) and (18) into

$$b_{1,2n-1} < b_{1,2n} < b_{1,2n+1}$$

the result is

$$\frac{\square}{2} \prod_{k=1}^{n} \frac{2k}{2k-1} < \prod_{k=1}^{n} \frac{2k+1}{2k} < \frac{\square}{2} \prod_{k=1}^{n+1} \frac{2k}{2k-1},$$

so rearrangement gives

$$\prod_{k=1}^{n} \frac{(2k)^2}{(2k-1)(2k+1)} < \frac{2}{\square} < \left[\prod_{k=1}^{n} \frac{(2k)^2}{(2k-1)(2k+1)} \right] \frac{2n+2}{2n+1}.$$

Since $2/\square = \pi/2$ and $(2n+2)/(2n+1)$ approaches one as $n\to\infty$, it finally follows that

$$\frac{\pi}{2} = \lim_{n\to\infty} \prod_{k=1}^{n} \frac{(2k)^2}{(2k-1)(2k+1)}$$
$$= \frac{2}{1} \cdot \frac{2}{3} \cdot \frac{4}{3} \cdot \frac{4}{5} \cdots$$

as desired.

Thus Wallis derived his famous infinite product on the basis of several unproved assumptions which, as we have indicated in the exercises, can be substantiated using elementary properties of the gamma function. This reliance upon reasoning "by analogy" did not escape the criticism of Wallis' contemporaries. Thomas Hobbes (1588–1679) attacked the *Arithmetica infinitorum* as "a scab of symbols," and objected to "the whole herd of them who apply their algebra to geometry." Fermat more specifically criticized Wallis' use of "incomplete" rather than "complete" (i.e. ordinary mathematical) induction. In his *Algebra* of 1685, Wallis replied that his purpose "was not so much to show a method Demonstrating things already known as to show a way of Investigation or finding out of things yet unknown. Thus I look upon [incomplete] induction [or analogy] as a very good Method of Investigation; as that which doth very often lead us to the easy discovery of a General Rule; or is, at least, a good preparative to such an one" (quoted by Nunn in [6], p. 385). Thus he alluded to the heuristic processes by which mathematical results are often discovered, prior to any attempt at rigorous proof.

For a translation of a pertinent part of the *Arithmetica infinitorum*, see Struik's source book ([7], pp. 244–253). Nunn [6] gives an English paraphrase that we have made use of. See also Whiteside's article ([8], pp. 236–241) for an outline of Wallis' interpolation method.

Quadrature of the Cissoid

In his *Tractatus duo de cycloide* (1659) Wallis applied his method of "interpolation by analogy" to the quadrature of the cissoid. The following set of exercises outlines his method (as described by Whiteside [8], pp. 242–243) and will provide the reader with first-hand practice.

EXERCISE 10. The *cissoid*, associated with the circle $y^2 = x(1-x)$ of diameter 1, is defined as the locus of the point B such that $BL/OL = OL/KL$ (Fig. 1). Show that its equation in rectangular coordinates is

$$y = x^{3/2}(1-x)^{-1/2}, \qquad x \in (0, 1).$$

Obviously $B\to O$ as $x\to 0$, and the line $x = 1$ is an asymptote. Hence the area under

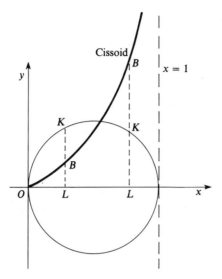

Figure 1

the cissoid is

$$A = \int_0^1 x^{3/2}(1-x)^{-1/2}\, dx$$

provided that this improper integral converges.

EXERCISE 11. Let

$$a_m = \int_0^1 x^{1/2}(1-x)^{m/2}\, dx = \int_0^1 x^{m/2}(1-x)^{1/2}\, dx$$

and note that $a_1 = \pi/8$, the area of the semi-circle of diameter one. Calculate directly the values

$$a_0 = \tfrac{2}{3}, \qquad a_2 = \tfrac{2}{3}\cdot\tfrac{2}{5}, \qquad a_4 = \tfrac{2}{3}\cdot\tfrac{2}{5}\cdot\tfrac{4}{7}, \qquad a_6 = \tfrac{2}{3}\cdot\tfrac{2}{5}\cdot\tfrac{4}{7}\cdot\tfrac{6}{9},$$

from which it appears that

$$a_m = \frac{m}{m+3} a_{m-2}$$

for m even. Assuming that this recursion relation holds for m odd as well, it follows that

$$a_3 = \frac{3a_1}{6} = \frac{\pi}{16}.$$

EXERCISE 12. Now let

$$b_n = \int_0^1 x^{3/2}(1-x)^{n/2}\, dx$$

and note that b_{-1} is the desired area under the cissoid. Calculate directly the values

$$b_0 = \tfrac{2}{5}, \qquad b_2 = \tfrac{2}{5}\cdot\tfrac{2}{7}, \qquad b_4 = \tfrac{2}{5}\cdot\tfrac{2}{7}\cdot\tfrac{4}{9},$$

from which it appears

$$b_n = \frac{n}{n+5} b_{n-2}$$

for n even. Assuming that this recursion relation holds for n odd as well, it follows that

$$b_{-1} = 6b_1 = 6a_3 = \frac{3\pi}{8},$$

so the area of the cissoid is three times the area of the generating semi-circle.

EXERCISE 13. Rigorously establish the results of the preceding exercise by noting that

$$b_n = B\left(\frac{5}{2}, \frac{n}{2}+1\right) = \frac{\Gamma(5/2)\Gamma(n/2+1)}{\Gamma(n/2+7/2)}$$

$$= \frac{n}{n+5} \frac{\Gamma(5/2)\Gamma(n/2)}{\Gamma(n+5/2)}.$$

The Discovery of the Binomial Series

The formulation of the binomial series in 1665, as a result of his reading of the *Arithmetica infinitorum*, was Newton's first mathematical discovery of lasting significance. He did not formally publish it, but described it in the two famous letters of 1676 that he sent to Henry Oldenburg, secretary of the Royal Society of London, for transmission to Leibniz. In the *epistola prior* dated June 3, 1676 (see Chapter 8 for references to Newton's correspondence) he states that

Extractions of roots are much shortened by this theorem,

$$(P+PQ)^{m/n} = P^{m/n} + \frac{m}{n}AQ + \frac{m-n}{2n}BQ + \frac{m-2n}{3n}CQ$$

$$+ \frac{m-3n}{4n}DQ + \text{etc.} \tag{19}$$

where $P+PQ$ signifies the quantity whose root or even any power, or the root of a power, is to be found; P signifies the first term of that quantity, Q the remaining terms divided by the first, and m/n the numerical index of the power of $P+PQ$, whether that power is integral or (so to speak) fractional, whether positive or negative.

Each of the symbols $A, B, C, \ldots$, denotes the immediately preceding term; that is, $A = P^{m/n}$, $B = (m/n)AQ$, etc.

EXERCISE 14. Show that formula (19) is equivalent to the binomial series in the more familiar form

$$(P+PQ)^{m/n} = P^{m/n}\left[1 + \sum_{k=1}^{\infty} \binom{m/n}{k} Q^k\right],$$

where the binomial coefficients are defined by

$$\binom{m/n}{k} = \frac{1}{k!}\left(\frac{m}{n}\right)\left(\frac{m}{n}-1\right)\cdot \cdots \cdot \left(\frac{m}{n}-k+1\right).$$

The binomial coefficients for positive integral powers had probably been known for hundreds of years. It is of independent significance that, in stating the generalization for positive or negative rational powers, Newton introduces for the first time the use of negative or fractional exponents for the purposes of routine algebraic computation. As we have seen, Wallis wrote $''1/a^2$, whose index is $-2''$ and $''\sqrt{a}$, whose index is $\frac{1}{2}''$, but never actually employed negative or fractional exponents. But Newton follows the above statement of the binomial series with the innocuously phrased remark that:

> For as analysts, instead of aa, aaa, etc., are accustomed to write a^2, a^3, etc., so instead of $\sqrt{a}$, $\sqrt{a^3}$, $\sqrt{c:a^5}$ (i.e. $\sqrt[3]{a^5}$), etc. I write $a^{1/2}$, $a^{3/2}$, $a^{5/3}$, and instead of $1/a$, $1/aa$, $1/a^3$, I write a^{-1}, a^{-2}, a^{-3}.

The statement without proof of the binomial series in the *epistola prior* is followed by nine illustrative examples, including the following.

$$(c^2 + x^2)^{1/2} = c + \frac{x^2}{2c} - \frac{x^4}{8c^3} + \frac{x^6}{16c^5}$$
$$- \frac{5x^8}{128c^7} + \frac{7x^{10}}{256c^9} + \text{etc.}$$

$$(d+e)^{4/3} = d^{4/3} + \frac{4ed^{1/3}}{3} + \frac{2e^2}{9d^{2/3}} - \frac{4e^3}{81d^{5/3}} + \text{etc.}$$

$$\frac{1}{d+e} = \frac{1}{d} - \frac{e}{d^2} + \frac{e^2}{d^3} - \frac{e^3}{d^4} + \text{etc.}$$

$$(d+e)^{-3} = \frac{1}{d^3} - \frac{3e}{d^4} + \frac{6e^2}{d^5} - \frac{10e^3}{d^6} + \text{etc.}$$

In return to Leibniz' request for information concerning the origin of the binomial series, Newton outlined in the *epistola posterior* of October 24, 1676 the steps by which he had been led to its discovery. The binomial formula for positive integral powers had not previously been known in a form that permitted the simple replacement of an integral exponent by a fractional one. Instead its discovery was based upon a complicated investigation using Wallis' method of tabular interpolation. As Newton tells it (in translation; see Newton's *Correspondence*, Vol. II, p. 130):

> At the beginning of my mathematical studies, when I had met with the works of our celebrated Wallis, on considering the series by the intercalation of which he himself exhibits the area of the circle and the hyperbola, the fact that, in the series of curves whose common base or axis is x and

the ordinates $(1-x^2)^{0/2}$, $(1-x^2)^{1/2}$, $(1-x^2)^{2/2}$, $(1-x^2)^{3/2}$, $(1-x^2)^{4/2}$, $(1-x^2)^{5/2}$, etc., if the areas of every other of them, namely x, $x-\frac{1}{3}x^3$, $x-\frac{2}{3}x^3+\frac{1}{5}x^5$, $x-\frac{3}{3}x^3+\frac{3}{5}x^5-\frac{1}{7}x^7$, etc. could be interpolated, we should have the areas of the intermediate ones, of which the first $(1-x^2)^{1/2}$ is the circle: in order to interpolate these series I noted that in all of them the first term was x and that the second terms $(0/3)x^3$, $(1/3)x^3$, $(2/3)x^3$, $(3/3)x^3$, etc., were in arithmetical progression, and hence that the first two terms of the series to be intercalated ought to be

$$x-\tfrac{1}{3}\left(\tfrac{1}{2}x^3\right), \quad x-\tfrac{1}{3}\left(\tfrac{3}{2}x^3\right), \quad x-\tfrac{1}{3}\left(\tfrac{5}{2}x^3\right), \quad \text{etc.}$$

To intercalate the rest I began to reflect that the denominators 1, 3, 5, 7, etc. were in arithmetical progression, so that the numerical coefficients of the numerators only were still in need of investigation. But in the alternately given areas these were the figures [i.e., digits] of powers of the number 11, namely of these 11^0, 11^1, 11^2, 11^3, 11^4, that is, first 1; then 1, 1; thirdly 1, 2, 1; fourthly 1, 3, 3, 1; fifthly 1, 4, 6, 4, 1, etc. And so I began to inquire how the remaining figures in these series could be derived from the first two given figures ... .

Thus Newton considered the sequence of *functions*

$$f_n(x) = \int_0^x (1-t^2)^{n/2}\, dt,$$

whereas Wallis had only considered the sequence of *numbers*

$$a_n = \int_0^1 (1-t^2)^{n/2}\, dt.$$

When n is even $f_n(x)$ can be evaluated explicitly, since he knows from Wallis that

$$\int_0^x t^p\, dt = \frac{x^{p+1}}{p+1}.$$

Thus

$$f_0(x) = 1(x),$$
$$f_2(x) = 1(x) + 1\left(-\tfrac{1}{3}x^3\right),$$
$$f_4(x) = 1(x) + 2\left(-\tfrac{1}{3}x^3\right) + 1\left(\tfrac{1}{5}x^5\right),$$
$$f_6(x) = 1(x) + 3\left(-\tfrac{1}{3}x^3\right) + 3\left(\tfrac{1}{5}x^5\right) + 1\left(-\tfrac{1}{7}x^7\right), \quad \text{etc.}$$

displaying the integral binomial coefficients as "figures of powers of the number 11." Newton wants to interpolate, for n odd, the coefficients of the infinite series

$$f_n(x) = \sum_{m=0}^{\infty} a_{mn}\left[(-1)^m \frac{x^{2m+1}}{2m+1}\right].$$

In particular, when $n=1$ this will give the area of a segment of the circle.

The details of his interpolation procedure are given in an unpublished manuscript composed in 1665 (see pages 126–134 of the first volume of

Table 4

m	0	1	2	3	4	5	6	$\cdots$	10
0	1	1	1	1	1	1	1	$\cdots$	1
1	0	$\frac{1}{2}$	1	$\frac{3}{2}$	2	$\frac{5}{2}$	3	$\cdots$	5
2	0	*	0	*	1	*	3	$\cdots$	10
3	0	*	0	*	0	*	1	$\cdots$	10
4	0	*	0	*	0	*	0	$\cdots$	5
5	0	*	0	*	0	*	0	$\cdots$	1

Newton's *Mathematical Papers*, referenced in Chapter 8). He first observes, as in the above quotation, that when n is even the first two coefficients are

$$a_{0n} = 1 \quad \text{and} \quad a_{1n} = \frac{n}{2},$$

and he assumes this to be true for n odd as well. The problem is to fill in the missing entries (asterisks) in the above tabular display of the coefficients a_{mn}.

He notices that, because of the familiar law of formation of the integral binomial coefficients,

$$a_{m, n+2} = a_{m-1, n} + a_{m, n} \tag{20}$$

in the present notation, the successive rows of the *subtabulation*, that consists only of the known columns (those for n even), are of the forms listed in Table 5.

He then assumes, in the fashion of Wallis, that the individual rows of the full tabulation (in Table 4) are of the same forms as in Table 5, but with the constants $a, b, c, \ldots$ to be determined separately for each row. It may be noted that this would follow from the assumption that, for each m, a_{mn} is a polynomial in n of degree m.

Equating the values given in Table 4 for a_{2n} with n even, with the literal expressions given in the row $m = 2$ of Table 5, we obtain the equations

$$0 = c$$
$$0 = a + 2b + c$$
$$1 = 6a + 4b + c$$

Table 5

m	0	2	4	6	8	10	12
0	a	a	a	a	a	a	a
1	b	$a+b$	$2a+b$	$3a+b$	$4a+b$	$5a+b$	$6a+b$
2	c	$b+c$	$a+2b+c$	$3a+3b+c$	$6a+4b+c$	$10a+5b+c$	$15a+6b+c$
3	d	$c+d$	$b+2c+d$	$a+3b+3c+d$	$4a+6b+4c+d$	$10a+10b+5c+d$	$20a+15b+6c+d$

which are readily solved for $a = \frac{1}{4}$, $b = -\frac{1}{8}$, $c = 0$. Therefore

$$a_{21} = b + c = -\frac{1}{8}, \qquad a_{23} = 3a + 3b + c = \frac{3}{8}, \quad \text{etc.}$$

Similarly, equating the known values of a_{3n} for n even with the above literal expressions for $m = 3$, we obtain the equations

$$0 = d$$
$$0 = b + 2c + d$$
$$0 = 4a + 6b + 4c + d$$
$$1 = 20a + 15b + 6c + d$$

which are readily solved for $a = \frac{1}{8}$, $b = -\frac{1}{8}$, $c = \frac{1}{16}$, $d = 0$. Therefore

$$a_{31} = c + d = \frac{1}{16}, \qquad a_{33} = a + 3b + 3c + d = -\frac{1}{16}, \quad \text{etc.}$$

EXERCISE 15. Write down enough terms in the row for $m = 4$, beginning with

$$e, \, d + e, \, c + 2d + e, \, b + 3c + 3d + e, \ldots,$$

to obtain in the above way 5 equations in the unknowns a, b, c, d, e. Solve these equations and substitute to obtain

$$a_{41} = -\frac{5}{128} \quad \text{and} \quad a_{43} = \frac{3}{128}.$$

The values of a_{m1} obtained thus far give

$$f_1(x) = \int_0^x (1 - t^2)^{1/2} \, dt = x - \frac{\frac{1}{2}x^3}{3} - \frac{\frac{1}{8}x^5}{5} - \frac{\frac{1}{16}x^7}{7} - \frac{\frac{5}{128}x^9}{9} + \text{etc.}$$

Newton actually continued in this manner to calculate the two additional coefficients $a_{51} = 7/256$ of $-x^{11}/11$ and $a_{61} = -21/1024$ of $x^{13}/13$. At this point he could discern the general form of the coefficients, namely that they arise from continued multiplication of terms of the product

$$\left(\frac{1}{2}\right) \times \frac{\left(\frac{1}{2}\right) - 1}{2} \times \frac{\left(\frac{1}{2}\right) - 2}{3} \times \frac{\left(\frac{1}{2}\right) - 3}{4} \times \cdots.$$

That is,

$$-\frac{1}{8} = \left(\frac{1}{2}\right) \times \frac{\left(\frac{1}{2}\right) - 1}{2}$$

$$\frac{1}{16} = \left(\frac{1}{2}\right) \times \frac{\left(\frac{1}{2}\right) - 1}{2} \times \frac{\left(\frac{1}{2}\right) - 2}{3}$$

$$-\frac{5}{128} = \frac{1}{2} \times \frac{\left(\frac{1}{2}\right) - 1}{2} \times \frac{\left(\frac{1}{2}\right) - 2}{3} \times \frac{\left(\frac{1}{2}\right) - 3}{4} \quad \text{etc.,}$$

or

$$\binom{\frac{1}{2}}{k} = \frac{\frac{1}{2}\left(\frac{1}{2} - 1\right)\left(\frac{1}{2} - 2\right) \cdots \cdots \left(\frac{1}{2} - k + 1\right)}{k!}$$

in modern notation.

Thus it was by means of Wallisian quadrature by interpolation that Newton discovered the general binomial coefficient! But, as he continues in the *epistola posterior*,

> When I had learnt this, I immediately began to consider that the terms
>
> $$(1-x^2)^{0/2}, \quad (1-x^2)^{2/2}, \quad (1-x^2)^{4/2}, \quad (1-x^2)^{6/2}, \quad \text{etc.,}$$
>
> could be interpolated in the same way as the areas generated by them: and that nothing else was required for this purpose but to omit the denominators 1, 3, 5, 7, etc., which are in the terms expressing the areas; this means that the coefficients of the terms of the quantity to be intercalated $(1-x^2)^{1/2}$, or $(1-x^2)^{3/2}$, or in general $(1-x^2)^m$, arise by the continued multiplication of the terms of this series
>
> $$m \times \frac{m-1}{2} \times \frac{m-2}{3} \times \frac{m-3}{4} \quad \text{etc.}$$

That is, termwise differentiation of his quadrature result gives

$$(1-x^2)^m = 1 + \sum_{k=1}^{\infty} (-1)^k \binom{m}{k} x^{2k},$$

where

$$\binom{m}{k} = \frac{m(m-1)\cdots(m-k+1)}{k!}.$$

As examples he lists

$$(1-x^2)^{1/2} = 1 - \tfrac{1}{2}x^2 - \tfrac{1}{8}x^4 - \tfrac{1}{16}x^6 + \cdots,$$

$$(1-x^2)^{3/2} = 1 - \tfrac{3}{2}x^2 + \tfrac{3}{2}x^4 + \tfrac{1}{16}x^6 + \cdots,$$

and

$$(1-x^2)^{1/3} = 1 - \tfrac{1}{3}x^2 - \tfrac{1}{9}x^4 - \tfrac{5}{81}x^6 + \cdots.$$

"So then the general reduction of radicals into infinite series by that rule, which I laid down at the beginning of [the *epistola prior*] became known to me, *and that before I was acquainted with the extraction of roots.*"

Of course Newton was aware that his interpolatory investigation did not constitute a proof. In order to test his results, he squared the series for $(1-x^2)^{1/2}$, "and it became $1 - x^2$, the remaining terms vanishing by the continuation of the series to infinity. And even so $1 - (1/3)x^2 - (1/9)x^4 - (5/81)x^6$, etc., multiplied twice into itself also produced $1 - x^2$."

EXERCISE 16. Square the quantity

$$P(x) = 1 - \tfrac{1}{2}x^2 - \tfrac{1}{8}x^4 - \tfrac{1}{16}x^6 - \tfrac{5}{128}x^8 + R(x),$$

where x^{10} is the lowest power occurring in $R(x)$, to obtain

$$[P(x)]^2 = 1 - x^2 + Q(x).$$

What is the lowest power of x occurring in $Q(x)$?

This verification of the binomial expansion motivated Newton to "try whether, conversely, these series, which it thus affirmed to be roots of the quantity $1 - x^2$, might not be extracted out of it in an arithmetical manner." As indicated by his notes on his early studies of contemporary mathematical texts, he was familiar with the following method of Viète for the extraction of the square root of a number N.

Given an estimate A of $\sqrt{N}$, denote by E the error, so $\sqrt{N} = A + E$. Then

$$N - A^2 = (A + E)^2 - A^2 = 2AE + E^2 \cong 2AE,$$

so

$$E \cong \frac{N - A^2}{2A}$$

if A is a sufficiently good estimate of $\sqrt{N}$ that E^2 is small compared with $2AE$. This provides the basis for the computation of successive approximations to $\sqrt{N}$, as follows. Starting with an initial estimate x_1 of $\sqrt{N}$, define

$$x_{n+1} = x_n + \frac{N - x_n^2}{2x_n} = x_n + e_n, \tag{21}$$

where

$$e_n = \frac{N - x_n^2}{2x_n}.$$

Note that $N - x_{n+1}^2 = (N - x_n^2) - 2x_n e_n - e_n^2$, so the numerator of e_{n+1} is obtained by simple subtractions from the numerator of e_n.

EXERCISE 17. Suppose that the sequence $\{x_n\}_1^\infty$ defined inductively by (21) converges to $x_* \neq 0$. Then take the limit as $n \to \infty$ in Equation (21) to show that $x_*^2 = N$.

Prior to the onslaught of the so-called "new math," Viète's method was frequently taught to school children in a rote form illustrated by the following computation.

$$
\begin{array}{rcl l}
N & = & 54{,}756 & x_1 = 200 \\
x_1^2 & = & 40{,}000 & e_1 = 10\left[\frac{14{,}756}{10(400)}\right] = 30 \\
\hline
N - x_1^2 & = & 14{,}756 & x_2 = 230 \\
2x_1 e_1 & = & 12{,}000 & e_2 = \left[\frac{1{,}856}{460}\right] = 4 \\
e_1^2 & = & 900 & x_3 = 234 \\
\hline
N - x_2^2 & = & 1856 & \\
2x_2 e_2 & = & 1840 & \\
e_2^2 & = & 16 & \\
\hline
N - x_3^2 & = & 0 &
\end{array}
$$

Thus $\sqrt{54{,}756} = 234$.

EXERCISE 18. Calculate the square root of 461,041 by Viète's method, as above.

In order to check his binomial expansion of $(1 - x^2)^{1/2}$, Newton applied a literal version of Viète's method. Starting with $y_1 = 1$, he iteratively calculated

$$y_{n+1} = y_n + e_n,$$

where e_n is the term of lowest degree in

$$\frac{(1 - x^2) - y_n^2}{2},$$

noting that

$$(1 - x^2) - y_{n+1}^2 = (1 - x^2) - 2y_n e_n - e_n^2$$

for the purpose of the successive computations. As he continues in the *epistola posterior*, "the matter turned out well. This was the form of the working in square roots."

$$1 - x^2 \qquad (1 - \tfrac{1}{2}x^2 - \tfrac{1}{8}x^4 - \tfrac{1}{16}x^6 \quad \text{etc.}$$
$$1$$
$$\overline{}$$
$$-x^2$$
$$-x^2 + \tfrac{1}{4}x^4$$
$$\overline{}$$
$$-\tfrac{1}{4}x^4$$
$$-\tfrac{1}{4}x^4 + \tfrac{1}{8}x^6 + \tfrac{1}{64}x^8$$
$$\overline{}$$
$$-\tfrac{1}{8}x^6 - \tfrac{1}{64}x^8$$

EXERCISE 19. Work through the above calculation, as well as the following extraction of $\sqrt{1 + x^2}$, identifying y_n and e_n at each stage of the computations.

$$1 + x^2 \qquad (1 + \frac{x^2}{2} - \frac{x^4}{8} + \frac{x^6}{16} - \frac{5x^8}{128} + \frac{7x^{10}}{256} - \frac{21x^{12}}{1024} \cdots$$
$$1$$
$$\overline{}$$
$$x^2$$
$$x^2 + x^4/4$$
$$\overline{}$$
$$-x^4/4$$
$$-x^4/4 - x^6/8 + x^8/64$$
$$\overline{}$$
$$x^6/8 - x^8/64$$
$$x^6/8 + x^8/16 - x^{10}/64 + x^{12}/256$$
$$\overline{}$$
$$-5x^8/64 + x^{10}/64 - x^{12}/256.$$

Continue through two more steps to obtain the terms in x^{10} and x^{12} listed above. Also verify that this is the same result that the binomial series gives.

Newton discovered the process of algebraic "long division" of polynomials in a similar way. His binomial series with $m = -1$ gave the geometric series

$$\frac{1}{1+x} = 1 - x + x^2 - x^3 + x^4 \cdots,$$

and he subsequently noted that the same result is obtained "by operating in general variables in the same way as arithmeticians in decimal numbers divide." In analogy with the common process for numerical long division, he sets out the computation in the following way.

$$1 + x)1 \qquad\qquad\qquad (1 - x + x^2 - x^3 \cdots$$
$$\underline{1+x}$$
$$-x$$
$$\underline{-x-x^2}$$
$$x^2$$
$$\underline{x^2+x^3}$$
$$-x^3$$
$$\underline{-x^3-x^4}$$
$$x^4 \cdots$$

Actually, Newton had first discovered the geometric series in his early binomial work when he investigated by Wallis' method of tabular interpolation the area

$$A = \int_0^x \frac{dt}{1+t}$$

under the hyperbola $y = 1/(1+x)$ over the interval $[0, x]$.

EXERCISE 20. Calculate the coefficients a_{mn} in

$$\int_0^x (1+t)^n \, dt = \sum_{m=1}^\infty a_{mn} \frac{x^m}{m}$$

for $n = 0, 1, 2, 3, 4$, so as to verify the entries in Table 6. Note the familiar Pascal triangle relation

$$a_{m-1,\,n} + a_{m,\,n} = a_{m,\,n+1}$$

for $n \geqslant 0$. Assuming that this relation holds for $n = -1$ as well, and that $a_{1,\,-1} = 1$, conclude that

$$a_{m,\,-1} \doteq (-1)^{m+1}.$$

Consequently the result of the "interpolation" is

$$\int_0^x \frac{dt}{1+t} = x - \frac{x^2}{2} + \frac{x^3}{3} - \frac{x^4}{4} + \cdots.$$

Table 6

m	−1	0	1	2	3	4
1		1	1	1	1	1
2		0	1	2	3	4
3		0	0	1	3	6
4		0	0	0	1	4
5		0	0	0	0	1
6		0	0	0	0	0

"Believe it or not," the computation indicated in Exercise 20 was Newton's original derivation of Mercator's series! He then noticed that this is likewise the result of termwise integration of the geometric series.

Obviously, the most extraordinary feature of Newton's remarkable binomial investigations is the *sequence of invention*. He began with the quadrature of circular and hyperbolic segments by Wallis' method of tabular interpolation. Next, by termwise differentiation of the results of these quadratures, he discovered the binomial series. Finally, the need to verify the binomial series led him to the algebraic versions of the familiar numerical processes of long division and root extraction.

The final result of this sequence of investigations—the application to infinite series of the familiar procedures of simple arithmetic—was of greater importance than any single example such as the binomial series. As Boyer puts it, Newton "had found that analysis by infinite series had the same inner consistency, and was subject to the same general laws, as the algebra of finite quantities. Infinite series were no longer to be regarded as approximating devices only; they were alternative forms of the functions they represented" ([2], p. 432). As Newton himself somewhat optimistically phrased it, "whatever common analysis [i.e., ordinary algebra] performs by equations made up of a finite number of terms [i.e., polynomials] (whenever it may be possible), this method may always perform by infinite equations [i.e., infinite series]." (See page 241 of Volume II of Newton's *Mathematical Papers*).

Thus was banished forever the "horror of the infinite" that had impeded the Greeks, and was set loose the torrent of infinite series expansions that were to play a central role in the development and applications of the new calculus.

References

[1] C. B. Boyer, Fractional indices, exponents, and powers. *Natl Math Mag* **17**, 81–86, 1943.
[2] C. B. Boyer, *A History of Mathematics*. New York: Wiley, 1968.

[3] F. Cajori, History of the exponential and logarithmic concepts. *Am Math Mon* **20**, 5–14, 35–47, 1913.

[4] F. Cajori, Controversies on mathematics between Wallis, Hobbes, and Barrow. *Math Teacher* **22**, 146–151, 1929.

[5] J. L. Coolidge, The story of the binomial theorem. *Am Math Mon* **56**, 147–157, 1949.

[6] T. P. Nunn, The arithmetic of infinites. *Math Gaz* **5**, 345–356, 377–386, 1909–1911.

[7] D. J. Struik, *A Source Book in Mathematics 1200–1800*. Cambridge, MA: Harvard University Press, 1969.

[8] D. T. Whiteside, Patterns of mathematical thought in the later 17th century. *Arch Hist Exact Sci* **1**, 179–388, 1960–1962.

[9] D. T. Whiteside, Newton's discovery of the general binomial theorem. *Math Gaz* **45**, 175–180, 1961.

See also the citations of Newton's collected *Correspondence* and *Mathematical Papers* in the References to Chapter 8.

The Calculus According to Newton

8

The Discovery of the Calculus

When we say that the calculus was discovered by Newton and Leibniz in the late seventeenth century, we do *not* mean simply that effective methods were then discovered for the solution of problems involving tangents and quadratures. For, as we have seen in preceding chapters, such problems had been studied with some success since antiquity, and with conspicuous success during the half century preceding the time of Newton and Leibniz.

The previous solutions of tangent and area problems invariably involved the application of special methods to particular problems. As successful as were, for example, the different tangent methods of Fermat and Roberval, neither developed them into general algorithmic procedures. Between these special techniques for the solution of individual problems, and the general methods of the calculus for the solution of whole classes of related problems, we today may see only a moderate gap, but it was one that Fermat and Roberval and their early seventeenth century contemporaries saw no reason to attempt to bridge.

What is involved here is the difference between the mere discovery of an important fact, and the recognition that it *is* important—that is, that it provides the basis for further progress. In mathematics, the recognition of the significance of a concept ordinarily involves its embodiment in new terminology or notation that facilitates its application in further investigations. As Hadamard remarks, "the creation of a word or a notation for a class of ideas may be, and often is, a scientific fact of very great importance, because it means connecting these ideas together in our subsequent thought" ([2], p. 38).

For example, we have seen that Fermat constructed the difference $f(A + E) - f(A)$, noted that (for the polynomial functions he dealt with) it contained E as a factor, divided by E, and finally cancelled every term still containing E as a factor, thereby obtaining the quantity

$$\left. \frac{f(A + E) - f(A)}{E} \right|_{E=0}.$$

Of course we now call this quantity the derivative, and denote it by $f'(A)$. But Fermat did not call it anything, nor introduce any particular notation for it. If he had, the way would have been open for general applications, and he might have been (as he has been erroneously called) at least a co-discoverer of the differential calculus.

Perhaps the most clear-cut example in the history of calculus, between discovery and the recognition of significance, is provided by the "fundamental theorem of calculus," which explicitly states the inverse relationship between tangent and area problems (or, in modern terminology, between differentiation and integration). This relationship was implicit (if not conspicuous) in the results of the early seventeenth century area computations that we have discussed in preceding chapters—e.g., the area under the curve $y = x^n$ over the interval $[0, x]$ is $x^{n+1}/n + 1$, while the slope of the tangent line to the curve $y = x^{n+1}/n + 1$ is x^n. Indeed, Barrow stated and proved (as we have seen in Chapter 5) a geometric theorem that clearly enunciated the inverse relationship between tangents and quadratures. However, he failed to recognize that his "fundamental theorem" provided the basis for "a new subject characterized by a distinctive method of procedure" (Boyer [1], p. 187). The contribution of Newton and Leibniz, for which they are properly credited as the discoverers of the calculus, was not merely that they recognized the "fundamental theorem of calculus" as a mathematical fact, but that they employed it to distill from the rich amalgam of earlier infinitesimal techniques a powerful algorithmic instrument for systematic calculation.

Isaac Newton (1642–1727)

Newton was born on Christmas Day in 1642. Nothing that is known about his youth and early education heralded the fact that his life and work would mark a new age in the intellectual history of mankind. He entered Cambridge in the summer of 1661 and received his B.A. early in 1665. Upon Barrow's retirement as Lucasian Professor in 1669, Newton was elected as his successor, and remained at Cambridge until 1696, when he left for London to serve as Warden of the Mint. Upon his death in 1727 he was buried in Westminster Abbey with such pomp that Voltaire remarked, "I have seen a professor of mathematics, only because he was great in his vocation, buried like a king who had done good to his subjects."

Apparently Newton did not begin his serious study of mathematics—beginning with Euclid's *Elements* and Descartes' *Geometrie*—until the summer of 1664. During the two years of 1665 and 1666 when

Cambridge closed because of the plague, he returned to his country home in Lincolnshire, and there laid the foundations for the three towering achievements of his scientific career—the calculus, the nature of light, and the theory of gravitation. Of this *biennium mirabilissimum* he later wrote that "in those days I was in the prime of my age of invention and minded mathematics and philosophy more than at any time since."

Newton's *Principia Mathematica* of 1687 and *Opticks* of 1704 detailed his contributions to mechanics and optics. However, his contributions to pure mathematics (including the calculus) remained largely unpublished during his lifetime. Mathematical discoveries at that time were not usually announced by means of prompt journal publication, because journals devoted to mathematics did not yet exist, but were often communicated in the form of personal letters and privately circulated manuscripts (and even sometimes proposed as riddles). At his death Newton left behind a mass of approximately 5000 sheets ([6], p. 70) of unpublished mathematical manuscripts, some of which had circulated amongst his contemporaries or served as a basis for his infrequent mathematical publications. This huge corpus of unpublished manuscript defied efforts directed towards its systematic organization for almost three centuries, until the appearance (in eight volumes from 1967) of the monumental Cambridge edition of *The Mathematical Papers of Isaac Newton*, edited by D. T. Whiteside. Throughout this chapter this edition is referenced as [NP]; for example, [NP III:2] refers to Newton's 1671 treatise on methods of series and fluxions in Part 2 of Volume III. Similarly, ([NC II], pp. 32–41) refers to Newton's first letter to Leibniz, in the second volume of *The Correspondence of Isaac Newton*, of which seven volumes have appeared since 1959.

The Introduction of Fluxions

In October of 1666 Newton gathered together and organized the results of his calculus research during the previous two years into a manuscript later referred to as "The October 1666 Tract on Fluxions" ([NP I], pp. 400–448). This was the first of his formal papers on the calculus. Although unpublished until recently, apparently copies of the manuscript were seen by a few English mathematicians during Newton's lifetime and after his death.

Beginning in late 1665, Newton had studied the tangent problem by the method of combining the velocity components of a moving point in a suitable coordinate system. This approach was (as we have seen in Chapter 5) previously developed by Roberval, but this earlier work was probably unknown to Newton. This investigation of tangents by means of component motions provided both the motivation for the new method of fluxions, and the key to its geometric applications.

Newton regarded the curve $f(x, y) = 0$ as the locus of the intersection of two moving lines, one vertical and the other horizontal. The x and y coordinates of the moving point are then functions of the time t, specifying the locations of the vertical and horizontal lines, respectively. The motion is then the composition of a horizontal motion with velocity vector having length $\dot{x}$ and a vertical motion with velocity vector having length $\dot{y}$. By the parallelogram law for the addition of velocity vectors (then well-known in the case of constant velocity vectors, and here applied to instantaneous velocity vectors), the tangent velocity vector is the parallelogram sum of these horizontal and vertical vectors. It follows that the slope of the tangent line to the curve is $\dot{y}/\dot{x}$ (Fig. 1).

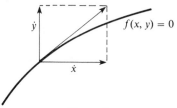

Figure 1

On this basis Newton considers the geometrical model of two (or more) points A and B traveling distances x and y along different straight lines in equal periods of time, such that $f(x, y) = 0$ at all times, with speeds $\dot{x}$ and $\dot{y}$, respectively, at a given time (Fig. 2).

He does not attempt to define the "fluxional speeds" of the points A and B, the *fluxions* $\dot{x}$ and $\dot{y}$ of x and y, with which the two points move with the "flux" (or flowing) of time. Instead, the concept of the speed of a point moving along a straight line is regarded as intuitively apparent on physical grounds. In modern terms, the fluxions $\dot{x}$ and $\dot{y}$ are simply the derivatives of x and y with respect to t,

$$\dot{x} = \frac{dx}{dt} \quad \text{and} \quad \dot{y} = \frac{dy}{dt},$$

and their ratio is the derivative of y with respect to x,

$$\frac{\dot{y}}{\dot{x}} = \frac{dy}{dx}.$$

It is with some irony that we use the differential notation of Leibniz to relate Newton's work in contemporary terms. It may also be noted that, in his early work, Newton generally used other letters, such as p and q instead of $\dot{x}$ and $\dot{y}$, for the fluxions of x and y. The dot notation, now regarded as characteristically Newtonian, was not consistently adopted by him until the early 1690s.

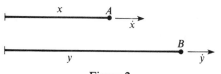

Figure 2

Newton's first problem is that of finding the relationship between the fluxions $\dot{x}$ and $\dot{y}$, given the relationship $f(x, y) = 0$ between x and y. For the case $f(x, y) = \Sigma a_{ij} x^i y^j$, a polynomial, he provides the following solution ([NP I], p. 402).

Set all y^e termes on one side of y^e equation that they become equal to nothing. And first multiply each terme by so many times $\dot{x}/x$ as x hath dimensions in that terme. Secondly multiply each terme by so many times $\dot{y}/y$ as y hath dimensions in it. . . . the summe of all these products shall bee equall to nothing. W^{ch} Equation gives y^e relation of y^e velocitys.

In other words, if $f(x, y) = \Sigma a_{ij} x^i y^j = 0$, his solution is

$$\Sigma \left(\frac{i\dot{x}}{x} + \frac{j\dot{y}}{y} \right) a_{ij} x^i y^j = 0. \tag{1}$$

EXERCISE 1. Show that (1) is equivalent to

$$\dot{x} \frac{\partial f}{\partial x} + \dot{y} \frac{\partial f}{\partial y} = 0$$

in terms of modern partial derivatives. Hence

$$\frac{\dot{y}}{\dot{x}} = - \frac{\partial f / \partial x}{\partial f / \partial y}.$$

In proof of (1), Newton first observes that if two bodies move with *uniform* (constant) velocities, then the distances traversed are proportional to their velocities. He continues, "And though they move not uniformly yet are y^e infinitely little lines wch each moment they describe, as their velocities wch they have while they describe y^m." His idea is that, during an "infinitely short" time interval o, the situation is the same as that during a finite time interval for the case of uniform motion—arbitrary motion is essentially uniform during an infinitely short time interval. "Soe y^t if y^e described lines bee x and y, in one moment, they will bee $x + \dot{x}o$ and $y + \dot{y}o$ in y^e next."

Illustrating the procedure by example ([NP I], p. 414), he therefore substitutes $x + \dot{x}o$ for x and $y + \dot{y}o$ for y in the equation $f(x, y) = 0$,

$$\Sigma a_{ij}(x + \dot{x}o)^i (y + \dot{y}o)^j = 0.$$

Binomial expansion then gives

$$\Sigma a_{ij} x^i y^j + \Sigma a_{ij} x^i (jy^{j-1}\dot{y}o + \text{terms in } o^2)$$

$$+ \Sigma a_{ij} y^j (ix^{i-1}\dot{x}o + \text{terms in } o^2)$$

$$+ \Sigma a_{ij}(ix^{i-1}\dot{x}o + \cdots)(jy^{j-1}\dot{y}o + \cdots) = 0.$$

Applying the fact that $\Sigma a_{ij}x^iy^j=0$, and dropping all terms involving o^2, the result is

$$\Sigma a_{ij}(ix^{i-1}y^j\dot{x}o + jx^iy^{j-1}\dot{y}o) = 0.$$

Division by o then gives (1) as desired. In justification of this procedure, Newton makes the following observations.

First y^t those termes ever vanish wch are not multiplyed by o, they being y^e propounded equation. Secondly those termes also vanish in wch o is of more y^n one dimension, because they are infinitely lesse y^n those in wch o is but of one dimension. Thirdly y^e still remaining termes, being divided by o will have [the desired form].

Thus, on the basis of plausible consequences of an intuitive physical conception of fluxions as instantaneous velocity components, he is able to compute (from (1) as in Exercise 1) the slope $\dot{y}/\dot{x}$ of the tangent line to an algebraic curve.

EXERCISE 2. Write $y=x^n$ in the form $f(x,y)=y-x^n=0$, and conclude from (1) that

$$\frac{\dot{y}}{\dot{x}} = nx^{n-1}.$$

The Fundamental Theorem of Calculus

Having calculated $dy/dx=\dot{y}/\dot{x}$ from the polynomial equation $f(x,y)=0$, Newton poses the converse problem: To find y in terms of x, given an equation expressing the relationship between x and the ratio $\dot{y}/\dot{x}$ of their fluxions. In case this equation is of the simple form

$$\frac{\dot{y}}{\dot{x}} = \phi(x)$$

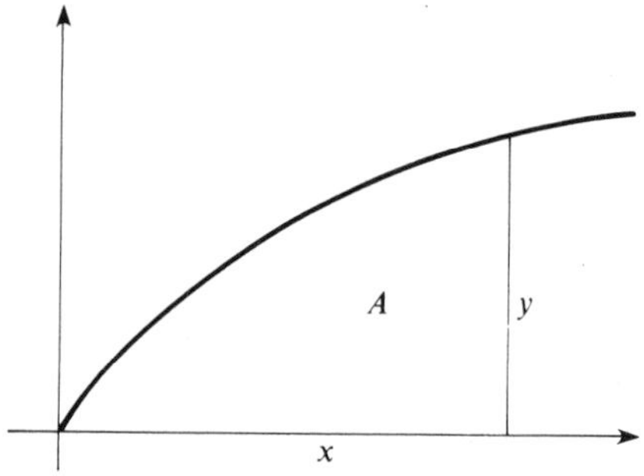

Figure 3

this is simply the problem of what we now call antidifferentiation, while the general case $g(x, \dot{y}/\dot{x}) = 0$ is a differential equation. An indication of Newton's insight (as early as 1666) is his remark that "could this ever bee done all problems whatever might bee resolved" ([NP I], p. 403).

In the fifth and seventh of his list of illustrative problems in the October 1666 tract, Newton discusses the computation of areas by means of antidifferentiation. This is the first historical appearance of the fundamental theorem of calculus in the explicit form

$$\frac{dA}{dx} = y$$

where A denotes the area under the curve $y = f(x)$, providing the basis for an algorithmic approach to the computation of areas. As we have seen, areas and tangents had been extensively calculated by ad hoc techniques throughout the early seventeenth century; it was the introduction and exploitation of general algorithmic techniques, by which these computations could be systematized, that constituted Newton's "discovery of the calculus."

Whereas previous infinitesimal techniques had been based, in principle, on the determination of an area as a limit of a sum (or, more crudely, as a sum of infinitesimal or indivisible elements of area), Newton introduced here the technique of first determining the rate of change of the desired area (with respect to x), and then calculating the area by antidifferentiation. In combination with his fluxional approach to tangents and rates of change, this made clear for the first time the precise nature of the inverse relationship between tangent problems and area problems, and the fact that both types of calculation are aspects of a single mathematical subject, one that is characterized by distinct and generally applicable algorithmic procedures.

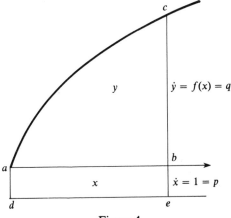

Figure 4

For Newton's fluxional formulation and derivation of the fundamental theorem of calculus, let y denote the area abc under the curve $q = f(x)$ (Fig. 4), and regard this area as being swept out by the vertical segment bc moving to the right with unit velocity $\dot{x} = 1$. If $p = 1$ (in the figure) then the area of the rectangle is x. Now, says Newton, "supposing y^e line cbe by parallel motion to describe y^e two [areas] x and y; The velocity wth wch they increase will bee, as be to bc: y^t is, y^e motion by wch x increaseth being $be = p = 1$, y^e motion by wch y increaseth will be $bc = q$." ([NP I], p. 427). Thus he takes it as obvious that the time rate of change of the area y is $q = f(x)$, with $\dot{x} = 1$, so

$$\frac{\dot{y}}{\dot{x}} = f(x).$$

His crucial insight consisted in the observation of this fact, rather than in a rigorous proof in modern terms. No doubt, he thought in terms of an increase in the area from y to $y + oq$, corresponding to an increase from x to $x + o$ during an "infinitely small" time interval o.

As an immediate application, this explained the reciprocal relationship between the fact that the slope of the curve with ordinate $x^{n+1}/n+1$ is x^n, and the fact that the area under the curve with ordinate x^n is $x^{n+1}/n+1$. For if

$$y = \frac{x^n + 1}{n + 1} \qquad \text{(the area)}$$

then the computational algorithm (1) gives

$$\frac{\dot{y}}{\dot{x}} = x^n,$$

and conversely. Taking $\dot{x} = p = 1$, corresponding to a unit time rate of increase of x, and $\dot{y} = q$, as in Fig. 4, this is

$$q = x^n \qquad \text{(the curve)}.$$

It may be noted that Newton habitually ignored the "constant of integration", taking all of his curves to pass through the origin.

The Chain Rule and Integration by Substitution

The tangent and area problems emphasize the importance of systematic procedures for differentiation (the calculation of $\dot{y}/\dot{x}$, given $f(x, y) = 0$) and antidifferentiation (the converse). Newton exploited the facility for differentiation and antidifferentiation by substitution methods—equivalent to what we call the chain rule and integration by substitution—that is essentially "built into" the calculus of fluxions.

As an example of the "built in" chain rule, suppose we want to calculate $\dot{y}/\dot{x}$ if

$$y = (1 + x^n)^{3/2}.$$

Newton would introduce a new variable $z = 1 + x^n$ with fluxion

$$\dot{z} = nx^{n-1}\dot{x}. \tag{2}$$

Then $y^2 = z^3$, so it follows that

$$2y\dot{y} = 3z^2\dot{z}. \tag{3}$$

EXERCISE 3. Apply the computational algorithm (1) to verify (2) and (3). But from (2) and (3) we conclude that

$$\frac{\dot{y}}{\dot{x}} = \frac{\dot{y}/\dot{z}}{\dot{x}/\dot{z}} = \frac{3z^2/2y}{1/nx^{n-1}}$$

$$= \frac{3nx^{n-1}(1+x^n)^2}{2(1+x^n)^{3/2}} = \frac{3}{2}nx^{n-1}\sqrt{1+x^n}\ .$$

This is an illustration of the following general procedure that Newton employed (in specific examples) to differentiate

$$y = [f(x)]^{m/n},$$

where $f(x)$ is a polynomial. First introduce the new variable $z = f(x)$ with fluxion $\dot{z} = f'(x)\dot{x}$. Then $y^n = z^m$, so it follows that

$$ny^{n-1}\dot{y} = mz^{m-1}\dot{z}$$

by application of (1). Finally

$$\frac{\dot{y}}{\dot{x}} = \frac{\dot{y}/\dot{z}}{\dot{x}/\dot{z}} = \frac{mz^{m-1}/ny^{n-1}}{1/f'(x)}$$

$$= \frac{m}{n}\frac{f'(x)[f(x)]^{m-1}}{[f(x)]^{(m/n)(n-1)}} = \frac{m}{n}[f(x)]^{m/n-1}f'(x),$$

the familiar "power formula" result of elementary calculus.

In a similar manner he is able to differentiate products and quotients, although he does not at this time formally state the product and quotient rules as explicit algorithms. Instead he illustrates by means of examples the following techniques. If

$$y = f(x)g(x),$$

let $u = f(x)$ and $v = g(x)$ have fluxions $\dot{u}$ and $\dot{v}$, respectively. If $f(x)$ and $g(x)$ are polynomials, then the basic rule (1) for the computation of the ratio of fluxions gives

$$\dot{y} = u\dot{v} + \dot{u}v,$$
$$\dot{u} = f'(x)\dot{x}, \quad \dot{v} = g'(x)\dot{x}. \qquad (4)$$

Hence

$$\frac{\dot{y}}{\dot{x}} = \frac{u\dot{v} + \dot{u}v}{\dot{x}} = f(x)g'(x) + f'(x)g(x).$$

EXERCISE 4. Apply (1) to verify (4).

If $y = f(x)/g(x) = u/v$, then $yv = u$ so

$$\dot{y}v + y\dot{v} = \dot{u}. \qquad (5)$$

Hence

$$\frac{\dot{y}}{\dot{x}} = \frac{\dot{u} - y\dot{v}}{v\dot{x}} = \frac{(\dot{u}/\dot{x}) - y(\dot{v}/\dot{x})}{v}$$
$$= \frac{f'(x) - g'(x)[f(x)/g(x)]}{g(x)}$$
$$= \frac{f'(x)g(x) - f(x)g'(x)}{[g(x)]^2}.$$

EXERCISE 5. Apply (1) to verify (5).

Newton describes this basic substitution process in the following language ([NP I], p. 411):

> Note y^t if there happen to bee in any Equation either a fraction or surde quantity . . . To find in what proportion the unknowne quantitys increase or decrease doe thus. 1. Take two letters y^e one (as ξ) to signify y^t quantity, y^e other (as $\dot{\xi}$) its motion of increase or decrease: And making an Equation betwixt y^e letter ξ & y^e quantity signifyed by it, find thereby [by (1)] y^e valor of the other letter $\dot{\xi}$. 2. Then substituting y^e letter ξ signifying y^t quantity, into its place in y^e maine Equation esteeme y^t letter ξ as an unknowne quantity & performe y^e worke of [(1)]; & into y^e resulting Equation instead of those letters ξ & $\dot{\xi}$ substitute theire valors. And soe you have y^e equation required.

For an example similar to one of Newton's, consider

$$y^2 = \frac{x}{\sqrt{a^2 - x^2}}.$$

First let

$$\xi = \frac{x}{\sqrt{a^2 - x^2}}, \quad \text{so} \quad a^2\xi^2 - x^2\xi^2 = x^2.$$

EXERCISE 6. Apply (1) to obtain

$$2a^2\xi\dot{\xi} - 2x\xi^2\dot{x} - 2x^2\xi\dot{\xi} = 2x\dot{x},$$

so

$$\dot{\xi} = \frac{x(1+\xi^2)\dot{x}}{(a^2-x^2)\xi}.$$

This is the first of Newton's three steps.

Now "y^e maine Equation" is $y^2 = \xi$, so

$$2y\dot{y} = \dot{\xi} = \frac{x(1+\xi^2)\dot{x}}{(a^2-x^2)\xi}$$

by Exercise 6. Hence

$$\frac{\dot{y}}{\dot{x}} = \frac{x(1+\xi^2)}{2y(a^2-x^2)\xi}$$

$$= \frac{x[1+x^2/(a^2-x^2)]}{2x^{1/2}(a^2-x^2)^{-1/4}(a^2-x^2)x(a^2-x^2)^{-1/2}}$$

$$= \tfrac{1}{2}a^2x^{-1/2}(a^2-x^2)^{-5/4}.$$

After several such examples, he concluded that "how to proceed in other cases (as when there are cube rootes, surde denominators, rootes within rootes (as $\sqrt{ax} + \sqrt{aa - xx}$) etc. in the equation may bee easily deduced from what hath been already said" ([NP I], p. 413).

He applied a similar substitution technique to construct a fairly extensive table of antiderivatives [NP I: pp. 405-410]. To start with, he showed that if

$$\frac{\dot{y}}{\dot{x}} = \frac{cx^{n-1}}{a+bx^n}, \quad \text{then} \quad y = \square\frac{c}{nab+nbz}. \tag{6}$$

(Read "area under" for $\square$, Newton's usual integral notation). To see this, substitute $z = bx^n$, so $\dot{z} = nbx^{n-1}\dot{x}$ as usual. Then

$$\frac{\dot{y}}{\dot{z}} = \frac{\dot{y}/\dot{x}}{\dot{z}/\dot{x}} = \frac{cx^{n-1}}{a+bx^n} + nbx^{n-1}$$

$$= \frac{c}{nb}\frac{1}{a+z},$$

so that (6) follows from the fundamental theorem of calculus.

EXERCISE 7. If $\dot{y}/\dot{x} = cx^{n-1}\sqrt{a+bx^n+cx^{2n}}$, substitute $z = x^n$ in this manner to show that

$$y = \square\frac{c}{n}\sqrt{a+bz+cz^2}.$$

EXERCISE 8. If $\dot{y}/\dot{x} = (c/x)\sqrt{ax^n + bx^{2n}}$, substitute $z^2 = x^n$, so $2z\dot{z} = nx^{n-1}\dot{x}$, to show that $y = \square\, (2c/n)\sqrt{a + bz^2}$.

In some cases, such substitutions lead to exact quadratures. For example, if

$$\frac{\dot{y}}{\dot{x}} = \frac{cx^{n-1}}{\sqrt{a + bx^n}} \, ,$$

then the substitution $z = x^n$ leads to

$$\frac{\dot{y}}{\dot{z}} = \frac{c}{n}\, \frac{1}{\sqrt{a + bz}} \, ,$$

so

$$y = \frac{2c}{nb}\sqrt{a + bz} = \frac{2c}{nb}\sqrt{a + bx^n}$$

(as may be verified by straightforward differentiation).

In cases such as Exercises 7 and 8, where an exact quadrature (in algebraic terms) seemed impossible, Newton's general goal was to reduce the quadrature to the calculation of the area under the graph of a circular or hyperbolic function, such as

$$\square\frac{a}{b + cx} \quad \text{or} \quad \square\sqrt{a^2 - x^2} \, .$$

These areas he calculated by binomial expansion followed by termwise integration. For example, if

$$\frac{\dot{y}}{\dot{x}} = \frac{a}{b + cx} = \frac{a}{b} - \frac{acx}{b^2} + \frac{ac^2x^2}{b^3} - \cdots ,$$

then

$$y = \frac{ax}{b} - \frac{acx^2}{2b^2} + \frac{ac^2x^3}{3b^2} - \cdots . \tag{7}$$

Similarly, if

$$\frac{\dot{y}}{\dot{x}} = \sqrt{a^2 - x^2} = a - \frac{x^2}{2a} + \frac{x^4}{8a^3} - \cdots ,$$

then

$$y = ax - \frac{x^3}{6a} + \frac{x^5}{40a^3} - \cdots . \tag{8}$$

Applications of Infinite Series

In 1668 Mercator's *Logarithmotechnia* appeared, containing his famous series for log $(1 + x)$, obtained by long division of $1 + x$ into 1, followed by the equivalent of termwise integration of the resulting infinite series.

Spurred by this publication of an important result contained in his own unpublished work of several years earlier, Newton wrote up in early summer 1669 a brief compendium of his results under the title *De Analysi per Aequationes Numero Terminorum Infinitas* (On Analysis by Equations Unlimited in the Number of Their Terms) ([NP II], pp. 206–247). Although the *De Analysi* itself remained unpublished until 1711, and contained only a fraction of Newton's 1664–1666 work, it was privately circulated and served to introduce him and his work to certain members of the English mathematical community.

The *De Analysi* opens with a description, by means of rules stated without proof, of his general method (by application of the fundamental theorem) for computing the area under a curve $y = f(x)$ (as Newton says, "rather briefly explained than narrowly demonstrated").

The first rule recalls that if $y = ax^{m/n}$ then the desired area is

$$\frac{a}{(m/n) + 1} x^{m/n+1} = \frac{na}{m+n} x^{(m+n)/n}.$$

The second rule—"If the value of y is compounded of several terms of that kind the area also will be compounded of the areas which arise separately from each of those terms"—asserts the validity of termwise integration.

The third rule states that "if the value of y or any of its terms be more compounded than the foregoing (that is, is not a polynomial), it must be reduced to simpler terms, by operating in general terms in the same way as arithmeticians in decimal numbers divide, extract roots or solve affected equations."

His examples "by division" and "by root extraction" are essentially the same as the computations by series of circular and hyperbolic areas mentioned at the end of the previous section (Equations (7) and (8)).

Newton's Method

In preparation for the computation of areas by "the resolution of affected equations," Newton introduces by example the technique for approximating solutions of equations that is now known as "Newton's method." In order to "resolve" the equation

$$y^3 - 2y - 5 = 0, \tag{9}$$

he starts with the approximation 2 to its root. Substitution of $y = p + 2$ into (9) yields the equation

$$p^3 + 6p^2 + 10p - 1 = 0$$

for p. Neglecting the nonlinear terms in p, he solves $10p - 1 = 0$ for the approximation $p = 0.1$, so 2.1 is his second approximation to the root. He then substitutes $y = q + 2.1$ into (9) and obtains the equation

$$q^3 + 6.3q^2 + 11.23q + 0.061 = 0$$

for q. Again neglecting the higher degree terms, he solves $11.23q + 0.061 = 0$ for $q = -0.0054$. This yields his third approximation 2.0946 to the actual root (2.09455148 to 8 decimal places).

This method for solving the polynomial equation

$$f(x) = \sum_{i=0}^{k} a_i x^i = 0 \tag{10}$$

may be described as follows. Given an approximation x_n to the actual root x_*, we substitute $x_* = x_n + p$ into (10), obtaining

$$0 = \sum_{i=0}^{k} a_i x_*^i$$

$$= \sum_{i=0}^{k} a_i (x_n + p)^i$$

$$= \sum_{i=0}^{k} a_i \left(x_n^i + i x_n^{i-1} p + \cdots \right)$$

$$= \sum_{i=0}^{k} a_i x_n^i + p \sum_{i=0}^{k} i a_i x_n^{i-1} + \cdots$$

$$0 = f(x_n) + p f'(x_n) + \cdots,$$

where the dots indicate higher degree terms in p. Neglecting these higher degree terms, we obtain

$$p \cong -\frac{f(x_n)}{f'(x_n)}$$

so

$$x_* \cong x_n - \frac{f(x_n)}{f'(x_n)} = x_{n+1}, \tag{11}$$

the familiar formula for the $(n+1)$st approximation using Newton's method.

It is interesting to note that Newton nowhere mentions the standard geometrical derivation of (11), that is, from the fact that the derivative $f'(x_n)$ is the slope of the hypotenuse of the right triangle in Figure 5.

Newton next sketches by example a generalization of the above method for solving an equation of the form

$$f(x, y) = 0$$

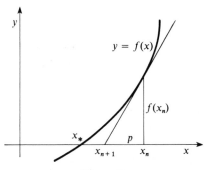

Figure 5

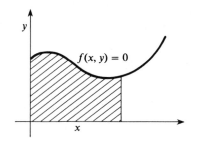

Figure 6

for y as a power series in x. His example is the equation

$$y^3 + a^2y + axy - x^3 - 2a^3 = 0,$$

EXERCISE 9. Let $a = 1$ in this equation, so

$$y^3 + y + xy - x^3 - 2 = 0. \tag{12}$$

When $x = 0$ the real solution is evidently $y = 1$, the first term of the power series for y. Substitute $y = 1 + p$ into (12), and discard the nonlinear terms in p and x in the resulting equation (in p and x) to obtain

$$p = -\tfrac{1}{4}x.$$

Next substitute $p = -\tfrac{1}{4}x + q$ into the preceding equation in p and x, and discard all terms of degree higher than 2 to obtain $q = x^2/64$. Thus the first three terms of the power series for y are

$$y = 1 - \frac{1}{4}x + \frac{x^2}{64} + \cdots .$$

Newton's idea (not exploited extensively in the *De Analysi*) is to calculate the area (Fig. 6) under a curve $f(x, y) = 0$ by "resolving the affected equation" for y as a power series in x, and then integrating this power series termwise (applying his first two rules).

The Reversion of Series

In the *De Analysi* Newton applies his method of successive approximations mainly to the "reversion of series". For example, given his series

$$z = x - \tfrac{1}{2}x^2 + \tfrac{1}{3}x^3 - \tfrac{1}{4}x^4 + \tfrac{1}{5}x^5 - \cdots$$

for the area z under the hyperbola $y = 1/(1+x)$, he wants to solve for x in terms of z (Fig. 7).

He decides to solve only for the first five terms in the series for x, and therefore drops all terms of degree higher than five, obtaining

$$\tfrac{1}{5}x^5 - \tfrac{1}{4}x^4 + \tfrac{1}{3}x^3 - \tfrac{1}{2}x^2 + x - z = 0. \tag{13}$$

Deletion of all nonlinear terms gives the first approximation

$$x \cong z.$$

Substitution of $x = z + p$ into (13) yields

$$\left(-\tfrac{1}{2}z^2 + \tfrac{1}{3}z^3 - \tfrac{1}{4}z^4 + \tfrac{1}{5}z^5\right) + p(1 - z + z^2 - z^3 + z^4)$$
$$+ p^2\left(-\tfrac{1}{2} + z - \tfrac{3}{2}z^2 + 2z^3\right) + \cdots = 0. \tag{14}$$

Neglecting nonlinear terms in p we obtain

$$p \cong \frac{\tfrac{1}{2}z^2 - \tfrac{1}{3}z^3 + \tfrac{1}{4}z^4 - \tfrac{1}{5}z^5}{1 - z + z^2 - z^3 + z^4} = \tfrac{1}{2}z^2 + \cdots,$$

so our second approximation is

$$x \cong z + \tfrac{1}{2}z^2.$$

Substitution of $p = \tfrac{1}{2}z^2 + q$ into (14) yields

$$\left(-\tfrac{1}{6}z^3 + \tfrac{1}{8}z^4 - \tfrac{1}{20}z^5\right) + q\left(1 - z + \tfrac{1}{2}z^2\right) + \cdots = 0$$

so

$$q \cong \frac{\tfrac{1}{6}z^3 - \tfrac{1}{8}z^4 + \tfrac{1}{20}z^5}{1 - z + \tfrac{1}{2}z^2} = \tfrac{1}{6}z^3 + \cdots .$$

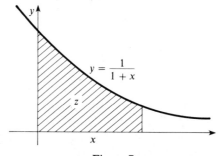

$$y = \frac{1}{1+x}$$

Figure 7

Thus the third approximation for x is

$$x \cong z + \tfrac{1}{2}z^2 + \tfrac{1}{6}z^3.$$

Continuing in this fashion Newton derives

$$x = z + \tfrac{1}{2}z^2 + \tfrac{1}{6}z^3 + \tfrac{1}{24}z^4 + \tfrac{1}{120}z^5 + \cdots . \tag{15}$$

Since $z = \log(1+x)$ is equivalent to $x = e^z - 1$, Newton has in (15) derived the exponential series

$$e^z = 1 + z + \tfrac{1}{2}z^2 + \tfrac{1}{6}z^3 + \cdots \tag{16}$$

for the first time.

In brief, the above method for the reversion of the series

$$z = a_1 x + a_2 x^2 + \cdots ,$$

to solve for

$$x = b_1 z + b_2 z^2 + \cdots ,$$

is as follows. Having found $b_1 (= 1/a_1)$, $b_2, \ldots, b_{n-1}$, substitute

$$x = b_1 z + \cdots + b_{n-1} z^{n-1} + r$$

into the original series and collect terms in the form

$$(Az^n + Bz^{n+1} + \cdots) + r(A' + B'z + \cdots) + \cdots = 0$$

Then

$$r \cong -\frac{Az^n + Bz^{n+1} + \cdots}{A' + B'z + \cdots} = -\frac{A}{A'}z^n + \cdots ,$$

so we take $b_n = -A/A'$.

Discovery of the Sine and Cosine Series

Newton proceeds to apply these techniques to obtain the power series for $\sin x$ and $\cos x$ for the first time. He derives first the series

$$\sin^{-1} x = x + \tfrac{1}{6}x^3 + \tfrac{3}{40}x^5 + \tfrac{5}{112}x^7 + \cdots \tag{17}$$

as follows. Consider the circle $x^2 + y^2 = 1$, as in Fig. 8. Then the angle $\theta = \sin^{-1} x$ (measured in radians) is twice the area of the circular sector OQR. But Newton knew, as a result of having integrated $\sqrt{1 - x^2}$ termwise after binomial expansion (see Equation (8)), the area of the segment $OPQR$ to be

$$x - \tfrac{1}{6}x^3 - \tfrac{1}{40}x^5 - \tfrac{1}{112}x^7 + \cdots .$$

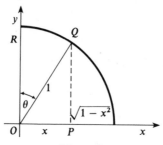

Figure 8

It therefore followed that

$$\theta = 2\left(x - \tfrac{1}{6}x^3 - \tfrac{1}{40}x^5 - \tfrac{1}{112}x^7 + \cdots\right) - x\sqrt{1-x^2}$$

$$= 2\left(x - \tfrac{1}{6}x^3 - \tfrac{1}{40}x^5 - \tfrac{1}{112}x^7 + \cdots\right)$$

$$- x\left(1 - \tfrac{1}{2}x^2 - \tfrac{1}{8}x^4 - \tfrac{1}{16}x^6 + \cdots\right)$$

$$\theta = x + \tfrac{1}{6}x^3 + \tfrac{3}{40}x^5 + \tfrac{5}{112}x^7 + \cdots$$

as desired (where $\sqrt{1-x^2}$ has been expanded by the binomial series).

Newton then obtained his series for the sine by reversion of series (17)!

EXERCISE 10. Apply Newton's method of successive approximations to invert the above series for $\sin^{-1}x$ and obtain

$$x = \sin\theta = \theta - \tfrac{1}{6}\theta^3 + \tfrac{1}{120}\theta^5 - \tfrac{1}{5040}\theta^7 + \cdots. \qquad (18)$$

EXERCISE 11. Calculate the square root to obtain the series for the cosine,

$$\cos\theta = \sqrt{1 - \sin^2\theta} = 1 - \tfrac{1}{2}\theta^2 + \tfrac{1}{24}\theta^4 - \tfrac{1}{720}\theta^6 + \cdots. \qquad (19)$$

This can be done by either (i) direct algebraic root extraction, (ii) application of the binomial series, or (iii) by application of Newton's method of successive approximations to solve the equation

$$y^2 + \left(\theta - \tfrac{1}{6}\theta^3 + \tfrac{1}{120}\theta^5 - \cdots\right)^2 = 1$$

for $y = \cos\theta$.

Having carried out these computations, Newton finally noticed the "obvious" factorial pattern of the coefficients,

$$\sin\theta = \sum_{k=0}^{\infty} \frac{(-1)^k \theta^{2k+1}}{(2k+1)!}, \qquad \cos\theta = \sum_{k=0}^{\infty} \frac{(-1)^k \theta^{2k}}{(2k)!}$$

In the *De Analysi*, Newton actually obtained (18) by inverting a series for the arclength $\stackrel{\frown}{QR}$. The above computation, based on area rather than arclength, is taken from his original investigation of 1665—see ([NP I],

p. 110) where he exuberantly writes out the series for $\sin^{-1}x$ through the term in x^{21}.

Newton next applied his trigonometric series to calculate the areas under the cycloid and the quadratrix ([NP II], pp. 239–241). He takes the cycloid ADG (Fig. 9) defined with respect to the circle of diameter $AH = 1$ by the relation

$$BD = BK + \overset{\frown}{AK},$$

that is, DK is equal to the length of the circular arc $\overset{\frown}{AK}$. If $AB = x$, then

$$BK = \sqrt{x - x^2}$$
$$= x^{1/2} - \tfrac{1}{2}x^{3/2} - \tfrac{1}{8}x^{5/2} - \tfrac{1}{16}x^{7/2}\cdots$$

by the binomial formula, and the circular arc from A to K is

$$\overset{\frown}{AK} = \frac{1}{2}\theta = \sin^{-1}\frac{AK}{AH}$$
$$= \sin^{-1}\sqrt{x}$$
$$= x^{1/2} + \tfrac{1}{6}x^{3/2} + \tfrac{3}{40}x^{5/2} + \tfrac{5}{112}x^{7/2}\cdots$$

by the above series for the arcsine. Consequently the Cartesian equation of the cycloid is

$$y = \sqrt{x - x^2} + \sin^{-1}\sqrt{x}$$
$$= 2x^{1/2} - \tfrac{1}{3}x^{3/2} - \tfrac{1}{20}x^{5/2} - \tfrac{1}{56}x^{7/2}\cdots.$$

By termwise integration the area ABD under the cycloid is therefore

$$\tfrac{4}{3}x^{3/2} - \tfrac{2}{15}x^{5/2} - \tfrac{1}{70}x^{7/2} - \tfrac{1}{252}x^{9/2}\cdots.$$

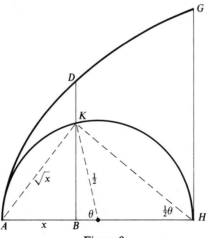

Figure 9

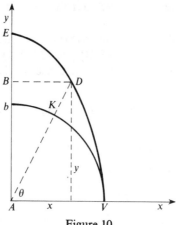

Figure 10

The quadratrix, a "mechanical" curve that was employed by the Greeks to square the circle and trisect an angle, is defined as follows (Fig. 10). Suppose the point B starts at the point $E(0, \pi/2)$ on the y-axis and moves to the origin A with constant speed during one unit of time, so its height at time t is

$$y = \frac{\pi}{2}(1-t).$$

At the same time suppose the point K starts at $b(0, 1)$ and moves down the unit quarter-circle bKV to V with constant angular speed in one unit of time, so its angle at time t is

$$\theta = \frac{\pi}{2}(1-t).$$

Then the typical point D on the quadratix is the intersection of the horizontal line through B and the radial ray through K. From the two expressions above we have

$$y = \theta = \tan^{-1}\frac{y}{x}$$

or

$$x = y \cot y$$

as the Cartesian equation of the quadratrix.

Now Newton has x and y interchanged (Fig. 11), so his equation is $y = x \cot x$. Substituting the series for the sine and cosine, he obtains

$$y = x\frac{\cos x}{\sin x} = x\frac{1-(1/2)x^2+(1/24)x^4-(1/720)x^6 \cdots}{x-(1/6)x^3+(1/120)x^5-(1/5040)x^7 \cdots}$$

$$= \frac{1-(1/2)x^2+(1/24)x^4-(1/720)x^6 \cdots}{1-(1/6)x^2+(1/120)x^4-(1/5040)x^6 \cdots}$$

$$y = 1-\tfrac{1}{3}x^2-\tfrac{1}{45}x^4-\tfrac{2}{945}x^6 \cdots$$

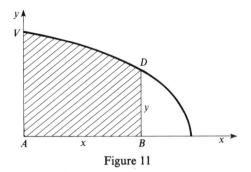

Figure 11

by division, so termwise-integration gives

$$x - \tfrac{1}{9}x^3 - \tfrac{1}{225}x^5 - \tfrac{2}{6615}x^7 \cdots$$

for the area $ABDV$.

Methods of Series and Fluxions

In 1671 Newton collected and organized his calculus investigations of the preceding half dozen years in the comprehensive treatise that is his mathematical chef-d'oeuvre, *De Methodis Serierum et Fluxionum* (Of the Methods of Series and Fluxions) ([NP III], pp. 32–353). His first attempts to publish this major work were unsuccessful, and then abandoned, but it was used throughout his life as a primary source of his early results, as (for example) in writing his two famous letters to Leibniz in 1676. It was loaned on occasion to interested parties, but did not appear in print until 1736 (after his death).

The first part of the *De Methodis* is an augmented version of the *De Analysi*, and includes an elaborate discussion of infinite series techniques for the solution of both algebraic and differential equations (the method of undetermined coefficients). This is followed by a richly detailed compilation, under the heading of twelve formally stated problems, of applications of Newton's series and fluxional methods.

For example, under "Problem 3—To Determine Maxima and Minima" ([NP III], pp. 117–121), he gives the following directions.

> When a quantity is greatest or least, at that moment its flow neither increases nor decreases: for if it increases, that proves that it was less and will at once be greater than it now is, and conversely so if it decreases. Therefore seek its fluxion [by previously described methods] and set it equal to nothing.

That is, Newton says to find the points at which $f(x)$ may attain its maximum or minimum values by solving the equation $f'(x) = 0$. He includes a list of nine geometrical problems that can be solved by this technique that is now a staple of introductory calculus courses.

In the *De Methodis* Newton's conception of fluxions has advanced from his original explanation in terms of velocities of moving bodies to the mature state that he many years later (see [NP III], p. 17 for reference) described as follows.

> I consider time as flowing or increasing by continual flux & other quantities as increasing continually in time & from ye fluxion of time I give the name of fluxions to the velocitys wth wch all other quantities increase. . . . I expose time by any quantity flowing uniformly & represent its fluxion by an unit, & the fluxions of other quantities I represent by any other fit symbols . . . This Method is derived immediately from Nature her self.

It is important to note that not only the time t but "any quantity flowing uniformly" can be chosen as the independent variable x. With $\dot{x} = 1$, the fluxion of any other variable is then its derivative with respect to x.

Applications of Integration by Substitution

Under Problem 8 ([NP III], pp. 119-209) Newton introduces his formal technique of integration by substitution. Let $v = f(x)$ and $y = g(z)$ describe the curves *FDH* and *GEI*, respectively (Fig. 12). Then, he says, "imagine their ordinates $DB = v$ and $EC = y$ to advance erect upon the bases $AB = x$ and $AC = z$: The increments, and so the fluxions, of the areas s and t thus traversed will then be as those ordinates multiplied into their speeds of advance, that is, into the fluxions of the bases." Therefore

$$\frac{\dot{s}}{\dot{t}} = \frac{v\dot{x}}{y\dot{z}}. \tag{20}$$

If we take $\dot{x} = 1$ so $\dot{s} = v$, it follows that $y = \dot{t}/\dot{z}$. If we further assume that $s = t$, then $\dot{s} = \dot{t} = v$, so

$$y = \frac{v}{\dot{z}}. \tag{21}$$

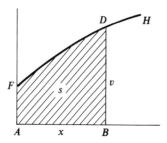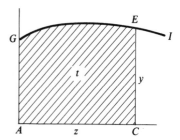

Figure 12

If, finally, z is given as a function of x, $z = \phi(x)$ or $x = \psi(z)$, then (21), with the right-hand side expressed as a function of z, defines the function $y = g(z)$ for which the two areas are equal. In particular

$$y = \frac{v}{\dot{z}} = \frac{f(x)}{\phi'(x)} = \frac{f(\psi(z))}{\phi'(\psi(z))}$$

$$y = f(\psi(z))\psi'(z),$$

so Newton is simply saying that

$$\int f(x)\,dx$$

transforms to

$$\int f(\psi(z))\psi'(z)\,dz$$

under the substitution $x = \psi(z)$.

As a first example, he takes $v = \sqrt{ax - x^2}$ and $z = \sqrt{ax}$ ($v^2 = ax - x^2$ and $z^2 = ax$). Then $2z\dot{z} = a$, so (21) gives

$$y = \frac{v}{\dot{z}} = \frac{2vz}{a}$$

$$= \frac{2z}{a}\sqrt{ax - x^2}$$

$$y = \frac{2z^2}{a^2}\sqrt{a^2 - z^2}$$

as the equation of the second curve, so

$$\int \sqrt{ax - x^2}\, dx = \int \frac{2z^2}{a^2}\sqrt{a^2 - z^2}\, dz.$$

Newton concludes the section with elegant quadratures of the cissoid, cycloid, and Archimedean spiral. These quadratures utilize a powerful technique for the transformation of integrals that generalizes the simple method of substitution discussed above.

Instead of assuming that the areas s and t under the curves $v = f(x)$ and $y = g(z)$, respectively, are equal, we assume a more general relation of the form

$$t = F(x, v) + ks \qquad (k = \text{constant}). \tag{22}$$

Upon substitution of

$$\dot{t} = \dot{x}F_x + \dot{v}F_v + k\dot{s}$$
$$= F_x + \dot{v}F_v + kv$$

into

$$y = \frac{\dot{t}}{\dot{z}},$$

the substitution $x = \psi(z)$ can be used to express y as a function of z, thereby obtaining $y = g(z)$ such that (22) holds. In particular

$$y = \frac{F_x(\psi(z), f(\psi(z))) + f'(\psi(z)) F_v(\psi(z), f(\psi(z))) + kf(\psi(z))}{\phi'(\psi(z))}.$$

It may be noted that this transformation incorporates the integration by parts technique as well as substitution. For if we take $v = f(x) = u(x)w'(x)$ and

$$t = u(x)w(x) - s, \tag{23}$$

then

$$\begin{aligned} \dot{t} &= u(x)w'(x) + u'(x)w(x) - u(x)w'(x) \\ &= u'(x)w(x), \end{aligned}$$

so

$$y = \frac{\dot{t}}{\dot{z}} = u'(\psi(z))w(\psi(z))\psi'(z) = g(z).$$

Since

$$s = \int f(x)\,dx, \qquad t = \int g(z)\,dz,$$

(23) therefore gives

$$\int u'(\psi(z))w(\psi(z))\psi'(z)\,dz = u(x)w(x) - \int u(x)w'(x)\,dx$$

or

$$\int w\,du = uw - \int u\,dw,$$

the familiar integration by parts formula.

For example, for his quadrature of the cissoid $v = x^2/\sqrt{ax - x^2}$, Newton takes $z = \sqrt{ax - x^2}$ and

$$t = \tfrac{1}{3}x\sqrt{ax - x^2} + \tfrac{2}{3}s. \tag{24}$$

He then computes $\dot{t} = ax/2\sqrt{ax - x^2}$, so $y = \dot{t}/\dot{z} = \sqrt{a^2 - z^2}$, the equation of a circle. Therefore (24) reduces the quadrature of a segment of the cissoid to the quadrature of a segment of the circle.

EXERCISE 12. Verify that $\dot{t} = ax/2\sqrt{ax - x^2}$ in the above computation.

Newton's Integral Tables

Newton's 1671 treatise contains two tables of integrals ([NP III], pp. 237–255). The first is entitled "A catalog of some curves related to rectilinear figures." It consists of a list of curves $y = f(z)$ such that the corresponding area $t = F(z)$ can be calculated explicitly by means of direct or inverse differentiation.

The second table, entitled "A catalogue of some curves related to conic sections," first appeared in print in an appendix ("De Quadratura Curvarum") to the 1704 *Opticks*. It consists of a list of curves $y = f(x)$ such that the corresponding area can be expressed in terms of the area under an appropriate conic section, i.e. in terms of an integral such as

$$\int \frac{dx}{a+bx} \quad \text{or} \quad \int \sqrt{a+bx+cx^2}\, dx.$$

These reductions are obtained by means of Newton's technique of comparing the area t under the curve $y = f(z)$ with the area s under the curve $v = g(x)$, given a "substitution" $x = \psi(z)$.

For example, given $y = z^{\eta-1}/(e+fz^{\eta})$, he substitutes $x = z^{\eta}$ and considers the hyperbola

$$v = \frac{1}{e+fx}.$$

Taking $\dot{z} = 1$, it follows that

$$\dot{t} = y\dot{z} = \frac{z^{\eta-1}}{e+fz^{\eta}}$$

and

$$\dot{s} = v\dot{x} = \frac{\dot{x}}{e+fx} = \frac{\eta z^{\eta-1}}{e+fx},$$

where

$$t = \int y\, dz \quad \text{and} \quad s = \int v\, dx.$$

Hence $\dot{t} = \dot{s}/\eta$ so

$$t = \frac{s}{\eta},$$

or

$$\int \frac{z^{\eta-1}dz}{e+fz^{\eta}} = \frac{1}{\eta}\int \frac{dx}{e+fx} \tag{25}$$

(taking lower limits so that the constant of integration is zero).

Given (25), Newton calculates the integrals

$$t_k = \int \frac{z^{k\eta-1}dz}{e+fz^{\eta}} \qquad (k \geqslant 1)$$

by recursion, as follows. Taking $\dot{z} = 1$,

$$\dot{t}_k = \frac{z^{k\eta - 1}}{e + fz^{\eta}}$$

$$= \frac{1}{f} z^{(k-1)\eta - 1} - \frac{e}{f} \frac{z^{(k-1)\eta - 1}}{e + fz^{\eta}}$$

$$\dot{t}_k = \frac{1}{f} z^{(k-1)\eta - 1} - \frac{e}{f} \dot{t}_{k-1},$$

so

$$t_k = \frac{z^{(k-1)\eta}}{(k-1)f\eta} - \frac{e}{f} t_{k-1}.$$

For $k = 2$ and $k = 3$, this yields

$$\int \frac{z^{2\eta - 1} dz}{e + fz^{\eta}} = \frac{z^{\eta}}{f\eta} - \frac{e}{f\eta} \int \frac{dx}{e + fx}$$

and

$$\int \frac{z^{3\eta - 1} dz}{e + fz^{\eta}} = \frac{z^{2\eta}}{2f\eta} - \frac{ez^{\eta}}{f^2\eta} + \frac{e^2}{f^2\eta} \int \frac{dx}{e + fx}$$

where $x = z^{\eta}$ as before.

The above integrals constitute the "first order" (or subsection) of Newton's table. The third of the ten "orders" in the table includes integrals of the form

$$\int \frac{z^{k\eta - 1} dz}{e + fz^{\eta} + gz^{2\eta}} \qquad (k \geqslant 1).$$

For example,

$$\int \frac{z^{\eta - 1} dz}{e + fz^{\eta} + gz^{2\eta}} = \frac{xv}{\eta} - \frac{2}{\eta} \int v \, dx, \qquad (26)$$

where

$$x^2 = \frac{1}{e + fz^{\eta} + gz^{2\eta}} \quad \text{and} \quad v^2 = \frac{4g + (f^2 - 4eg)x^2}{4g^2}.$$

The verification of (26) by means of this substitution turns out to be somewhat tedious; the following exercise gives a simpler approach to the same integral.

EXERCISE 13. Provide the details for Newton's alternative evaluation of

$$t = \int \frac{z^{\eta - 1} dz}{e + fz^{\eta} + gz^{2\eta}},$$

as follows. First substitute $x = z^\eta$ to obtain

$$t = \frac{1}{\eta} \int \frac{dx}{e + fx + gx^2}.$$

Then use the factorization

$$4g(e + fx + gx^2) = (f + p + 2gx)(f - p + 2gx),$$

where $p^2 = f^2 - 4eg$, to conclude that

$$t = \frac{2g}{p\eta} \int \left(\frac{1}{f - p + 2gx} - \frac{1}{f + p + 2gx} \right) dx.$$

Newton includes a number of area computations illustrating the applications of his tables of integrals. The first such example is the quadrature of the *versiera* (later called "Agnesi's witch") that he describes as follows. Let *AHQ* be a semicircle of diameter 1 on the *y*-axis as pictured (Fig. 13). Given a point *C* on the horizontal *z*-axis, let *E* be the point on *CI* at the same height as the intersection *H* of the semicircle *AHQ* and the diagonal *AI* of the rectangle *ACIQ*. Then *E* is the typical point on the *versiera*. It is easy to calculate

$$y = \frac{1}{1 + z^2},$$

so Newton needs to compute the integral

$$t = \int_0^z \frac{dz}{1 + z^2},$$

which of course equals $\tan^{-1} z$.

Consulting one of the entries in his tables, he notes that the area *s* under the circle $v = \sqrt{1 - x^2}$ transforms to the above area *t* under the transformation

$$z = \frac{1}{x} \sqrt{1 - x^2} = \phi(x) \quad \text{or} \quad x = \frac{1}{\sqrt{1 + z^2}} = \psi(x),$$

$$t = xv - 2s. \tag{27}$$

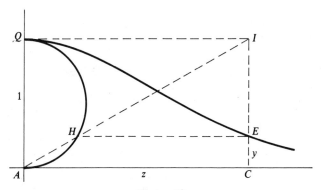

Figure 13

This may be verified as follows.

$$\dot{t} = v + x\dot{v} - 2\dot{s} = x\dot{v} - v \qquad (\dot{s} = v)$$

$$= \frac{-x^2}{\sqrt{1-x^2}} - \sqrt{1-x^2} = -\frac{1}{\sqrt{1-x^2}}$$

$$\dot{t} = -\frac{1}{z}\sqrt{1+z^2} \, ,$$

and

$$\dot{z} = \frac{-1}{\sqrt{1-x^2}} - \frac{1}{x^2}\sqrt{1-x^2}$$

$$= -\left(z + \frac{1}{z}\right)\sqrt{1+z^2} \, ,$$

so

$$y = -\frac{\dot{t}}{\dot{z}} = \frac{1/z}{z + (1/z)} = \frac{1}{1+z^2}.$$

The limits $z = 0$ and $z = z$ correspond to $x = 1$ and $x = 1/\sqrt{1+z^2}$, so (27) gives

$$\int_0^z \frac{dz}{1+z^2} = x\sqrt{1-x^2} - 2\int_1^{1/\sqrt{1+z^2}} \sqrt{1-x^2} \, dx.$$

$$= x\sqrt{1-x^2} + 2\int_{1/\sqrt{1+z^2}}^1 \sqrt{1-x^2} \, dx. \qquad (28)$$

Newton expresses this result geometrically as follows. The proportion

$$\frac{AH}{AI} = \frac{CE}{CI}$$

gives $AH = 1/\sqrt{1+z^2} = x$, so (28) simply says that the area t of $ACEQ$ (Fig. 13) is twice the area of the circular sector ABD in Figure 14. Since $\tan \alpha = z$, this means that

$$\int_0^z \frac{dz}{1+z^2} = \tan^{-1}z \quad \text{as desired.}$$

EXERCISE 14. Newton's next example is the quadrature of the "kappa curve" which he describes as the locus of the corner point of a right-angled rule AEF with unbounded leg AE passing through the origin A, as F moves along the horizontal y-axis, with the finite leg EF being of unit length. Then the similar triangles AEH and EFH (Fig. 15) give

$$y = \frac{z^2}{\sqrt{1-z^2}} \, ,$$

so the area $AGEC$ is

$$t = \int_0^z \frac{z^2 dz}{\sqrt{1-z^2}}.$$

Show by the transformation method of the previous section that, if s is the area

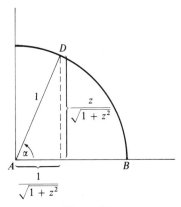

Figure 14

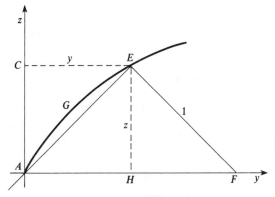

Figure 15

under the circle $v = \sqrt{1 - x^2}$, and $x = z$, then

$$t = s - xv.$$

That is,

$$\int_0^z \frac{z^2 dz}{\sqrt{1 - z^2}} = \int_0^x \sqrt{1 - x^2}\ dx - x\sqrt{1 - x^2}$$

$$= \tfrac{1}{2} \sin^{-1}x - \tfrac{1}{2}x\sqrt{1 - x^2}\ .$$

$$= \tfrac{1}{2} \sin^{-1}z - \tfrac{1}{2}z\sqrt{1 - z^2}\ .$$

Arclength Computations

Under Problem 12: To Determine the Lengths of Curves ([NP III], pp. 315–329) of "Methods of Series and Fluxions," Newton applies the basic fluxional technique for the computation of arclength that he describes as follows (see Fig. 16).

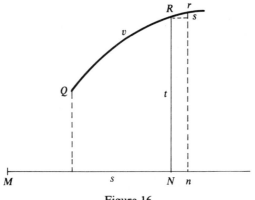

Figure 16

The fluxion of the length is determined by setting it equal to the square root of the sum of the squares of the fluxions of the base and the perpendicular ordinate. For let RN be the line perpendicularly ordinate upon the base MN, and QR the proposed curve at which RN terminates. Calling $MN = s$, $NR = t$ and $QR = v$, and their respective fluxions $\dot{s}$, $\dot{t}$, $\dot{v}$, imagine the line NR to move forward to the next closest possible place nr and then, when the perpendicular Rs is let fall to nr, will Rs, sr and Rr be contemporaneous moments of the lines MN, NR and QR, by whose addition they come to be Mn, nr and Qr. Since these are to one another as the fluxions of the same lines and, because of the right angle Rsr, it is $\sqrt{Rs^2 + sr^2} = Rr$, therefore $\sqrt{\dot{s}^2 + \dot{t}^2} = \dot{v}$.

That is, he deduces from the "characteristic triangle" Rsr the arclength relation that takes the familiar form

$$\frac{ds}{dt} = \sqrt{\left(\frac{dx}{dt}\right)^2 + \left(\frac{dy}{dt}\right)^2}$$

in terms of rectangular coordinates x and y and arclength s. However, in the examples that follow, he denotes by z and y the horizontal and vertical coordinates, and by t the arclength. Setting $\dot{z} = 1$, he therefore has $\dot{t} = \sqrt{1 + \dot{y}^2}$. For example, if $y = (z^3/a^2) + (a^2/12z)$, then $\dot{t} = [1 + [(3z^2/a^2) - (a^2/12z^2)]^2]^{1/2} = (3z^2/a^2) + (a^2/12z^2)$, so $t = (z^3/a^2) - (a^2/12z)$. Newton points out that, if $Ab = \frac{1}{2}a$ (Fig. 17), substitution of $z = \frac{1}{2}a$ gives $t = -a/24$, so the length of dD (where $AB = z$) is

$$\frac{z^3}{a^2} - \frac{a^2}{12z} - \left(-\frac{a}{24}\right) = \frac{z^3}{a^2} - \frac{a^2}{12z} + \frac{a}{24}.$$

This is an example of "choosing the constant of integration."

In order to rectify the semicubical parabola $z^3 = ay^2$, Newton calculates

$$\dot{t} = \sqrt{1 + \dot{y}^2} = \sqrt{1 + \frac{9z}{4a}},$$

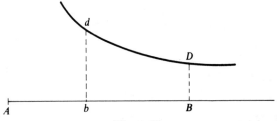

Figure 17

whence

$$t = \frac{8a}{27}\left(1 + \frac{9z}{4a}\right)^{3/2}$$

from his table of integrals.

EXERCISE 15. In order to rectify the curve $z^4 = ay^3$, Newton first calculates

$$\dot{i} = \sqrt{1 + \dot{y}^2} = \sqrt{1 + \frac{16z^{2/3}}{9a^{2/3}}},$$

that is,

$$\frac{\dot{i}}{\dot{z}} = \sqrt{1 + \frac{16z^{2/3}}{9a^{2/3}}}.$$

Now show that the area s under the hyperbola

$$v = \sqrt{x + \frac{16x^2}{9a^{2/3}}}$$

and the area t under the curve

$$y = \sqrt{1 + \frac{16z^{2/3}}{9a^{2/3}}}$$

correspond under the transformation

$$z = x^{3/2}, \qquad t = \tfrac{3}{2}s.$$

Consequently the length (from $z = 0$ to $z = z$) of the curve $z^4 = ay^3$ is equal to a hyperbolic area,

$$t = \int_0^z \sqrt{1 + \frac{16z^{2/3}}{9a^{2/3}}}\ dz = \int_0^{z^{2/3}} \sqrt{x + \frac{16x^2}{9a^{2/3}}}\ dx.$$

In his Example 5 Newton gives the first rectification of the "cissoid of the ancients," described in rectangular coordinates by

$$y = \frac{(a - z)^2}{\sqrt{z(a - z)}}.$$

Then

$$\dot{y} = -\tfrac{1}{2}z^{-3/2}(a-z)^{1/2}(a+2z)$$

so

$$\dot{t} = \sqrt{1+\dot{y}^2} = \tfrac{1}{2}az^{-3/2}\sqrt{a+3z} .$$

In order to find t, he refers to an entry in his catalogue of integrals for the fact that the area s under the hyperbola $v = \sqrt{a^2 + 3x^2}$ corresponds to the area t under the curve

$$y = \tfrac{1}{2}az^{-3/2}\sqrt{a+3z}$$

under the transformation

$$x^2 = az, \qquad t = \frac{6s}{a} - \frac{v^3}{ax} .$$

To see this, we calculate

$$\dot{t} = \frac{6\dot{v}}{a} - \frac{3v^2\dot{v}}{ax} + \frac{v^3}{ax^2}$$

$$= \frac{a}{x^2}\sqrt{a^2+3x^2} .$$

Upon substitution of $v = \sqrt{a^2 + 3x^2}$ and simplification, it follows that

$$y = \frac{\dot{t}}{\dot{z}} = \frac{a}{2x}\cdot\frac{a}{x^2}\sqrt{a^2+3x^2}$$

$$= \frac{a}{2}z^{-3/2}\sqrt{a+3z} ,$$

as desired, since $x^2 = az$.

Thus Newton has expressed the length

$$t = \int_z^a \tfrac{1}{2}az^{-3/2}\sqrt{a+3z}\ dz$$

of the cissoid over the interval $[z, a]$ in terms of the area s (Fig. 18) under the hyperbola $v = \sqrt{a^2 + 3x^2}$ over the interval $[x, a]$,

$$t = \frac{6}{a}\int_{\sqrt{az}}^a \sqrt{a^2+3x^2}\ dx - \left[\frac{(a^2+3x^2)^{3/2}}{ax}\right]_{\sqrt{az}}^a .$$

In his Example 8 Newton derives an infinite series for the arclength of the ellipse

$$y^2 + bz^2 = a^2.$$

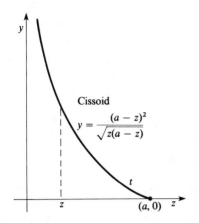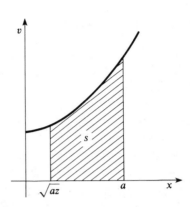

Figure 18

Differentiation gives

$$\dot{y} = \frac{bz}{\sqrt{a^2 + bz^2}}$$

so

$$\dot{t} = \sqrt{1 + \dot{y}^2} = \sqrt{\frac{a^2 + (b + b^2)z^2}{a^2 + bz^2}}$$

$$= \left[1 + \frac{b^2}{a^2}z^2 + \frac{b^3}{a^4}z^4 + \frac{b^4}{a^6}z^6 + \cdots \right]^{1/2}$$

$$\dot{t} = 1 + \frac{b^2}{2a^2}z^2 + \frac{4b^3 - b^4}{8a^4}z^4 + \frac{8b^4 - 4b^5 + b^6}{16a^6}z^6 + \cdots$$

by division and root extraction. Termwise antidifferentiation then gives

$$t = z + \frac{b^2}{6a^2}z^3 + \frac{4b^3 - b^4}{40a^4}z^5 + \frac{8b^4 - 4b^5 + b^6}{112a^6}z^7 + \cdots$$

for the length of the arc of the ellipse over the interval $[0, z]$.

In his final arclength example in the *De Methodis*, Newton uses his series

$$\sin x = x - \frac{x^3}{3!} + \frac{x^5}{5!} - \frac{x^7}{7!} + \cdots$$

(from the *De Analysi*) to rectify the quadratrix

$$y = x \cot \frac{x}{a} = \frac{x\sqrt{1 - \sin^2(x/a)}}{\sin(x/a)}.$$

By root extraction and division of series he obtains

$$y = a - \frac{x^2}{3a} - \frac{x^4}{45a^3} - \frac{2x^6}{945a^5} \cdots ,$$

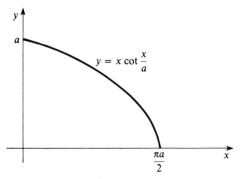

Figure 19

so

$$\dot{y} = -\frac{2x}{3a} - \frac{4x^3}{45a^3} - \frac{4x^5}{315a^5} \cdots .$$

Then

$$\dot{t} = \sqrt{1 + \dot{y}^2}$$

$$= 1 + \frac{2}{9}\frac{x^2}{a^2} + \frac{14}{405}\frac{x^4}{a^4} + \frac{604}{127,575}\frac{x^6}{a^6} + \cdots ,$$

so

$$t = x + \frac{2}{27}\frac{x^3}{a^2} + \frac{14}{2025}\frac{x^5}{a^4} + \frac{604}{893,025}\frac{x^7}{a^6} + \cdots$$

is an infinite series expansion for the arc of the quadratrix above the interval $[0, x]$ (Fig. 19).

EXERCISE 16. Set $a = 1$ and carry out the above series computations to verify Newton's result.

The Newton–Leibniz Correspondence

As we saw in Chapter 7, Newton announced his binomial series and described its discovery in the two letters of 1676 that he sent to Henry Oldenburg, secretary of the Royal Society of London, for transmission to Leibniz. These famous letters, the *epistola prior* dated June 13, 1676 ([NC II], pp. 32–41) and the *epistola posterior* dated October 24, 1676 ([NC II], pp. 130–149), served to establish Newton's priority for many of his early calculus applications that we have recounted from the *De Analysi* and the *De Methodis*. While concealing his basic fluxional techniques by means of anagrams—an unfortunate device that was not altogether unknown in the seventeenth century—he listed in these letters a large number of the infinite series, quadrature, and rectification results whose derivations we have described in this chapter.

An additional item of interest in this correspondence concerns quadratures of the circle that result in infinite series representations of the number π. In his reply to the *epistola prior*, Leibniz presented the alternating series

$$\frac{\pi}{4} = 1 - \frac{1}{3} + \frac{1}{5} - \frac{1}{7} + \cdots \tag{29}$$

that now bears his name ([NC II], pp. 65–71). Contrary to common assumption, Leibniz did not obtain this result directly from the integral

$$\frac{\pi}{4} = \int_0^1 \frac{dx}{1 + x^2}$$

by termwise integration after expansion of the integrand as a geometric series. As will be explained in Chapter 9, Leibniz' derivation of (29) was an application of his general "transmutation" method for the transformation of integrals.

In reply in the *epistola posterior*, Newton mentions his own table of integrals for "comparison of curves with conic sections," including in particular the integral of

$$\frac{z^{\eta - 1}}{e + fz^{\eta} + gz^{2\eta}}$$

(see Exercise 13). He first remarks that with $f = 0$ and $\eta = 1$ this gives Leibniz' series (29) (evidently by the inverse tangent approach mentioned above), and then presents the series

$$1 + \tfrac{1}{3} - \tfrac{1}{5} - \tfrac{1}{7} + \tfrac{1}{9} + \tfrac{1}{11} - \tfrac{1}{13} - \tfrac{1}{15} + \cdots \tag{30}$$

"for the length of the quadrantal arc of which the chord is unity (i.e. for $\pi/2\sqrt{2}$), or, what is the same thing, this:

$$\tfrac{1}{2} + \tfrac{1}{15} - \tfrac{1}{63} + \tfrac{1}{143} - \cdots$$

for the half of its length. And these perhaps, since they are just as simple as the others and converge more rapidly, you and your friends will not disdain. But I for my part regard the matter otherwise. For that is better which is more useful and solves the problem with less labor" ([NC II], 138–139).

EXERCISE 17. Provide, as follows, the details in Newton's likely derivation of the series

$$\frac{\pi}{4\sqrt{2}} = \frac{1}{2} + \sum_1^{\infty} \frac{(-1)^{n+1}}{16n^2 - 1} = \frac{1}{2} + \frac{1}{15} - \frac{1}{63} + \frac{1}{143} - \frac{1}{255} + \cdots .$$

(a) Show that

$$\int_0^1 \frac{1 + x^2}{1 + x^4} \, dx = 1 + \frac{1}{3} - \frac{1}{5} - \frac{1}{7} + \frac{1}{9} + \frac{1}{11} - \cdots$$

by expanding $1/(1 + x^4)$ as a geometric series and integrating termwise.

(b) Noting that

$$2\frac{1+x^2}{1+x^4} = \frac{1}{1+\sqrt{2}\,x+x^2} + \frac{1}{1-\sqrt{2}\,x+x^2},$$

conclude that

$$\int_0^1 \frac{1+x^2}{1+x^4}\,dx = \frac{1}{2}\int_{-1}^1 \frac{dx}{1+\sqrt{2}+x^2}.$$

Substitute $x+(\sqrt{2}\,/2)=(1/\sqrt{2}\,)\tan\theta$ to obtain

$$\int_0^1 \frac{1+x^2}{1+x^4}\,dx = \frac{1}{\sqrt{2}}\left[\tan^{-1}(1+\sqrt{2}\,)-\tan^{-1}(1-\sqrt{2}\,)\right]$$

$$= \frac{1}{\sqrt{2}}\left[\left(\frac{3\pi}{8}\right)-\left(-\frac{\pi}{8}\right)\right]$$

$$= \frac{\pi}{2\sqrt{2}}$$

as desired.

(c) Finally group terms as

$$1+\left(\tfrac{1}{3}-\tfrac{1}{5}\right)-\left(\tfrac{1}{7}-\tfrac{1}{9}\right)+\left(\tfrac{1}{11}-\tfrac{1}{13}\right)-\cdots$$

and find common denominators.

The Calculus and the *Principia Mathematica*

The magisterial *Philosophiae Naturalis Principia Mathematica* (Mathematical Principles of Natural Philosophy) was published in 1687. This founding document of modern exact science sets forth in comprehensive detail Newton's system of mechanics and theory of gravitation.

The *Principia* bristles with infinitesimal considerations and limit arguments, and is therefore sometimes regarded as Newton's first published account of the calculus. However, its exposition is couched almost entirely in the language and form of classical synthetic geometry, and makes little or no significant use of the algorithmic computational machinery of Newton's calculus of fluxions. The traditional view is that Newton first discovered the basic propositions of the *Principia* by means of fluxional analyses and computations, and later clothed them in the accepted dress of synthetic geometry, presumably in an effort to avoid controversy ("to avoid being baited by little smatterers in Mathematicks"). However, according to Whiteside, "it is futile to plough laboriously through the voluminous mass of Newton's extant papers (containing 10–15 million words at a conservative estimate) in search of manuscripts bearing dotted fluxional arguments which reappear, suitably recast in geometrical mould, in the pages of the first edition of the *Principia*. . . . Nowhere, let me repeat, are there to be found extant autograph manuscripts of Newton's, preceding the *Principia* in time, which could conceivably buttress the

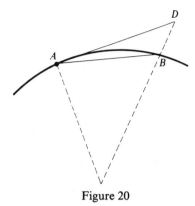

Figure 20

conjecture that he first worked the proofs in that book by fluxions before remoulding them in traditional geometrical form. Nor in all the many thousands of such sheets relating to the composition and revision of the *Principia* is there the faint trace of a suggestion that such papers ever existed" ([7], pp. 9–10).

What Newton did, in fact, need and use throughout the *Principia* was a facility for dealing with limits of ratios of geometrical quantities. For example, Lemma VII in Section I of Book I states that, given a chord $\overline{AB}$ of the arc $\overparen{AB}$ of a curve, and a corresponding tangent segment $\overline{AD}$ (Fig. 20), "if the points A and B approach one another and meet," then "the ultimate ratio of the arc, chord, and tangent, any one to any other, is the ratio of equality." The Scholium to Section I specifies that

> By the ultimate ratio of evanescent quantities (i.e., ones that are approaching zero) is to be understood the ratio of the quantities not before they vanish, nor afterwards, but with which they vanish. . . . Those ultimate ratios with which quantities vanish are not truly the ratios of ultimate quantities, but limits towards which the ratios of quantities decreasing without limit do always converge; and to which they approach nearer than by any given difference, but never go beyond, nor in effect attain to, till the quantities are diminished *in infinitum*.

Thus the "ultimate ratio of evanescent quantities" is simply the limit of their ratio. Lemma I of Book I of the *Principia* is, in effect, Newton's attempted definition of the limit concept: "Quantities, and the ratios of quantities, which in any finite time converge continually to equality, and before the end of that time approach nearer to each other than by any given difference, become ultimately equal." In modern notation, we would say that if, given $\epsilon > 0$, it follows that $f(t)$ and $g(t)$ differ by less than ϵ for t sufficiently close to a, then

$$\lim_{t \to a} f(t) = \lim_{t \to a} g(t).$$

Although Newton did not make explicit and systematic use of the fluxional calculus in the *Principia*, he provided in these passages his clearest exposition of the limit concept on which that calculus is based.

Newton's apparent use of infinitesimals or indivisibles in subsequent portions of the Principia has frequently been a source of confusion or misunderstanding. However, he warned in the Scholium to Section I of Book I that, "if hereafter I should happen to consider quantities as made up of particles, or should use little curved lines for right (straight) ones, I would not be understood to mean indivisibles but evanescent divisible quantities; not the sums and ratios of determinate parts, but always the limits of sums and ratios; and that the force of such demonstrations always depends on the method laid down in the foregoing Lemmas." In the introduction to the *De Quadratura* (discussed in the next section) he similarly stressed that

> By like ways of arguing, and by the method of Prime and Ultimate Ratios, may be gathered the Fluxions of Lines, whether Right or Crooked in all cases whatever, as also the Fluxions of Surfaces, Angles and other Quantities. In Finite Quantities so to frame a Calculus, and thus to investigate the Prime and Ultimate Ratios of Nascent or Evanescent Finite Quantities, is agreeable to the Geometry of the Ancients; and I was willing to shew, that in the Method of Fluxions there's no need of introducing Figures infinitely small into Geometry. For this Analysis may be performed in any Figures whatsoever, whether finite or infinitely small, so they are but imagined to be similar to the Evanescent Figures; as also in Figures which may be reckoned as infinitely small, if you do but proceed cautiously.

In other words, Newton says, exposition in terms of indivisibles or infinitesimals is simply a convenient shorthand (but not a substitute) for rigorous mathematical proof in terms of ultimate ratios (limits).

Newton's Final Work on the Calculus

Of Newton's several treatises on the calculus, the last written but first published was the *De Quadratura Curvarum* (On the Quadrature of Curves). This severely technical exposition of Newton's mature calculus of fluxions was written in 1691–1693 (rather than 1676 as stated in most histories of mathematics) and appeared as a mathematical appendix to the 1704 edition of his *Opticks*.

In the *epistola posterior* Newton had stated without proof the following "prime theorem" concerning the squaring of curves. The area under the curve

$$y = x^\theta (e + fx^\eta)^\lambda$$

is

$$Q\left\{ \frac{x^{\pi}}{s} - \frac{r-1}{s-1}\frac{eA}{fx^{\eta}} + \frac{r-2}{s-2}\frac{eB}{fx^{\eta}} - \frac{r-3}{s-3}\frac{eC}{fx^{\eta}} + \cdots \right\} \qquad (31)$$

where

$$Q = \frac{(e+fx^{\eta})^{\lambda+1}}{\eta f}, \qquad r = \frac{\theta+1}{\eta}, \qquad s = \lambda + r, \qquad \pi = \eta(r-1),$$

and the letters $A, B, C, \ldots,$ denote the immediately preceding terms, that is,

$$A = \frac{x^{\pi}}{s}, \qquad B = -\frac{r-1}{s-1}\frac{eA}{fx^{\eta}}, \qquad C = \frac{r-2}{s-2}\frac{eB}{fx^{\eta}}, \qquad \text{etc.}$$

In equivalent summation notation,

$$\int x^{\theta}(e+fx^{\eta})^{\lambda}\,dx = \frac{Qx^{\pi}}{s}\left[1 + \sum_{k=1}^{\infty} (-1)^{k}\frac{(r-1)(r-2)\cdots(r-k)}{(s-1)(s-2)\cdots(s-k)}\frac{e^{k}}{f^{k}x^{k\eta}} \right].$$
$$(31')$$

If r is a positive integer, then this is a finite sum with r terms; otherwise it is an infinite series whose convergence requires discussion (which Newton does not provide).

For example, if $y = x^{n} = x^{0}(0+x^{1})^{n}$, then $\theta = 0$, $f = \eta = 1$, $\lambda = n$ and $Q = x^{n+1}$, $r = 1$, $s = n+1$, $\pi = 0$, so (31) reduces to a single term,

$$\int x^{n}\,dx = \frac{x^{n+1}}{n+1}.$$

If $y = x/(1-2x^{2}+x^{4}) = x(1-x^{2})^{-2}$, then $\theta = 1$, $f = -1$, $\eta = 2$, $\lambda = -2$ and $Q = -\frac{1}{2}(1-x^{2})^{-1}$, $r = 1$, $s = -1$, $\pi = 0$, so (31) gives

$$\int \frac{x\,dx}{1-2x^{2}+x^{4}} = \frac{1}{2(1-x^{2})}.$$

Alternatively, if we write $y = x^{-3}(-1+x^{-2})^{-2}$, then $\theta = -3$, $f = 1$, $\eta = -2$ $= \lambda$ and $Q = -\frac{1}{2}(-1+x^{-2})^{-1}$, $r = 1$, $s = -1$, $\pi = 0$, so (31) gives

$$\int \frac{x\,dx}{1-2x^{2}+x^{4}} = \frac{x^{2}}{2(1-x^{2})}.$$

The two antiderivatives, obtained by application of (31) to different expressions for the same integrand function, differ by a "constant of integration."

The following two exercises correspond to additional examples that Newton gives in the *epistola posterior*.

EXERCISE 18. Write $\sqrt{x + x^2}\,/x^5 = x^{-4}(1 + x^{-1})^{1/2}$ and apply (31) to obtain

$$\int \frac{1}{x^5}\sqrt{x + x^2}\ dx = \frac{-16x^2 + 24x - 30}{105x^4}(x + 1)\sqrt{x^2 + x}\ .$$

EXERCISE 19. Write

$$\frac{x^{1/3}}{\sqrt[5]{1 - 3x^{2/3} + 3x^{4/3} - x^2}} = x^{1/3}(1 - x^{2/3})^{-3/5}$$

and apply (31) to obtain

$$\int \frac{x^{1/3}\ dx}{\sqrt[5]{1 - 3x^{2/3} + 3x^{4/3} - x^2}} = -\frac{30x^{2/3} + 75}{28}(1 - x^{2/3})^{2/5}.$$

The following exercise outlines a proof of the Newton's "prime theorem" that is suggested by the methods of the *De Quadratura* (see [NP VII], p. 28, note (21)).

EXERCISE 20. If

$$I_i = \int x^{\theta - i\eta}(e + fx^\eta)^\lambda\ dx \qquad i = 0, 1, 2, \ldots,$$

then (31') gives the value of I_0.
   (a) Show that the derivative of $x^{(r - i)\eta}(e + fx^\eta)^{\lambda + 1}$ is

$$\eta(e + fx^\eta)^\lambda[(r - i)ex^{\theta - i\eta} + (s - i + 1)fx^{\theta - (i - 1)\eta}].$$

Conclude by antidifferentiation that

$$I_{i-1} = \frac{Qx^{(r - i)\eta}}{s - i + 1} - \frac{(r - i)e}{(s - i + 1)f}I_i. \tag{32}$$

   (b) Show by repeated application of the recursion formula (32) that

$$I_0 = \frac{Qx^\eta}{s}\left[1 + \sum_{k=1}^{n-1}(-1)^k\frac{(r - 1)(r - 2)\cdots(r - k)}{(s - 1)(s - 2)\cdots(s - k)}\frac{e^k}{f^k x^{k\eta}}\right]$$

$$+ (-1)^n\frac{(r - 1)(r - 2)\cdots(r - n)e^n}{s(s - 1)(s - 2)\cdots(s - n + 1)f^n}I_n. \tag{33}$$

This provides the remainder term that must be investigated in order to establish convergence of (31') in case r is not a positive integer.
   (c) Similarly, obtain the following expansion in ascending powers of x^η:

$$I_0 = \frac{Qfx^{\eta + \eta}}{re}\left[1 + \sum_{k=1}^{n-1}(-1)^k\frac{(s + 1)(s + 2)\cdots(s + k)}{(r + 1)(r + 2)\cdots(r + k)}\frac{f^k x^{k\eta}}{e^k}\right]$$

$$+ (-1)^n\frac{(s + 1)(s + 2)\cdots(s + n)f^n}{r(r + 1)(r + 2)\cdots(r + n - 1)e^n}I_{-n}. \tag{34}$$

The *De Quadratura* contains the following generalization of (34)—see ([NW I], p. 145 or [NP VII], p. 521). Let

$$R = e + fx^\eta + gx^{2\eta} + hx^{3\eta} + \cdots,$$

$$S = a + bx^\eta + cx^{2\eta} + dx^{3\eta} + \cdots,$$

$$r = \theta/\eta, \qquad s = r + \lambda, \qquad t = s + \lambda, \qquad v = t + \lambda, \qquad \cdots.$$

Then

$$\int x^{\theta-1}R^{\lambda-1}S\,dx = x^\theta R^\lambda\left[\frac{a/\eta}{re} + \frac{b/\eta - sfA}{(r+1)e}x^\eta\right.$$

$$+ \frac{c/\eta - (s+1)fB - tgA}{(r+2)e}x^{2\eta}$$

$$\left. + \frac{d/\eta - (s+2)fC - (t+1)gB - vhA}{(r+3)e}x^{3\eta} + \cdots\right],$$

$$(35)$$

where each $A, B, C, \ldots$, is the coefficient of the preceding power of x, that is,

$$A = \frac{a/\eta}{re}, \qquad B = \frac{b/\eta - sfA}{(r+1)e}, \qquad C = \frac{c/\eta - (s+1)fB - tgA}{(r+2)e}, \qquad \cdots.$$

EXERCISE 21. Write

$$\frac{x^5 + x^4 - 8x^3}{(x-1)^3(x+2)^2} = \frac{x^6 - 9x^4 + 8x^3}{(x-1)^4(x+2)^2}$$

$$= x^3(2 - 3x + x^3)^{-2}(8 - 9x + x^3)$$

and apply (35) to obtain

$$\int \frac{x^5 + x^4 - 8x^3}{(x-1)^3(x+2)^2}\,dx = \frac{x^4}{x^3 - 3x + 2}.$$

EXERCISE 22. Apply (35) to obtain

$$\int \frac{3 + 2x + 8x^2 + 8x^3 - 7x^4 - 6x^5}{(1 + x - x^2 - x^3)^{4/3}}\,dx = \frac{3x(1 + x^2)}{(1 + x - x^2 - x^3)^{1/3}}.$$

EXERCISE 23. Show that the infinite series corresponding to (34) is a special case of (35).

As Hadamard has remarked, the *De Quadratura* "brings the integration of rational functions to a state hardly inferior to what it is now" ([2], p. 41).

In 1676 Newton himself had written, in a letter to John Collins ([NC II], p. 179),

> There is no curve line exprest by any equation of three terms—but I can in less then half a quarter of an hower tell whether it may be squared or what are ye simplest figures it may be compared wth, be those figures Conic sections or others. And then by a direct & short way (I dare say ye shortest ye nature of ye thing admits of for a general one) I can compare them. And so if any two figures exprest by such equations be propounded I can by ye same rule compare them if they may be compared. This may seem a bold assertion because it's hard to say a figure may or may not be squared or compared with another, but it's plain to me by ye fountain I draw it from, though I will not undertake to prove it to others.

And in the same dozen years from 1664 to 1676 he had discovered the law of universal gravitation, explained the color spectrum of the rainbow, invented and built reflecting telescopes, and devoted inordinate amounts of time to smoky chemical experiments!

References

Primary References

[NP] D. T. Whiteside (ed), *The Mathematical Papers of Isaac Newton*. Cambridge University Press, 1967–76, 7 volumes.

[NW] D. T. Whiteside (ed), *The Mathematical Works of Isaac Newton*. New York: Johnson Reprint, 1964. 2 volumes.

[NC] H. W. Turnbull et. al. (eds), *The Correspondence of Isaac Newton*. Cambridge University Press, 1959–78. 7 volumes.

[PM] F. Cajori (ed), *Newton's Mathematical Principles of Natural Philosophy*, A. Motte's Translation Revised. University of California Press, 1934.

Secondary References

[1] C. B. Boyer, *The History of the Calculus and its Conceptual Development*. New York: Dover (reprint), Chapter V, 1959.

[2] J. Hadamard, Newton and the Infinitesimal Calculus, in *Newton Tercentennary Celebrations*. Cambridge: The Royal Society, 1947.

[3] P. Kitcher, Fluxions, limits, and infinite littleness—A study of Newton's presentation of the calculus. *Isis* **64**, 33–49, 1973.

[4] L. T. More, *Isaac Newton, a Biography*. New York: Dover (reprint), 1962.

[5] C. J. Scriba, The inverse method of tangents: A dialogue between Leibniz and Newton. *Arch Hist Exact Sci* **2**, 113–137, 1962.

[6] D. T. Whiteside, Sources and strengths of Newton's early mathematical thought, in R. Palter, (ed), *The Annus Mirabilis of Sir Isaac Newton 1666–1966*. M.I.T. Press, 1970.

[7] D. T. Whiteside, The mathematical principles underlying Newton's *Principia Mathematica*. *J Hist Astron* **1**, 116–138, 1970.

The Calculus According to Leibniz $\quad$ 9

Gottfried Wilhelm Leibniz (1646–1716)

In the century of Kepler, Galileo, Descartes, Pascal, and Newton, the most versatile genius of all was Gottfried Wilhelm Leibniz. He was born at Leipzig, entered the university there at the age of fifteen, and received his bachelor's degree at seventeen. He continued his studies in logic, philosophy and law, and at twenty completed a brilliant thesis on the historical approach to teaching law. When the University of Leipzig denied his application for a doctorate in law because of his youth, he transferred to the University of Altdorf in Nuremberg, and received his doctorate in philosophy there in 1667.

Upon the completion of his academic work, Leibniz entered the political and governmental service of the Elector of Mainz. His serious study of mathematics did not begin until 1672 (at the age of twenty-six) when he was sent to Paris on a diplomatic mission. The following four years that he spent in Paris were Leibniz' "prime age of invention" in mathematics (similar to Newton's 1664–66 period). During his stay in Paris he conceived the principal features of his own version of the calculus, an approach that he elaborated during the balance of his life, and which during the eighteenth century was dominant over Newton's approach. In 1676 he returned to Germany, and served for the next forty years as librarian and councillor to the Elector of Hanover. Although his professional career was devoted mainly to law and diplomacy, the breadth of his fundamental contributions—to diverse areas of mathematics, philosophy, and science—is probably not matched by the work of any subsequent scholar.

In regard to the calculating machine that he built during the Paris years, Leibniz remarked, "It is unworthy of excellent men to lose hours like slaves in the labor of calculation which could safely be relegated to anyone else if machines were used." A lifelong project was his search for a universal language or symbolic logic that would standardize and mechanize not only numerical computations but all processes of rational human thought, and would eliminate the mental labor of routine and repetitive steps. His goal was the creation of a system of notation and terminology that would codify and simplify the essential elements of logical reasoning so as to

> furnish us with an Ariadne's thread, that is to say, with a certain sensible and palpable medium, which will guide the mind as do the lines drawn in geometry and the formulas for operations, which are laid down for the learner in arithmetic (quoted by Baron [1], p. 9).

Such a universal "characteristic" or language, he hoped, would provide all educated people—not just the fortunate few—with the powers of clear and correct reasoning.

Apparently the formulation of this far-reaching goal antedated Leibniz' serious interest in or detailed knowledge of mathematics. But, as Hofmann remarks, "A man who places such thoughts into the forefront of his mind has mathematics in his blood even if he is still ignorant of its detail" ([7], p. 2). Indeed, it was precisely (and only) in mathematics that Leibniz fully accomplished his goal. His infinitesimal calculus is the supreme example, in all of science and mathematics, of a system of notation and terminology so perfectly mated with its subject as to faithfully mirror the basic logical operations and processes of that subject. It is hardly an exaggeration to say that the calculus of Leibniz brings within the range of an ordinary student problems that once required the ingenuity of an Archimedes or a Newton. Perhaps the best measure of its triumph is the fact that today we can scarcely discuss the results of Leibniz' predecessors without restating them in his differential notation and terminology (as in our discussion of Newton's work in Chapter 8).

A few examples will indicate what Leibniz meant by symbolic notation as a "sensible and palpable medium, which will guide the mind" to correct conclusions. In the functional notation introduced much later by Lagrange, the chain rule says that, if $h(x) = f(g(x))$, then

$$h'(x) = f'(g(x))g'(x). \tag{1}$$

Nothing about the notation in Formula (1) suggests why it is true, nor how to prove it. But in differential notation, with $z = f(y)$ and $y = g(x)$, Formula (1) becomes

$$\frac{dz}{dx} = \frac{dz}{dy} \cdot \frac{dy}{dx}. \tag{2}$$

This formula, by contrast, conspicuously suggests its own validity, by

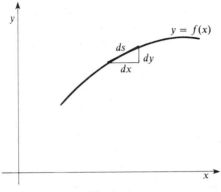

Figure 1

cancellation of the right-hand side differential dy's as though they were real numbers. This symbolic cancellation of differentials also suggests a logical proof of the formula—by replacing the differentials dx, dy, dz by the finite increments Δx, Δy, Δz and proceeding to the limit.

The integral version of the chain rule is the formula for integration by substitution,

$$\int f(g(x))g'(x)\, dx = \int f(u)\, du. \tag{3}$$

The symbolic substitution $u = g(x)$, $du = g'(x)\, dx$ makes Formula (3) seem inevitable, whatever its proof may be. This amounts to the invariance of the differential form $f(u)\, du$ with respect to arbitrary changes of variable —one of Leibniz' most important discoveries.

Now consider a surface that is generated by revolving the curve $y = f(x)$ around the x-axis. Thinking of an infinitesimal segment ds of the curve as the hypotenuse of the "characteristic triangle" with sides dx and dy (see Fig. 1), the Pythagorean theorem gives

$$ds = \sqrt{(dx)^2 + (dy)^2} = \sqrt{1 + \left(\frac{dy}{dx}\right)^2}\, dx.$$

When this segment ds is revolved around the x-axis in a circle of radius y, it generates an infinitesimal area

$$dA = 2\pi y\, ds = 2\pi y\sqrt{1 + \left(\frac{dy}{dx}\right)^2}\, dx.$$

Adding up the infinitesimal areas, we obtain

$$A = \int dA = \int 2\pi y\sqrt{1 + \left(\frac{dy}{dx}\right)^2}\, dx. \tag{4}$$

for the area of the surface. Thus we "discover" the correct Formula (4) by a quite routine and plausible manipulation of Leibniz' symbols. By contrast, its rigorous justification would require a detailed discussion and definition of the concept of surface area, followed by a proof (perhaps in terms of Riemann sums) that Formula (4) agrees with this definition.

These simple examples illustrate the principal features of the analytical or symbolic calculus of Leibniz—the central role of infinitesimal differences (differentials) and sums (integrals), and the inverse relationship between them; the characteristic triangle as a link between tangent (differential) and quadrature (integral) problems; the transformation of integrals by means of substitutions—and the manner in which this calculus does, indeed, guide the mind in the formal derivation of correct results.

In this chapter we outline the stages by which Leibniz gradually discovered and elaborated his calculus. The first crucial steps were taken during his Paris years, 1672–76, eight or ten years after Newton's formative period. However, Leibniz' first publication of the calculus was in 1684, twenty years prior to the publication of Newton's *De Quadratura* in 1704. In the final section of the chapter we discuss briefly the chief differences between the Newtonian and Leibnizian approaches to the calculus, and the unfortunate priority dispute between their respective followers that took place in the early eighteenth century.

In 1714, two years before his death, Leibniz composed the essay *Historia et origo calculi differentialis* (History and Origin of the Differential Calculus), opening with the lines (in the English translation of this extract provided by Weil [12])

> It is most useful that the true origins of memorable inventions be known, especially of those which were conceived not by accident but by an effort of meditation. The use of this is not merely that history may give everyone his due and others be spurred by the expectation of similar praise, but also that the art of discovery may be promoted and its method become known through brilliant examples. One of the noblest inventions of our time has been a new kind of mathematical analysis, known as the differential calculus; but while its substance has been adequately explained, its source and original motivation have not been made public. It is almost forty years now that its author invented it . . . .

English translations of the complete *Historia et origo*, and of a number of letters and manuscripts supplying additional details, are available in the volume of J. M. Child [5], to which frequent reference will be made in this chapter. In addition, Struik's source book [11] contains English translations of three of Leibniz' earliest published papers on the calculus.

The Beginning—Sums and Differences

In the *Historia et origo* and elsewhere, Leibniz always traced his inspiration for the calculus back to his early work with sequences of sums and differences of numbers. As a young student he had been interested in simple number properties, and in 1666 had published an essay entitled *De arte combinatoria* (On the Art of Combinations) that dealt with elementary properties of combinations and permutations.

Shortly after his arrival in Paris in 1672, he noticed an interesting fact about the sum of the differences of consecutive terms of a sequence of numbers. Given the sequence

$$a_0, a_1, a_2, \ldots, a_n$$

consider the sequence

$$d_1, d_2, \ldots, d_n$$

of differences, $d_i = a_i - a_{i-1}$. Then

$$d_1 + d_2 + \cdots + d_n = (a_1 - a_0) + (a_2 - a_1) + \cdots + (a_n - a_{n-1})$$
$$= a_n - a_1. \tag{5}$$

Thus the *sum of the consecutive differences equals the difference of the first and last terms of the original sequence.*

As an example, he observed that the "difference sequence" of the sequence of squares,

$$0, 1, 4, \ldots, n^2$$

is the sequence of consecutive odd numbers,

$$1, 3, 5, \ldots, 2n - 1,$$

because $i^2 - (i-1)^2 = 2i - 1$. It follows that the sum of the first n odd numbers is n^2,

$$1 + 3 + 5 + \cdots + (2n - 1) = n^2. \tag{6}$$

EXERCISE 1. (a) By adding $2 + 4 + \cdots + 2n$ to both sides of (6), show that

$$1 + 2 + 3 + \cdots + 2n = \frac{2n}{2}(2n + 1).$$

(b) By adding $(2n + 1)$ to both sides of the result of (a), show that

$$1 + 2 + 3 + \cdots + 2n + (2n + 1) = \frac{2n + 1}{2}(2n + 2).$$

Note that (a) and (b) together yield the familiar result

$$1 + 2 + 3 + \cdots + n = \frac{n}{2}(n + 1)$$

for all positive integers n.

EXERCISE 2. Apply (5) to the sequence of cubes

$$0, 1, 8, \ldots, n^3$$

to obtain

$$3 \sum_{i=1}^{n} i^2 - 3 \sum_{i=1}^{n} i + n = n^3.$$

Solve this equation, using the result of Exercise 1, for

$$\sum_{i=1}^{n} i^2 = 1^2 + 2^2 + \cdots + n^2 = \frac{n}{6}(n+1)(2n+1).$$

His result on sums of differences also suggested to Leibniz the possibility of summing an *infinite* series of numbers. Suppose the numbers

$$b_1, b_2, \ldots, b_n, \ldots$$

are the differences of consecutive terms of the sequence

$$a_1, a_2, \ldots, a_n, \ldots,$$

that is, $b_i = a_i - a_{i+1}$. Then

$$b_1 + b_2 + \cdots + b_n = a_1 - a_{n+1}.$$

If, in addition, $\lim_{n \to \infty} a_n = 0$, then it follows that

$$\sum_{n=1}^{\infty} b_n = a_1. \tag{7}$$

EXERCISE 3. Apply (7) with $\{a_n\}_1^\infty$ being the sequence $1, \frac{1}{3}, \frac{1}{5}, \ldots,$ of reciprocals of the odd integers to show that

$$\frac{1}{1 \cdot 3} + \frac{1}{3 \cdot 5} + \frac{1}{5 \cdot 7} + \cdots = \frac{1}{2}.$$

EXERCISE 4. Noting that the sequence of differences of the terms of the geometric progression

$$1, a, a^2, \ldots, a^n, \ldots$$

is

$$(1-a), (1-a)a, (1-a)a^2, \ldots, (1-a)a^n, \ldots,$$

apply Formula (7) to show that

$$\sum_{n=0}^{\infty} a^n = \frac{1}{1-a}$$

if $0 < a < 1$.

Not long after his arrival in Paris, Leibniz called on Christiaan Huygens (1629–1695), who was completing his comprehensive treatise *Horologium oscillatorium* (1673) on the theory of the pendulum clock, and was probably the most renowned scientist on the continent. When Leibniz described his results on sums of differences, Huygens suggested that he try to find the sum of the series

$$\frac{1}{1} + \frac{1}{3} + \frac{1}{6} + \frac{1}{10} + \cdots + \frac{1}{n(n+1)/2} + \cdots \tag{8}$$

of reciprocals of the triangular numbers. This problem had risen somewhat earlier in a discussion with Hudde on computing probabilities for certain games of chance.

Huygen's problem was an especially propitious one for Leibniz to consider at that time. From his earlier work he was familiar with the combinatorial or figurate numbers as they appear in Pascal's "arithmetic triangle." Let this triangle be written in the form

1	1	1	1	1	1	$\cdots$
1	2	3	4	5	6	$\cdots$
1	3	6	10	15	21	$\cdots$
1	4	10	20	35	56	$\cdots$
1	5	15	35	70	126	$\cdots$
$\vdots$	$\vdots$	$\vdots$	$\vdots$	$\vdots$	$\vdots$	

The nth element of each row is the sum of the first n elements of the preceding row. Thus the nth triangular number (or figurate number of type 1), the sum of the first n integers, is the nth element of the third row. Since the nth figurate number of type k is the sum of the first n figurate numbers of type $k-1$ (see Formula (18) and Exercise 12 of Chapter 4), the $(k+2)$th row consists of the figurate numbers of type k. Hence the arithmetic triangle exhibits the triangular numbers as sums of integers, the pyramidal numbers as sums of triangular numbers, etc. Conversely, the triangular numbers are differences of consecutive pyramidal numbers, etc.

Leibniz saw that questions of the sort asked by Huygens could be answered by starting with the sequence of reciprocals of the integers, instead of the integers themselves, and constructing subsequent rows by taking differences rather than sums. In this way he obtained the following array, which he called his "harmonic triangle".

$\frac{1}{1}$	$\frac{1}{2}$	$\frac{1}{3}$	$\frac{1}{4}$	$\frac{1}{5}$	$\frac{1}{6}$	$\frac{1}{7}$	$\cdots$	
$\frac{1}{2}$	$\frac{1}{6}$	$\frac{1}{12}$	$\frac{1}{20}$	$\frac{1}{30}$	$\frac{1}{42}$	$\cdot$	$\cdot$	$\cdot$
$\frac{1}{3}$	$\frac{1}{12}$	$\frac{1}{30}$	$\frac{1}{60}$	$\frac{1}{105}$	$\cdot$	$\cdot$		
$\frac{1}{4}$	$\frac{1}{20}$	$\frac{1}{60}$	$\frac{1}{140}$	$\cdot$	$\cdot$			
$\frac{1}{5}$	$\frac{1}{30}$	$\frac{1}{105}$	$\cdot$	$\cdot$				
$\cdot$	$\cdot$	$\cdot$						
$\cdot$	$\cdot$	$\cdot$						
$\cdot$	$\cdot$	$\cdot$						

Thus each row of the harmonic triangle is the sequence of differences of consecutive terms of the preceding row. Therefore Formula (7) implies that the sum of the terms of each row is equal to the first element of the preceding row. In particular,

$$\tfrac{1}{2} + \tfrac{1}{6} + \tfrac{1}{12} + \tfrac{1}{20} + \cdots = 1, \tag{9}$$

$$\tfrac{1}{3} + \tfrac{1}{12} + \tfrac{1}{30} + \tfrac{1}{60} + \cdots = \tfrac{1}{2}, \tag{10}$$

$$\tfrac{1}{4} + \tfrac{1}{20} + \tfrac{1}{60} + \tfrac{1}{140} + \cdots = \tfrac{1}{3}. \tag{11}$$

The nth element of the second row of the harmonic triangle is

$$\frac{1}{n} - \frac{1}{n+1} = \frac{1}{n(n+1)},$$

which is the half of the reciprocal of the nth triangular number $n(n+1)/2$. Hence multiplication of Equation (9) by 2 yields the sum asked for by Huygens,

$$\tfrac{1}{1} + \tfrac{1}{3} + \tfrac{1}{6} + \tfrac{1}{10} + \cdots = 2.$$

Similarly, multiplication of Equation (10) by 3 yields the sum of the reciprocals of the pyramidal numbers,

$$\tfrac{1}{1} + \tfrac{1}{4} + \tfrac{1}{10} + \tfrac{1}{20} + \cdots = \tfrac{3}{2}.$$

EXERCISE 5. Show that the nth element of the third row of the harmonic triangle is $2/n(n+1)(n+2)$, which is one-third of the reciprocal of the nth pyramidal number.

EXERCISE 6. Show by induction on k that the nth element of the $(k+1)$st row of the harmonic triangle is

$$\frac{k!}{n(n+1) \cdots (n+k)} = \frac{1}{(k+1)F(n, k)},$$

where $F(n, k)$ denotes the nth figurate number of type k (see Formula (18) of Chapter 4). Conclude that

$$\sum_{n=1}^{\infty} \frac{1}{F(n, k)} = \frac{k+1}{k}.$$

Pascal's arithmetic triangle and Leibniz' harmonic triangle enjoy a certain inverse relationship with respect to their manners of formation —involving sums in the former case and differences in the latter. In the arithmetic triangle each row consists of sums of the terms in the preceding row, and differences of terms in the following row. In the harmonic triangle, however, each row consists of differences of the terms in the preceding row.

These considerations implanted in Leibniz' mind a vivid conception that was to play a dominant role in his development of the calculus—the notion of an inverse relationship between the operation of taking differences and that of forming sums of the elements of a sequence.

EXERCISE 7. Given a sequence $\{a_n\}_1^{\infty}$ or $a_1, a_2, a_3, \ldots$, define the difference sequence

$$\Delta\{a_n\} = a_2 - a_1, \quad a_3 - a_2, \quad \ldots, \quad a_{n+1} - a_n, \quad \cdots$$

and the sum sequence

$$\sum\{a_n\} = a_1, \quad a_1 + a_2, \quad \ldots, \quad \sum_{i=1}^{n} a_i, \quad \cdots.$$

Then show that

$$\Delta \sum \{a_n\} = \{a_{n+1}\}$$

and

$$\sum \Delta \{a_n\} = \{a_{n+1} - a_1\}.$$

The Characteristic Triangle

At the time of his investigation of sum and difference sequences in late 1672, Leibniz was still largely ignorant of the mathematical work that was then contemporary. In a 1680 letter to Tschirnhaus ([5], p. 215), he told of a memorable conversation with Huygens in early 1673 that led to his mathematical self-education.

> The prime occasion from which arose my discovery of the method of the Characteristic Triangle, and other things of the same sort, happened at a time when I had studied geometry for not more than six months. Huygens, as soon as he had published his book on the pendulum, gave me a copy of it; and at that time I was quite ignorant of Cartesian algebra and also of the method of indivisibles, indeed I did not know the correct definition of the center of gravity. For, when by chance I spoke of it to Huygens, I let him know that I thought that a straight line drawn through the center of gravity always cut a figure into two equal parts; since that clearly happened in the case of a square, or a circle, an ellipse, and other figures that have a center of magnitude, I imagined that it was the same for all other figures. Huygens laughed when he heard this, and told me that nothing was further from the truth. So I, excited by this stimulus, began to apply myself to the study of the more intricate geometry, although as a matter of fact I had not at that time really studied the Elements. But I found in practice that one could get on without a knowledge of the Elements, if only one was master of a few propositions. Huygens, who thought me a better geometer than I was, gave me to read the letters of Pascal, published under the name of Dettonville; and from these I gathered the method of indivisibles and centers of gravity, that is to say the well-known methods of Cavalieri and Guldinus.

It was in his study of Pascal's work that Leibniz found his famous "characteristic triangle." In June 1658 Pascal had proposed a contest, with a closing date of 1 October 1658, for the solution of several problems concerning the cycloid—to find the area and centroid of an arbitrary segment of a cycloid, and to find the volumes and centroids of various solids of revolution obtained by revolving such a segment about either its base or its ordinate. In 1643 Roberval had shown that the area of the whole cycloid is three times that of its generating circle, and that the volume obtained by revolving the cycloid about its base is five-eighths that of the circumscribed cylinder (for this work by Roberval see Struik's source book ([11], pp. 232–238).

Most of the leading mathematicians of the day followed this contest with interest, and a number of them submitted proposed solutions. After none of the submitted solutions had been judged fully acceptable, Pascal published his own work on the cycloid and related problems in the form of *Lettres de A. Dettonville* (the pseudonym of Amos Dettonville being an anagram on Louis, or Lovis, de Montalte, the pseudonym under which Pascal's *Lettres provinciales* had appeared).

In the *Historia et origo* Leibniz, referring to himself in the third person, described his decisive discovery of the characteristic triangle as follows ([5], p. 38).

> From one example given by Dettonville, a light suddenly burst upon him, which strange to say Pascal himself had not perceived in it. For when he proves the theorem of Archimedes for measuring the surface of a sphere or parts of it, he used a method in which the whole surface of the solid formed by a rotation round any axis can be reduced to an equivalent plane figure. From it our young friend made out for himself the following general theorem. Portions of a straight line normal to a curve, intercepted between the curve and an axis, when taken in order and applied at right angles to the axis give rise to a figure equivalent to the moment of the curve about the axis.

The Pascal reference here is to the short "treatise on the sines of a quadrant of a circle" which is part of the first Dettonville letter (see Struik's source book ([11], pp. 239–241) for an English translation). In its Proposition 1 Pascal proved that "the sum of the sines [ordinates] of any arc of a quadrant [of a circle] is equal to the portion of the base between the extreme sines multiplied by the radius." The use of the word sine for ordinate connotes the fact that, in the sum referred to, each ordinate is to be multiplied by a corresponding infinitesimal arc ds of the circle (rather than by an infinitesimal segment dx of the base).

To prove the proposition, Pascal constructed the right triangle E_1E_2K with hypotenuse E_1E_2 tangent to the circle at a typical point D, and then noted that the triangles E_1E_2K and ADI are similar (see Fig. 2). Therefore

$$\frac{AD}{E_1E_2} = \frac{DI}{E_2K}, \quad \text{so} \quad DI \cdot E_1E_2 = AD \cdot E_2K = AD \cdot R_1R_2.$$

Thus, if $y = DI$, $a = AD$, $\Delta s = E_1E_2$, $\Delta x = R_1R_2$, then $y\Delta s = a\Delta x$. Regarding Δs and Δx as indivisibles and summing up, we see that Pascal's result is

$$\int y \, ds = \int a \, dx. \tag{12}$$

Because $2\pi y \, ds$ is the area of an infinitesimal zone on the hemisphere of

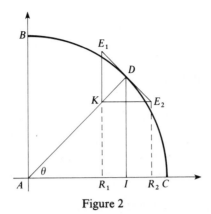

Figure 2

radius a that is obtained by revolving the quarter-circle around the x-axis, it follows that the area of the hemisphere is

$$A = \int 2\pi y \, ds = 2\pi a \int_0^a dx = 2\pi a^2.$$

Thus (12) provides an infinitesimal derivation of the area formula $A = 4\pi a^2$ for the sphere of radius a.

EXERCISE 8. Apply (12) to the arc of the quarter-circle corresponding to $\alpha \leqslant \theta \leqslant \beta$ to conclude that

$$\int_\alpha^\beta \sin \theta \, d\theta = \cos \alpha - \cos \beta.$$

Leibniz' "burst of light" consisted of noticing the quite general application of Pascal's infinitesimal triangle construction to an arbitrary curve, with the role of the radius of the circle played by the normal to the given curve. Thus, from the similarity of the triangles shown in Figure 3, it

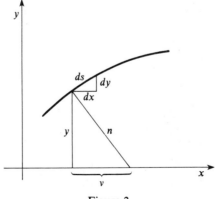

Figure 3

follows that

$$\frac{ds}{n} = \frac{dx}{y} \quad \text{or} \quad y \, ds = n \, dx.$$

Summation of infinitesimals then gives

$$\int y \, ds = \int n \, dx. \tag{13}$$

Since Leibniz did not invent his differential-integral notation until two years later in 1675, he had to express (13) in verbal form—the moment of the given curve about the x-axis is equal to the area under a second curve whose ordinate is the normal n to the given curve (see the *Historia et origo* ([5], pp. 38–41) for Leibniz' presentation of (13) together with Formulas (14) and (15) below). Multiplication of the moment by 2π gives the area $A = \int 2\pi y \, ds$ of the surface of revolution obtained by rotating the original curve around the x-axis. When Leibniz showed this result to Huygens the latter "confessed to him that by the help of this very theorem he had found the surface of parabolic conoids [paraboloids] and others of the same sort, stated without proof many years before" [in 1657].

EXERCISE 9. Consider the parabola $y = \sqrt{x}$, $0 \leqslant x \leqslant a$. Knowing that $D\sqrt{x} = \frac{1}{2}\sqrt{x}$, show that the normal to the parabola is $n = \frac{1}{2}\sqrt{4x+1}$. Hence apply Formula (13) to show that the area of the paraboloid obtained by revolving this parabola around the x-axis (see Fig. 4) is

$$A = \int 2\pi y \, ds = \pi \int_0^a \sqrt{4x+1} \; dx = \frac{\pi}{6}\left[(a+1)^{3/2} - 1\right].$$

At the same time, Leibniz saw how to apply the characteristic triangle method to rectification and quadrature problems. Given a curve whose arclength is sought, let t denote the length of the tangent line intercepted between the x-axis and a vertical ordinate of (constant) length a. Then

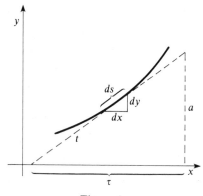

Figure 4

from the similarity of the triangles in Figure 4 it follows that

$$\frac{ds}{t} = \frac{dy}{a} \quad \text{or} \quad a \, ds = t \, dy.$$

Hence

$$\int a \, ds = \int t \, dy, \tag{14}$$

so the rectification of the given curve reduces to a quadrature problem—the calculation of the area of the region between the y-axis and a second curve whose abscissa x is the tangent t to the given curve.

EXERCISE 10. Consider the semi-cubical parabola $y = x^{2/3}$, $0 \leqslant x \leqslant 8$. Knowing that $Dx^{2/3} = 2x^{-1/3}/3$, show that the tangent (taking $a = 1$) is $t = \frac{1}{2}\sqrt{9y + 4}$. Hence apply Formula (14) to show that the arclength of this semi-cubical parabola is

$$s = \int ds = \int t \, dy$$

$$= \int_0^4 \tfrac{1}{2}\sqrt{9y + 4} \; dy = \tfrac{1}{27}\big[(40)^{3/2} - 8\big].$$

For Leibniz' third application of the characteristic triangle, note that the similarity of the triangles in Figure 3 implies that

$$\frac{dy}{v} = \frac{dx}{y} \quad \text{or} \quad v \, dx = y \, dy,$$

where v is the subnormal to the given curve. Hence

$$\int v \, dx = \int y \, dy. \tag{15}$$

Leibniz noted that, if the given curve passes through the origin and its base interval is $[0, b]$, then the right-hand integral in (15) is simply the area $\frac{1}{2}b^2$ of a triangle with base and height equal to b—"straight lines that continually increase from zero, when each is multiplied by its element of increase, form together a triangle."

> Thus, to find the area of a given figure, another figure is sought such that its subnormals are respectively equal to the ordinates of the given figure, and then this second figure is the quadratrix of the given one; and thus from this extremely elegant consideration we obtain the reduction of the areas of surfaces described by rotation to plane quadratures [Formula (13)], as well as the rectification of curves [Formula (14)]; at the same time we can reduce these quadratures of figures to an inverse problem of tangents [Formula (15)] (see [5], p. 41).

Since $v = y(dy/dx)$, Formula (15) says that

$$\int y\left(\frac{dy}{dx}\right)dx = \int y\,dy.$$

This was the first appearance of two ideas that were to play central roles in Leibniz' calculus—the transformation of integrals by means of substitutions, and the reduction of quadrature problems to inverse tangent problems, the latter being problems in which a curve is to be determined from a knowledge of its tangent line.

To illustrate the way in which Formula (15) reduces quadratures to inverse tangent problems, suppose we want to find the area $\int_0^a x^n dx$ under the curve $z = x^n$. If we can find a curve $y = f(x)$ with subnormal $v = x^n$, then Formula (15) will yield

$$\int_0^a x^n dx = \int y\,dy = \left[\tfrac{1}{2}y^2\right]_{x=0}^{x=a} = \tfrac{1}{2}f(a)^2,$$

assuming that $f(0) = 0$. Trying $y = bx^k$, we want

$$v = y\frac{dy}{dx} = bx^k \cdot bkx^{k-1} = b^2 kx^{2k-1} = x^n.$$

This requires that

$$k = \tfrac{1}{2}(n+1) \quad \text{and} \quad b = \left[\tfrac{1}{2}(n+1)\right]^{-1/2}.$$

It follows that

$$\int_0^a x^n dx = \frac{1}{2}(ba^k)^2 = \frac{a^{n+1}}{n+1}.$$

In these investigations of 1673, his first year of serious work in mathematics, Leibniz obtained few if any results that were actually new, that is, no specific quadratures or rectifications that had not been discovered previously by others. Even his touchstone, the characteristic triangle, was implicit in Pascal's work (and fairly explicit in Barrow's *Geometrical Lectures*). But he took significant first steps towards his real goal—the development of a general algorithmic method that would unify the diverse results and techniques that he found in the existing mathematical literature. Two decades later, in a letter to l'Hospital, he summarized these first steps as follows ([5], pp. 220–222).

> [With the] use of what I call the "characteristic triangle", formed from the elements of the coordinates and the curve, I thus found as it were in the twinkling of an eyelid nearly all the theorems that I afterward found in the works of Barrow and Gregory. Up to that time, I was not sufficiently versed in the calculus [algebra] of Descartes, and as yet did not make use of equations to express the nature of curved lines; but, on the advice of

Huygens, I set to work at it, and I was far from sorry that I did so: for it gave me the means almost immediately of finding my differential calculus. This was as follows. I had for some time previously taken a pleasure in finding the sums of series of numbers, and for this I had made use of the well-known theorem, that, in a series decreasing to infinity, the first term is equal to the sum of all the differences. From this I had obtained what I call the "harmonic triangle," as opposed to the "arithmetical triangle" of Pascal · · · Recognizing from this the great utility of differences and seeing that by the calculus of M. Descartes the ordinates of the curve could be expressed numerically, I saw that to find quadratures or the sums of the ordinates was the same thing as to find an ordinate (that of the quadratrix), of which the difference is proportional to the given ordinate. I also recognized almost immediately that to find tangents is nothing else but to find differences, and that to find quadratures is nothing else but to find sums, provided that one supposes that the differences are incomparably small.

Transmutation and the Arithmetical Quadrature of the Circle

In late 1673 or early 1674 Leibniz discovered a general "transmutation" or transformation method with which he could derive essentially all of the previously known plane quadrature results. He described its advantages in his reply to Newton's *Epistola prior* (see [9], pp. 65–66).

My method is but a corollary of a general theory of transformations, by the help of which any given figure whatever, by whatever equation it may be accurately stated, is reduced to another analytically equivalent figure ... Furthermore, the general method of transformations itself seems to me proper to be counted among the most powerful methods of analysis, for not merely does it serve for infinite series and approximations, but also for geometrical solutions and endless other things that are scarcely manageable otherwise · · · The basis of the transformation is this: that a given figure, with innumerable lines [ordinates] drawn in any way (provided they are drawn according to some rule or law), may be resolved into parts, and that the parts—or others equal to them—when reassembled in another position or another form compose another figure, equivalent to the former or of the same area even if the shape is quite different; whence in many ways the quadratures can be attained · · · These steps are such that they occur at once to anyone who proceeds methodically under the guidance of Nature herself; and they contain the true method of indivisibles as most generally conceived and, as far as I know, not hitherto expounded with sufficient generality. For not merely parallel and convergent straight lines, but any other lines also, straight or curved, that are constructed by a definite law can be applied to the resolution [of the original figure into parts that are to be reassembled to

246 The Calculus According to Leibniz

form another figure]; but he who has grasped the universality of the
method will judge how great and how abstruse are the results that can
thence be obtained: For it is certain that all squarings hitherto known,
whether absolute or hypothetical, are but limited specimens of this.

Here Leibniz describes in very general terms the following principle, to
which the term "transmutation" was applied during the seventeenth
century. Let A and B be two plane (or space) regions, each subdivided into
"indivisibles", generally infinitely small rectangles (or prisms). If there is a
one-to-one correspondence between the indivisibles in A and those in B,
such that corresponding indivisibles have equal areas (or volumes), then it
is said that B is derived from A by a "transmutation", and we conclude
that A and B have equal areas (or volumes).

As we saw in Chapter 4, this principle was the basis for the computations of Cavalieri and others who used rectangular indivisibles. It enabled
them to accomplish (in a variety of special cases) what would be done now
by means of changes of variable and integration by parts. Although he
described the inherent possibilities somewhat more generally in his letter to
Newton, Leibniz' main innovation in practice was the use of *triangular*
indivisibles in a systematic transformation process.

Given neighboring points $P(x, y)$ and $Q(x + dx, y + dy)$ on the curve
$y = f(x)$, $x \in [a, b]$, Leibniz considers the infinitesimal triangle OPQ, where
O is the origin. Let the tangent line determined by the infinitesimal arc ds
joining P and Q intersect the y-axis at the point $T(0, z)$ (see Fig. 5), where

$$z = y - x\frac{dy}{dx},\qquad(16)$$

and denote by OS the perpendicular segment of length p from O to this
tangent line (extended). Then the triangle OST is similar to the characteristic triangle PRQ, so it follows that $dx/p = ds/z$. Hence the area of the

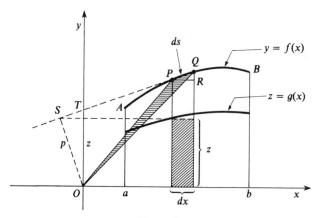

Figure 5

infinitesimal triangle OPQ is

$$a(OPQ) = \tfrac{1}{2}p \; ds = \tfrac{1}{2}z \; dx. \tag{17}$$

If we consider the sector OAB, bounded by the graph AB of $y = f(x)$ and the radii OA and OB, as being subdivided into infinitesimal triangles like OPQ, then it follows from (17) that

$$a(OAB) = \frac{1}{2}\int_a^b z \; dx, \tag{18}$$

where $z = g(x)$ is defined by (16). But

$$\int_a^b y \; dx = \tfrac{1}{2}bf(b) - \tfrac{1}{2}af(a) + a(OAB)$$

$$= \tfrac{1}{2}\big[xy\big]_a^b + a(OAB),$$

so it follows from (18) that

$$\int_a^b y \; dx = \frac{1}{2}\left(\big[xy\big]_a^b + \int_a^b z \; dx\right). \tag{19}$$

Formula (19) is Leibniz' "transmutation theorem." Its significance (like that of Formula (15)) is that it established an inverse relationship between the tangent problem (since z is defined in terms of the tangent) and the quadrature problem (of computing $\int_a^b y \; dx$). Moreover, a new curve $z = g(x)$ was introduced to serve as a "quadratrix" for the original curve $y = f(x)$, in case $\int_a^b z \; dx$ turned out to be a simpler integral in terms of which $\int_a^b y \; dx$ could be evaluated. Note also that the substitution of $z = y - x(dy/dx)$ into (19) yields the integration by parts formula

$$\int_a^b y \; dx = \big[xy\big]_a^b - \int_{f(a)}^{f(b)} x \; dy.$$

EXERCISE 11. Consider the "higher parabola" $y^q = x^p$, $q > p > 0$. Show that

$$\frac{q}{y}\frac{dy}{dx} = \frac{p}{x}, \quad \text{so} \quad z = \frac{q-p}{q}y.$$

Conclude from the transmutation formula that

$$\int_a^b x^{p/q}dx = \frac{q}{p+q}[xy]_a^b = \frac{q}{p+q}[x^{(p+q)/q}]_a^b.$$

Leibniz' most interesting application of the transmutation theorem was his so-called "arithmetical quadrature of the circle"—the derivation of the famous infinite series

$$\frac{\pi}{4} = 1 - \frac{1}{3} + \frac{1}{5} - \frac{1}{7} + \cdots \tag{20}$$

that now bears his name.

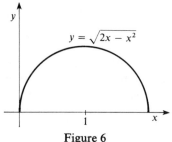

Figure 6

The upper half of the unit circle tangent to the y-axis at the origin (Fig. 6) is the graph of $y = \sqrt{2x - x^2}$. Since

$$\frac{dy}{dx} = \frac{1-x}{y},$$

we find that

$$z = y - x\frac{1-x}{y} = \sqrt{\frac{x}{2-x}},$$

or

$$x = \frac{2z^2}{1+z^2}.$$

Leibniz then applies the transmutation formula to compute the area of the quarter-circle as follows.

$$\frac{\pi}{4} = \int_0^1 y \, dx$$

$$= \frac{1}{2}\left(\left[x\sqrt{2x - x^2} \, \right]_0^1 + \int_0^1 z \, dx \right) \quad \text{(by (19))}$$

$$= \frac{1}{2}\left[1 + \left(1 - \int_0^1 x \, dz \right) \right] \quad \text{(Figure 7)}$$

$$= 1 - \int_0^1 \frac{z^2 dz}{1+z^2}$$

$$= 1 - \int_0^1 z^2(1 - z^2 + z^4 \cdots) \, dz \quad \text{(geometric series)}$$

$$= 1 - \left[\frac{1}{3}z^3 - \frac{1}{5}z^5 + \frac{1}{7}z^7 \cdots \right]_0^1 \quad \text{(termwise integration)}$$

$$\frac{\pi}{4} = 1 - \frac{1}{3} + \frac{1}{5} - \frac{1}{7} + \cdots,$$

ignoring the question of convergence when $z = 1$.

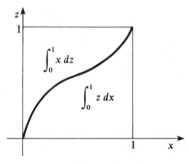

Figure 7

EXERCISE 12. Consider the portion of the rectangular hyperbola defined by $y = \sqrt{x^2 + 2x}$, $x \geqslant 0$. Show by Leibniz' transmutation method, i.e., by a computation analogous to his arithmetical quadrature of the circle, that the area of the shaded region in Figure 8 is

$$\int_0^x \sqrt{x^2 + 2x} \; dx = z + \frac{z^3}{3} + \frac{z^5}{5} + \cdots .$$

Leibniz was intrigued by the comparison between the series

$$\frac{\pi}{8} = \left(\frac{1}{2} - \frac{1}{6}\right) + \left(\frac{1}{10} - \frac{1}{14}\right) + \left(\frac{1}{18} - \frac{1}{22}\right) + \cdots$$

$$= \frac{1}{3} + \frac{1}{35} + \frac{1}{99} + \cdots$$

$$\frac{\pi}{8} = \frac{1}{1 \cdot 3} + \frac{1}{5 \cdot 7} + \frac{1}{9 \cdot 11} + \cdots \tag{21}$$

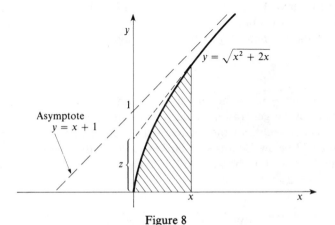

Figure 8

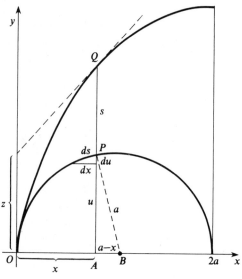

Figure 9

and Mercator's series in the form

$$\frac{1}{4}\log 2 = \frac{1}{4}\left(1 - \frac{1}{2} + \frac{1}{3} - \frac{1}{4} + \frac{1}{5} - \frac{1}{6} + \cdots\right)$$

$$= \left(\frac{1}{4} - \frac{1}{8}\right) + \left(\frac{1}{12} - \frac{1}{16}\right) + \left(\frac{1}{20} - \frac{1}{24}\right) + \cdots$$

$$\frac{1}{4}\log 2 = \frac{1}{2\cdot 4} + \frac{1}{6\cdot 8} + \frac{1}{10\cdot 12} + \cdots . \qquad (22)$$

Another impressive accomplishment of the transmutation method was Leibniz' quadrature of a general segment of a cycloid (the first of Pascal's contest problems). Figure 9 shows half of an arch of the cycloid that is generated by a circle of radius a rolling along the vertical line $x = 2a$. The length y of the ordinate AQ of the typical point Q on the cycloid is given by

$$y = u + s,$$

where $u = \sqrt{2ax - x^2}$ is the length of the ordinate AP of the corresponding point P on the generating circle, and s is the length of the circular arc $\widehat{OP}$. That is, the length of the segment PQ is equal to that of the arc $\widehat{OP}$; see Exercise 13 below for this standard property of the cycloid.

The similarity of the characteristic triangle for the circle and the triangle ABP implies that

$$\frac{du}{dx} = \frac{a - x}{u} \quad \text{and} \quad \frac{ds}{dx} = \frac{a}{u},$$

so

$$\frac{dy}{dx} = \frac{du}{dx} + \frac{ds}{dx} = \frac{2a - x}{u}.$$

Therefore

$$z = y - x\frac{dy}{dx} = (u + s) - \frac{2ax - x^2}{u} = (u + s) - u = s.$$

The transmutation theorem therefore gives

$$\int_0^{x_1} s\, dx = \int_0^{x_1} z\, dx = 2\int_0^{x_1} y\, dx - x_1 y_1$$

$$= 2\int_0^{x_1} (u + s)\, dx - x_1(u_1 + s_1). \tag{23}$$

Hence

$$\int_0^{x_1} s\, dx = x_1(u_1 + s_1) - 2\int_0^{x_1} u\, dx.$$

But subtraction of the triangle ABP from the circular sector OBP gives

$$\int_0^{x_1} u\, dx = \tfrac{1}{2}as_1 - \tfrac{1}{2}u_1(a - x_1).$$

It follows that

$$\int_0^{x_1} s\, dx = au_1 - s_1(a - x_1). \tag{24}$$

Finally, the area of the cycloidal segment over the interval $[0, x_1]$ is, from (24) and (25),

$$\int_0^{x_1} y\, dx = \tfrac{1}{2}x_1 y_1 + \frac{1}{2}\int_0^{x_1} s\, dx$$

$$= \tfrac{1}{2}x_1 y_1 + \tfrac{1}{2}au_1 - \tfrac{1}{2}s_1(a - x_1). \tag{25}$$

For example, with $x_1 = 2a$, $y_1 = \pi a$, $u_1 = 0$, $s_1 = \pi a$, Formula (25) gives $3\pi a^2/2$ for the area of the half-arch shown in Figure 9.

EXERCISE 13. Figure 10 shows the cycloid generated by a circle of radius a, with parametric equations

$$x = a(t - \sin t), \qquad y = a(1 - \cos t).$$

Show that the length s of the segment PQ is equal to the length $a(\pi - t)$ of the circular arc $\widehat{OP}$.

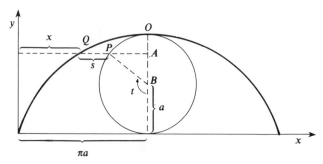

Figure 10

The Invention of the Analytical Calculus

Leibniz recorded the invention of his analytical calculus in a series of somewhat disjointed notes that he wrote during late October and early November of 1675. We will refer to the English translations of these crucial notes provided by Child [5].

Given a curve described in terms of its abscissa x and ordinate y, Leibniz envisions a discrete sequence of infinitely many values of y associated with the corresponding sequence of values of x. The sequence of ordinates is in some way analogous to an ordinary sequence of numbers, and the abscissas (like subscripts) determine the order of this sequence. However, the difference between two successive values of y is assumed to be infinitesimal, or "negligible" compared with the y values themselves.

At first Leibniz uses the letter ℓ to denote the infinitesimal difference between two successive values of y, and designates sums by writing omn. as an abbreviation of the Latin *omnia*. Thus, in the manuscript of October 29, he starts with his previous result $\frac{1}{2}y^2 = \int y\, dy$ written in the form

$$\frac{\overline{\text{omn. } \ell}^{\,2}}{2} = \text{omn.} \ \overline{\frac{\ell}{\overline{\text{omn. } \ell} \ a}}. \tag{26}$$

He uses the overbars in place of parentheses, and inserts the constant $a = 1$ to preserve dimensional homogeneity. Thus (26) means

$$\frac{1}{2}\left(\int dy\right)^2 = \int\left(\int dy\right) dy.$$

He remarks that "this is a very fine theorem, and one that is not at all obvious." Continuing, he says

> Another theorem of the same kind is
>
> $$\text{omn. } x\ell = x \text{ omn. } \ell - \text{omn. omn. } \ell \tag{27}$$
>
> where ℓ is taken to be a term of a progression [of differences], and x is the

number which expresses the position or order of the ℓ corresponding to it;
or x is the ordinal number and ℓ is the ordered thing.

Thus he is now talking about a sequence of *differences* of ordinates.
Equation (27) amounts to

$$\int x \, dy = x \left(\int dy \right) - \int \left(\int dy \right)$$
$$= xy - \int y \, dx.$$

In these early notes he often writes $\int y$, not making it clear whether $\int y \, dx$
or $\int y \, dy$ is intended.

It is actually at this point in the discussion that he introduces the
integral symbol with the innocuous-looking remark

It will be useful to write $\int$ for omn, so that $\int \ell = $ omn. ℓ, or the sum of
the ℓ's. Thus,

$$\frac{\int \bar{\ell}^{\,2}}{2} = \int \int \bar{\ell} \frac{\ell}{a}, \quad \text{and} \quad \int \overline{x\ell} = x \int \bar{\ell} - \int \int \ell. \qquad (28)$$

He adds that "all these theorems are true for series in which the differences
of the terms bear to the terms themselves a ratio that is less than any
assignable quantity" [i.e., is infinitesimal].

Having introduced the symbol $\int$ (evidently an elongated S for "sum"),
he proceeds to investigate its rules of operation. For example, with $\ell = dx$
in the first of Equations (28), he recovers $\int x \, dx = \frac{1}{2} x^2$. Then, with $\ell = x \, dx$
in the second of Equations (28), he obtains

$$\int x^2 \, dx = x \int x \, dx - \int \left(\int x \, dx \right)$$
$$= x \cdot \frac{x^2}{2} - \int \frac{x^2}{2},$$

from which it follows that

$$\int x^2 \, dx = \tfrac{1}{3} x^3.$$

Actually, in the October 29 manuscript he writes $\ell = y/d$, which becomes the now familiar dy three days later in his November 1 manuscript.
The difference notation y/d first appears in his discussion of the inverse
tangent problem:

Given ℓ, and its relation to x, to find $\int \ell$. This is to be obtained from the
contrary calculus, that is to say, suppose that $\int \ell = ya$. Let $\ell = ya/d$.
Then just as $\int$ will increase, so d will diminish the dimensions. But $\int$
means a sum, and d a difference. From the given y, we can always find
y/d or ℓ, that is, the difference of the y's. Hence one equation may be
transformed into the other.

In the manuscript of November 11, he poses the questions as to whether $d(uv) = (du)(dv)$ and $d(v/u) = (dv)/(du)$, and answers in the negative by noting that

$$d(x^2) = (x + dx)^2 - x^2 = 2x\,dx + (dx)^2 = 2x\,dx,$$

ignoring the higher-order infinitesimal, while

$$(dx)(dx) = (x + dx - x)(x + dx - x) = (dx)^2.$$

At this time Leibniz is still searching for the correct product and quotient rules. Nevertheless, he can already use his embryonic calculus to solve a non-trivial geometrical problem—to find the curve $y = f(x)$ whose subnormal v is inversely proportional to its ordinate, that is,

$$v = \frac{b}{y}.$$

He starts with his previous result

$$\int v\,dx = \tfrac{1}{2}y^2 \quad \text{(Formula (15))}.$$

Application of the inverse operator d gives

$$v\,dx = d\int v\,dx = d\left(\tfrac{1}{2}y^2\right) = y\,dy.$$

Substitution of $v = b/y$ then gives

$$\frac{b}{y}\,dx = y\,dy,$$

$$b\,dx = y^2\,dy,$$

$$\int b\,dx = \int y^2\,dy,$$

so

$$bx = \tfrac{1}{3}y^3$$

is the equation of the desired curve. He proceeds to check this result by use of Sluse's tangent rule (Chapter 5), thereby verifying in a non-trivial problem the validity of his calculus.

EXERCISE 14. Show by differentiation that the curve $bx = y^3/3$ has the subnormal property $v = b/y$.

By July of 1676, (see [4], pp. 118–122) Leibniz consistently includes the differential under the integral sign. In a manuscript dated November 1676 ([5], pp. 124–127), he states clearly the rules for differentiation and integration of powers,

$$dx^e = ex^{e-1} \quad \text{and} \quad \int x^e\,dx = \frac{x^{e+1}}{e+1},$$

with e not necessarily a positive integer. He adds the important remark that "this reasoning is general, and it does not depend upon what the progression for the x's may be." This is his way of saying that x may be a function of the independent variable, rather than the independent variable itself. This generality made possible the method of substitution for differentiating compositions of functions (i.e., what we now call the chain rule).

For example, to compute $d\sqrt{a + bz + cz^2}$ he substitutes $x = a + bz + cz^2$. Noting that

$$dVx = \frac{dx}{2Vx} \quad \text{and} \quad dx = (b + 2cz)\, dz,$$

it follows that

$$d\sqrt{a + bz + cz^2} = \frac{(b + 2cz)\, dz}{2\sqrt{a + bz + cz^2}}.$$

Previously, Leibniz had accepted Sluse's tangent rule without proof. In the November 1676 manuscript he shows how to derive it from his calculus. For example, given

$$z = ay^2 + byx + cx^2 + fx + gy + h = 0, \tag{29}$$

he substitutes $x + dx$ for x and $y + dy$ for y, obtaining

$$ay^2 + 2ay\, dy + a(dy)^2 + byx + by\, dx + bx\, dy + b\, dx\, dy$$
$$+ cx^2 + 2cx\, dx + c(dx)^2 + fx + f\, dx + gy + g\, dy + h = 0.$$

By (29) and the assumption that

$$a(dy)^2 + b\, dx\, dy + c(dx)^2 = 0,$$

there remains

$$2ay\, dy + by\, dx + bx\, dy + 2cx\, dx + f\, dx + g\, dy = 0,$$

so

$$\frac{dy}{dx} = -\frac{by + 2cx + f}{2ay + bx + g} = -\frac{\partial z/\partial x}{\partial z/\partial y},$$

in agreement with Sluse's rule.

In a manuscript dated 11 July 1677, and in an undated revision of it ([5], pp. 134–144), Leibniz gives statements and proofs of the product and quotient rules. To show that

$$d(xy) = x\, dy + y\, dx,$$

he writes

$$d(xy) = (x + dx)(y + dy) - xy$$
$$= x\, dy + y\, dx + dx\, dy,$$

and remarks that "the omission of the quantity $dx\, dy$, which is infinitely small in comparison with the rest, for it is supposed that dx and dy are

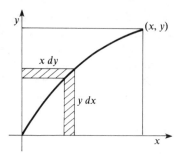

Figure 11

infinitely small, will leave $x\ dy + y\ dx$." To show that

$$d\frac{y}{x} = \frac{x\ dy - y\ dx}{x^2},$$

he writes

$$d\frac{y}{x} = \frac{y + dy}{x + dx} - \frac{y}{x} = \frac{x\ dy - y\ dx}{x^2 + x\ dx}$$

"which becomes (if we write x^2 for $x^2 + x\ dx$, since $x\ dx$ can be omitted as being infinitely small in comparison with x^2) equal to $(x\ dy - y\ dx)/x^2$."

Leibniz was careful to verify whenever possible agreement between the results of his evolving analytical or operational calculus and the results of familiar geometrical arguments. For example, he noted that the product rule $d(xy) = x\ dy + y\ dx$ agrees with the addition of areas in Figure 11,

$$\int x\ dy + \int y\ dx = xy.$$

Similarly, addition of moments about the x- and y-axes, respectively, gives

$$\int \tfrac{1}{2}y^2\ dx + \int xy\ dy = \tfrac{1}{2}xy^2$$

and

$$\int xy\ dx + \int \tfrac{1}{2}x^2\ dy = \tfrac{1}{2}x^2y.$$

However, as he remarked in the *Historia* ([5], pp. 55–56), "the calculus also shows this without reference to any figure, for $\tfrac{1}{2}d(x^2y) = xy\ dx + \tfrac{1}{2}x^2\ dy$; so that now there is need for no greater number of the fine theorems of celebrated men for Archimedean geometry, than at most those given by Euclid in his Book II or elsewhere, for ordinary geometry." The calculus has become a "sensible and palpable medium, which will guide the mind"!

In the revised 1677 manuscript the role of the infinitesimal characteristic triangle is made explicit in the new calculus. A curve is now a polygon with infinitely many angles and infinitesimal sides. The arclength element ds is a side of this infinite-angled polygon—an infinitesimal straight line seg-

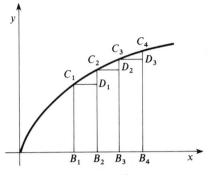

Figure 12

ment joining two adjacent vertices, so

$$ds = \sqrt{(dx)^2 + (dy)^2} = \sqrt{1 + \left(\frac{dy}{dx}\right)^2} \; dx,$$

where dx and dy are the differences of the x- and y-coordinates of these two adjacent vertices. Thus, for the parabola $y = \frac{1}{2}x^2$, the arclength is given by

$$s = \int ds = \int \sqrt{1 + x^2} \; dx,$$

so the rectification of this parabola depends on the quadrature of the hyperbola $y = \sqrt{1 + x^2}$.

In this manuscript the integral $\int y \, dx$ is clearly identified with a sum of infinitesimal rectangles with heights y and width dx. Referring to Figure 12, Leibniz says ([5], p. 13)

> I represent the area of a figure by the sum of all the rectangles contained by the ordinates and the differences of the abscissae, i.e., by $B_1D_1 + B_2D_2 + B_3D_3 +$ etc. For the narrow triangles $C_1D_1C_2$, $C_2D_2C_3$, etc., since they are infinitely small compared with the said rectangles, may be omitted without risk; and thus I represent in my calculus the area of the figure by $\int y \, dx$, or the rectangles contained by each y and the dx that corresponds to it.

Next he introduces what we now call the fundamental theorem of calculus—"we, now mounting to greater heights, obtain the area of a figure by finding the figure of its summatrix or quadratrix." Given a curve with ordinate x, whose area is sought, suppose it is possible to find a curve with ordinate y such that

$$\frac{dy}{dx} = \frac{z}{a}$$

where a is a constant (presumably included for the sake of dimensional homogeneity). Then

$$z \, dx = a \, dy,$$

so the area under the original curve is

$$\int z \, dy = a \int dy = ay, \qquad (30)$$

assuming (as usual with Leibniz) that the y-curve passes through the origin. Thus quadrature problems reduce in Leibniz' calculus to inverse tangent problems. That is, in order to find the area under the curve with ordinate z, it suffices to find a curve whose tangent satisfies the condition

$$\frac{dy}{dx} = z.$$

Subtracting the area over $[0, a]$ from that over $[0, b]$, and setting $a = 1$ in (30), it follows that

$$\int_a^b z \, dx = y(b) - y(a).$$

The First Publication of the Calculus

Leibniz' first published article on his differential calculus appeared in 1684 in the Leipzig periodical *Acta Eruditorum*. An English translation is included in Struik's source book ([11], pp. 272–280).

This first paper was entitled "A new method for maxima and minima as well as tangents, which is impeded neither by fractional nor by irrational quantities, and a remarkable type of calculus for this." Differentials are introduced without much indication of the infinitesimal considerations that had been their motivation. Given an arbitrary number dx, dy is defined to be that number dy such that the ratio dy/dx is equal to the slope of the tangent line. By modern standards, this is not so bad, except that no real definition of the tangent line is supplied—"We have only to keep in mind that to find a *tangent* means to draw a line that connects two points of the curve at an infinitely small distance, or the continued side of a polygon with an infinite number of angles, which for us takes the place of the *curve*."

The mechanical rules for computing differentials of powers, products, and quotients are stated without any explanation of their source. It is pointed out that dv is positive when the ordinate v increases with increasing x, while dv is negative when v is decreasing. Since "none of these cases happens $\cdots$ when v neither increases nor decreases, but is stationary," the necessary condition $dv = 0$ for a maximum or minimum, corresponding to a horizontal tangent line, is noted. The necessary condition $d(dv) = 0$ for an inflection point is explained similarly.

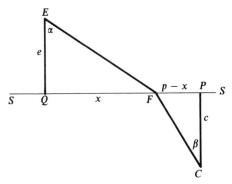

Figure 13

As a first application of his max-min method, Leibniz solves the following problem. "Let two points C and E [Fig. 13] be given and a line SS in the same plane. It is required to find a point F on SS such that when E and C are connected with F the sum of the rectangle [product] of CF and a given line h and the rectangle of FE and a given line r is as small as possible."

EXERCISE 15. With the notation indicated in Figure 13, the quantity to be minimized is

$$w = h\sqrt{(p-x)^2 + c^2} + r\sqrt{x^2 + e^2}.$$

Apply the condition $dw = 0$ to conclude that

$$\frac{h(p-x)}{\sqrt{(p-x)^2 + c^2}} = \frac{rx}{\sqrt{x^2 + e^2}},$$

or

$$\frac{\sin \alpha}{\sin \beta} = \frac{h}{r}.$$

Leibniz interprets this result as the law of refraction for a light ray passing from a medium of density r (with respect to the velocity of light) into one of density h, the line SS representing the interface between the two media. He adds that "other very learned men have sought in many devious ways what someone versed in this calculus can accomplish in these lines as by magic."

"And this is only the beginning of much more sublime Geometry, pertaining to even the most difficult and most beautiful problems of applied mathematics, which without our differential calculus or something similar no one could attack with any such ease." The 1684 paper concludes with the solution of a problem of De Beaune that Descartes had been unable to solve—to find the ordinate w of a curve whose subtangent τ is a constant, $\tau = a$. For such a curve (Fig. 14),

$$\frac{dw}{dx} = \frac{w}{\tau} = \frac{w}{a},$$

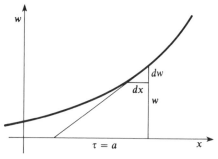

Figure 14

or

$$w = a\frac{dw}{dx}.\tag{31}$$

Leibniz considers a sequence of values of x with constant differences $dx = b$. Then

$$dw = \frac{b}{a}w,$$

so the corresponding sequence of ordinates w is proportional to its sequence of differences. Knowing that this is the characteristic property of a geometrical progression (see Exercise 4), he concludes that "if the x form an arithmetic progression, then the w form a geometric progression. In other words, if the w are numbers, the x will be logarithms, so the [desired curve] is logarithmic."

EXERCISE 16. Integrate Equation (31) to show that $w = e^{x/a}$, or $x = a \log w$, if $w = 1$ when $x = 0$.

EXERCISE 17. Sharpen Exercise 4 to prove that a series is geometric if and only if its terms are proportional to its differences.

The integral and the symbol $\int$ first appeared in print in a paper published by Leibniz in the *Acta Eruditorum* of 1686 (see [(11], pp. 281–282), where he presented the result expressed by Equation (15). The fundamental theorem of calculus, with the proof discussed in the preceding section, appeared in the *Acta Eruditorum* of 1693 (see [11], pp. 282–284).

Higher-Order Differentials

We have seen that Leibniz' infinitesimal calculus had its roots in a certain logical extrapolation—from the simple concepts of sum and difference sequences, for ordinary sequences of numbers, to the case of sequences of variables associated with a geometric curve. The curve is envisioned as an infinite-angled polygon with infinitely many infinitesimal sides, each of which is coincident with a tangent line to the curve. The basic sequences of variables associated with the curve are then the sequences of abscissas x and ordinates y of the infinitely many vertices of this polygon.

The difference of two successive values of x is the differential dx, and similarly for dy. It is assumed that the quantities dx and dy are non-zero but incomparably small, and therefore negligible, with respect to the values of the variables x and y. Similarly, it is assumed that a product of differentials, such as $(dx)(dy)$ or $(dx)^2$, is in turn negligible in comparison with the differentials dx and dy. On the basis of these assumptions, taken as operational rules, the standard differentiation formulas are derived.

It is important to note that the differentials dx are fixed non-zero quantities; they are neither variables approaching zero nor ones that are intended to eventually approach zero. There is actually a *sequence* of differentials dx (or dy) associated with the curve—it is simply the *difference sequence* of the sequence of abscissas x (or ordinates y). This difference sequence in turn has a difference sequence whose elements are the *second-order* differentials

$$d(dx) = d\,dx = d^2x.$$

Similarly, d^2y is the difference of successive differences of y values. By taking differences iteratively, the higher-order differentials $d^kx = d(d^{k-1}x)$ and $d^ky = d(d^{k-1}y)$ are obtained.

It is assumed that d^2y is incomparably small with respect to dy, and in general that d^ky is incomparably small with respect to $d^{k-1}y$. In addition, it is assumed that a kth-order differential d^ky is of the same order of magnitude as a kth power $(dx)^k$ of a first-order differential, in the sense that the quotient $d^ky/(dx)^k$ is a real number (except in singular cases). On the basis of these assumptions, the product and quotient rules can be used to compute differentials of differentials. For example,

$$d(x\,dy) = (dx)(dy) + x\,d^2y,$$
$$d^2(x^n) = d(nx^{n-1}dx)$$
$$= n(n-1)x^{n-2}(dx)^2 + nx^{n-1}d^2x, \qquad (32)$$
$$d\left(\frac{dy}{dx}\right) = \frac{(d^2y)(dx) - (d^2x)(dy)}{(dx)^2}. \qquad (33)$$

EXERCISE 18. (a) Show that

$$d^2(uv) = u\,d^2v + 2(du)(dv) + (d^2u)v$$
$$= (d^0u)(d^2v) + 2(du)(dv) + (d^2u)(d^0v)$$

where we write $d^0u = u$ and $d^0v = v$.

(b) Use induction on n to prove "Leibniz' rule"

$$d^n(uv) = \sum_{p=0}^{n} \binom{n}{p}(d^pu)(d^{n-p}v). \qquad (34)$$

In Leibnizian computations with differentials, the choice of x as the *independent variable* was effected by choosing the original sequence of x values or abscissas as an *arithmetic* progression, so that dx is constant and therefore $d^2x = 0$. Note that, with $d^2x = 0$, Formulas (32) and (33) become

$$d^2(x^n) = n(n-1)x^{n-2}(dx)^2 \quad \text{and} \quad d\left(\frac{dy}{dx}\right) = \frac{d^2y}{dx},$$

from which it follows that

$$\frac{d^2(x^n)}{(dx)^2} = n(n-1)x^{n-2}$$

and

$$\frac{d(dy/dx)}{dx} = \frac{d^2y}{(dx)^2}.$$

EXERCISE 19. Assuming that $d^2x = 0$, show by induction on n that

$$d\left(\frac{d^{n-1}y}{(dx)^{n-1}}\right) = \frac{d^n y}{(dx)^{n-1}},$$

so division by dx gives

$$\frac{d}{dx}\left(\frac{d^{n-1}y}{(dx)^{n-1}}\right) = \frac{d^n y}{(dx)^n}. \tag{35}$$

Whereas higher-order differentials, in contrast with first-order ones, no longer are with us in ordinary everyday calculus, their legacy survives in the notation

$$\frac{d^n y}{dx^n}$$

(Formula (35) without parentheses) for the nth derivative of the function y of the independent variable x.

In a problem where the choice of y as the independent variable was indicated, the sequence of ordinates y (rather than abscissas x) was taken as an arithmetic progression, so that $d^2y = 0$ (instead of $d^2x = 0$). This choice was referred to as the "specification of the progression of the variables." See the article by Bos ([2], pp. 25–35) for a discussion of the consequences of this choice. In particular, this freedom of choice was the basis for the method of integration by substitution. For example, if the sequence of abscissas x is taken to be a sequence of squares, one has the substitution $x = t^2$, $dx = 2t\, dt$, where t is the new independent variable with $d^2t = 0$. As Leibniz himself put it, "in this way I can transform the given quadrature into others in an infinity of ways, and thus find the one by means of the other."

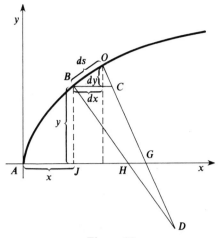

Figure 15

To illustrate the geometric applications of higher-order differentials, we include John Bernoulli's derivation of a formula for the radius of curvature at a point on a curve (as described by Bos ([2], pp. 36–37). The radii OD and BD in Figure 15 are perpendicular to the curve AB, and intersect at the center of curvature D. The radius of curvature at B is $r = BD$, and the arclength differential is $ds = BO$. From the fact that triangle BHJ is similar to the characteristic triangle, it follows that

$$AH = x + y\frac{dy}{dx}.$$

Taking x as the independent variable so that $d^2x = 0$, it follows that GH, the differential of AH, is given by

$$GH = d(AH) = d\left(x + y\frac{dy}{dx}\right)$$

$$GH = dx + \frac{(dy)^2 + yd^2y}{dx}. \tag{36}$$

Now the similarity of the triangles DGH and DCB gives the proportion

$$\frac{BC}{HG} = \frac{BD}{HD}, \tag{37}$$

in which

$$BC = \frac{(dx)^2 + (dy)^2}{dx}, \qquad BD = r,$$

and

$$HD = r - BH = r - \frac{y\sqrt{(dx)^2 + (dy)^2}}{dx}.$$

After substitution of these values and (36) into (37), the resulting equation is easily solved for

$$r = -\frac{\left[(dx)^2 + (dy)^2\right]^{3/2}}{(dx)(d^2y)} = -\frac{(ds)^3}{(dx)(d^2y)}. \tag{38}$$

Division of the numerator and denominator in (38) by $(dx)^3$ gives the familiar formula in terms of derivatives of y with respect to x,

$$r = \frac{(ds/dx)^3}{|d^2y/dx^2|} = \frac{\left[1 + (dy/dx)^2\right]^{3/2}}{|d^2y/dx^2|}.$$

The Meaning of Leibniz' Infinitesimals

In his publications on the calculus, Leibniz stressed the routine and formal character of his rules for the calculation and manipulation of differentials, and asserted that the proper application of these rules of operation would invariably lead to correct and meaningful results, even if uncertainty remained as to the precise meaning of the infinitesimals that appeared in the computations. Indeed, it was the correctness of the results obtained that had been his guide in the formulation of his algorithms, and had confirmed his confidence in their operational validity.

Mathematical tradition generally attributes to Leibniz a belief in the actual existence of infinitesimal quantities—an infinitesimal quantity being one that is non-zero, yet smaller than every positive real number—and allegations to this effect are sometimes found in discussions of twentieth century "non-standard analysis" (see Chapter 12). Nevertheless, Leibniz seems not to have committed himself on the question of the actual existence of infinitesimals, and he certainly expressed doubts on occasion (e.g. see [3], p. 219). At any rate, he recognized that the question of the existence of infinitesimals is independent of the question as to whether computations with infinitesimals, carried out in accordance with the operational rules of the calculus, lead to correct solutions of problems. Consequently, whether or not infinitesimals actually exist, they can serve as "fictions useful to abbreviate and to speak universally." Leibniz gave a comprehensive statement of this point of view in an unpublished manuscript probably written sometime after 1700, in reply to criticisms of the calculus advanced in 1694 by the Dutch physician and geometer Bernard Nieuwentijdt ([5], pp. 149–150):

> Whether infinite extensions [quantities] successively greater and greater, or infinitely small ones successively less and less, are legitimate considerations, is a matter that I own to be possibly open to question; but for him who would discuss these matters, it is not necessary to fall back upon metaphysical controversies, such as the composition of the continuum, or to make geometrical matters depend thereon. $\cdot\cdot\cdot$ It will be sufficient if,

when we speak of infinitely great (or more strictly unlimited), or of infinitely small quantities (i.e., the very least of those within our knowledge), it is understood that we mean quantities that are indefinitely great or indefinitely small, i.e., as great as you please, or as small as you please, so that the error that any one may assign may be less than a certain assigned quantity. Also, since in general it will appear that, when any small error is assigned, it can be shown that it should be less, it follows that the error is absolutely nothing ... If any one wishes to understand these [the infinitely great and infinitely small] as the ultimate things, or as truly infinite, it can be done, and that too without falling back upon a controversy about the reality of extensions, or of infinite continuums in general, or of the infinitely small, ay, even though he think that such things are utterly impossible; it will be sufficient simply to make use of them as a tool that has advantages for the purpose of the calculation, just as the algebraists retain imaginary roots with great profit. For they contain a handy means of reckoning, as can manifestly be verified in every case in a rigorous manner by the method already stated.

Thus Leibniz presents his calculus of infinitesimals as an abbreviated form of the rigorous Greek method of exhaustion, one whose more concise language is better adapted to the art of discovery. The basis for his argument is that, given an equality between two expressions involving differentials, that has been obtained by discarding higher-order differentials, it could have been established rigorously (and more tediously) by substituting for each differential the corresponding finite difference, and then proving that the difference between the resulting expressions could be made arbitrarily small by choosing the finite differences sufficiently small.

Finally, it should be mentioned that whereas Leibniz himself was somewhat circumspect regarding the actual existence of infinitesimals, this appropriate caution was generally not shared by his immediate followers (such as the Bernoulli brothers), who uncritically accepted infinitesimals as genuine mathematical entities. Indeed, this freedom from doubts about the foundations of the calculus probably promoted the rapid development of the subject and its applications.

Leibniz and Newton

In this chapter and the previous one we have detailed the separate approaches of Newton and Leibniz to the development of the calculus as a new and coherent mathematical discipline. It is instructive, finally, to compare and contrast their two approaches.

Leibniz' devotion to the advantages of appropriate notation was so wholehearted that one could ask whether he invented the calculus or merely a particularly felicitous system of notation for the calculus. Of course the answer is that he did both; indeed, his differential and integral notation so captured the essence of his calculus as to make notation and concept virtually inseparable. Newton, on the other hand, had little interest in notational matters; neither suggestive nor consistent notation was of great importance to him.

Leibniz' constant goal was the formulation of general methods and algorithms that could serve to unify the treatment of diverse problems. General methods are certainly implicit in all of Newton's work, but his greater enthusiasm for the solution of particular problems is evident. The difference is one of emphasis—Leibniz emphasizes general techniques that can be applied to specific problems, whereas Newton emphasizes concrete results that can be generalized.

In regard to the calculus itself, discrete infinitesimal differences of geometric variables played the central role in Leibniz' approach, while Newton's fundamental concept was the fluxion or time rate of change, based on intuitive ideas of continuous motion. As a consequence, Leibniz' notation and terminology effectively disguises the limit concept, which by contrast is fairly explicit in Newton's calculus.

For Leibniz, the separate differentials dx and dy are fundamental; their ratio dy/dx is "merely" a geometrically significant quotient. For Newton, however, especially in his later work, the derivative itself—as a ratio of fluxions or an "ultimate ratio of evanescent quantities"—is the heart of the matter. A second derivative is simply a fluxion of a fluxion, each fluxion involving only first-order infinitesimals, so Newton has no need of Leibniz' higher-order infinitesimals.

The integral of Newton is an indefinite integral, a fluent to be determined from its given fluxion; he solves area and volume problems by interpreting them as inverse rate of change problems. Leibniz' integral, by contrast, is an infinite sum of differentials. Of course, both ultimately compute their integrals by the process of antidifferentiation; the computational exploitation of the inverse relationship between quadrature and tangent problems was their key common contribution.

Whereas Leibniz had only a peripheral interest in infinite series (apart from their contribution to his early motivation), the expansion of functions in power series was for Newton an everyday working tool that he always regarded as an indispensable part of his "method" of analysis. For example, Newton was happy to evaluate an integral or solve a differential equation in terms of an infinite series for its solution, but Leibniz always preferred a "closed form" solution.

We have seen that Newton's formative work on the calculus dated from 1664–1666, while Leibniz' analogous period was 1672–1676. However, Leibniz' first publications on the calculus appeared in 1684 and 1686 (his *Acta Eruditorum* articles), whereas Newton, although he had shown manuscripts to colleagues in England, published nothing on the calculus until his *Principia* of 1687 and his *Opticks* of 1704 (with the *De Quadratura* as a mathematical appendix).

Beginning in the late 1690's Leibniz came under attack by followers of Newton who assumed that he had taken and used crucial suggestions (without acknowledging credit to Newton) from the letters of 1676, and that he had learned of Newton's work during his brief visits to London in 1673 and 1676 (although he and Newton never met). Eventually, inferences became public charges of plagiarism. Leibniz in 1711 appealed for redress from the Royal Society of London (of which he was a member and Newton the president). The Royal Society appointed a commission which ruled in 1712, in a decision that was evidently stage-managed by Newton, that Leibniz was essentially guilty as charged.

This unfortunate controversy had less to do with mathematics than with nationalistic rivalry between English and continental European mathematicians (see the article by Hofmann [8], pp. 164–165 for further details). Any serious study of the investigations of Newton and Leibniz makes it clear that their respective contributions were discovered independently.

An irony of the English "victory" in the Newton–Leibniz dispute was that English mathematicians, in steadfastly following Newton and refusing to adopt Leibniz' analytical methods, effectively closed themselves off from the mainstream of progress in mathematics for the next century. Although Newton's spectacular applications of mathematics to scientific problems inspired much of the eighteenth century progress in mathematics, these advances came mainly at the hands of continental mathematicians using the analytical machinery of Leibniz' calculus, rather than the methods of Newton.

References

[1] M. E. Baron, *The Origins of the Infinitesimal Calculus*. Oxford: Pergamon, 1969, Chapter 7.
[2] H. J. M. Bos, Differentials, higher-order differentials and the derivative in the Leibnizian calculus. *Arch Hist Exact Sci* **14**, 1 – 90, 1974–75.
[3] C. B. Boyer, *The History of the Calculus and Its Conceptual Development*. New York: Dover, 1959, Chapter V.
[4] F. Cajori, Leibniz, the master builder of notations. *Isis* **7**, 412–429, 1925.
[5] J. M. Child, *The Early Mathematical Manuscripts of Leibniz*. Chicago: Open Court, 1920.
[6] J. Earman, Infinities, infinitesimals, and indivisibles: the Leibnizian labyrinth. *Stud Leibnitiana* **7**, 236–251, 1975.
[7] J. E. Hofmann, *Leibniz in Paris 1672–1676*. Cambridge University Press, 1974.
[8] Article on Leibniz, *Dictionary of Scientific Biography*, Vol. VIII. New York: Scribners, 1973.
[9] H. W. Turnbull et al. (eds.), *The Correspondence of Isaac Newton*, Vol. 2. Cambridge University Press, 1959–1978.
[10] C. J. Scriba, The inverse method of tangents: A dialogue between Leibniz and Newton. *Arch Hist Exact Sci* **2**, 113–137, 1962.
[11] D. J. Struik, *A Source Book in Mathematics 1200–1800*. Cambridge, MA: Harvard University Press, 1969.
[12] A. Weil, Review of Hofmann (Reference 7 above). *Bull Am Math Soc* **81**, 676–688, 1975.

10 The Age of Euler

Leonhard Euler (1707–1783)

The eighteenth century was in mathematics a period of consolidation and exploitation of the great discoveries of the seventeenth century, and of their application to the investigation of scientific problems. The dominant figure of this period was Leonhard Euler, the most prolific mathematician of all time—his collected works amount to approximately seventy-five substantial volumes. The range and creativity of his fundamental contributions, to all branches of both pure and applied mathematics, would perhaps justify Euler's inclusion in the traditional short list—Archimedes, Newton, Gauss—of the incomparable giants of mathematics.

Euler's professional career was spent at the royal academies in St. Petersburg (1727–1741 and 1766–1783) and Berlin (1741–1766). He was born and educated at Basel in Switzerland, where he completed his university education at the age of fifteen. Although his father, a clergyman who had studied mathematics under James Bernoulli, preferred a theological career for his son, young Euler learned mathematics from John Bernoulli, and thereby found his true vocation.

The famous Bernoulli brothers—James (= Jacques = Jakob, 1654–1705) and John (= Jean = Johann, 1667–1748)—were frequent correspondents with Leibniz and, after his 1684–86 calculus publications, equal collaborators with him in the initial development of the Leibnizian calculus. It was James Bernoulli who introduced the word "integral" in suggesting the name *calculus integralis* instead of Leibniz' original *calculus summatorius* for the inverse of the *calculus differentialis*.

John Bernoulli wrote during 1691–1692 two small unpublished treatises on the differential and integral calculus. Shortly thereafter he agreed to teach the subject to the young Marquis de l'Hospital (1661–1704) and, in return for a regular salary, to communicate to l'Hospital his own mathematical discoveries, to be used as the Marquis saw fit. The result was the publication by l'Hospital in 1696 of the first differential calculus textbook, entitled *Analyse des infiniment petits pour l'intelligence des lignes courbes* (Analysis of the Infinitely Small for the Understanding of Curves). The book opens with two definitions—"variable quantities are those that continually increase or decrease," and "the infinitely small part whereby a variable quantity is continually increased or decreased is called the differential of that quantity"—and two postulates—"two quantities, whose difference is an infinitely small quantity, may be taken (or used) indifferently for each other," and "a curve may be considered as a polygon of an infinite number of sides, each of infinitely small length, which determine the curvature of the curve by the angles they make with each other." On this basis the basic formulas for differentials of algebraic functions are derived, and applied to problems involving tangents, maxima and minima, and curvature.

This first calculus text is now remembered mainly for its inclusion of a result of Bernoulli that is known as "l'Hospital's rule" for indeterminate forms—if $f(x)$ and $g(x)$ are differentiable functions with $f(a) = g(a) = 0$, then

$$\lim_{x \to a} \frac{f(x)}{g(x)} = \lim_{x \to a} \frac{f'(x)}{g'(x)}$$

provided that the right-hand limit exists. L'Hospital's argument, which is stated verbally without functional notation (see the English translation included in Struik's source book [12], pp. 313–316), amounts simply to the assertion that

$$\frac{f(a + dx)}{g(a + dx)} = \frac{f(a) + f'(a)\,dx}{g(a) + g'(a)\,dx} = \frac{f'(a)\,dx}{g'(a)\,dx} = \frac{f'(a)}{g'(a)}$$

provided that $f(a) = g(a) = 0$. He concludes that, if the ordinate y of a given curve "is expressed by a fraction, the numerator and denominator of which do each of them become 0 when $x = a$," then "if the differential of the numerator be found, and that is divided by the differential of the denominator, after having made $x = a$, we shall have the value of [the ordinate y when $x = a$]."

L'Hospital's was the first printed textbook on the new calculus, but it was Euler's two-volume *Introductio in analysin infinitorum* of 1748 that forged a new branch of mathematics—analysis, to stand alongside geometry and algebra—from the concept of a function and infinite processes (such as summation of series) for the representation and investigation of functions. In the *Introductio* we find, for the first time, systematic treatments of logarithms as exponents and of the trigonometric functions as numerical ratios rather than line segments, and a study of the functional properties of the elementary transcendental functions by means of their infinite series expansions. The *Introductio* is the earliest mathematics textbook that can be read with comparative ease by the modern student (even by one who has little knowledge of Latin). Euler's notation and terminology seem almost "modern" for the simple reason that he originally introduced so much of the notation and terminology that is still used.

Euler's *Introductio* was followed by his *Institutiones calculi differentialis* of 1755 and the three-volume *Institutiones calculi integralis* of 1768–1770. These great treatises on the differential and integral calculus provide the original source for much of the content and methods of modern courses and textbooks on calculus and differential equations.

The Concept of a Function

In modern mathematics courses a *function* from X to Y (where X and Y are sets of real or complex numbers) is defined to be a rule that assigns to each element x of the set X a unique element $y = f(x)$ of the set Y. Sometimes the function f is defined in terms of the set of all pairs $(x, f(x))$, a subset of the Cartesian product set $X \times Y$.

Euler's *Introductio* was the first work in which the function concept played a central and explicit role. It was the identification of functions, rather than curves, as the principal objects of study, that permitted the arithmetization of geometry, and the consequent separation of infinitesimal analysis from geometry proper.

In seventeenth century infinitesimal analysis geometrical curves were the principal objects of study, and this study was carried out largely within the framework of Cartesian geometry. The *variables* associated with a particular curve were exclusively geometrical quantities—abscissas, ordinates, subtangents and subnormals, arclengths of segments of the curve, areas between the curve and the coordinate axes, etc. Relationships between these quantities were often described by means of equations, except in the case of transcendental relationships—ones that "transcended" description by means of algebraic equations; these had to be described in terms of verbal descriptions of geometrical constructions. However, these geometrical variables were viewed primarily as being associated with the curve itself, rather than with each other.

In particular, the several variables associated with a curve were not generally viewed as depending upon some single "independent" variable. A partial exception was Newton's fluxional approach in which all of the geometrical variables were regarded (in effect) as functions of time. Indeed, Newton could force upon a given variable the role of independent variable by choosing it to play the role of the time variable. As he put it in *Methods of Series and Fluxions* ([NP III], p. 73),

We can, however, have no estimate of time except in so far as it is expounded and measured by an equable local motion, and furthermore quantities of the same kind alone, and so also their speeds of increase and decrease, may be compared one with another. For these reasons I shall, in what follows, have no regard to time, formally so considered, but from quantities propounded which are of the same kind shall suppose some one to increase with an equable flow: to this all the others may be referred as though it were time, and so by analogy the name of 'time' may not improperly be conferred upon it. And so whenever in the following you meet with the word 'time' (as I have, for clarity's and distinction's sake, on occasion woven it into my text), by that name should be understood not time formally considered but that other quantity through whose equable increase or flow time is expounded and measured.

Nevertheless, Newton's view in practice remained essentially geometric and kinematic rather than functional in character.

Leibniz introduced the word "function" into mathematics precisely as a term designating the various geometrical quantities associated with a curve; they were the "functions" of the curve. Then, as increased emphasis was placed on the formulas and equations relating the functions of a curve, attention came naturally to be focused on their roles as the symbols appearing in these equations, that is, as variables depending only on the values of other variables and constants in equations (and thus no longer depending explicitly on the original curve). This gradual shift of emphasis led ultimately to the definition of a function given by Euler at the beginning of the *Introductio* ([5], p. 185, § 4).

> A function of variable quantity is an analytical expression composed in any way from this variable quantity and from numbers or constant quantities.

Euler's admissible operations for "composing analytical expressions" were the standard algebraic operations (including the solution of algebraic equations) and various enumerated transcendental processes, including taking limits of sequences, sums of infinite series, infinite products, etc. On this basis his arithmetization of infinitesimal analysis was so complete that no pictures or drawings appear in Volume I of his *Introductio* (Volume II deals with analytic geometry) nor in his calculus treatises.

Euler later gave (in the preface to his *Institutiones calculi differentialis* [6], p. 4) a still broader definition that is virtually equivalent to modern definitions of functions.

> If some quantities so depend on other quantities that if the latter are changed the former undergo change, then the former quantities are called functions of the latter. This denomination is of broadest nature and comprises every method by means of which one quantity could be determined by others. If, therefore, x denotes a variable quantity, then all quantities which depend upon x in any way or are determined by it are called functions of it (translation quoted from [14], p. 70).

The article by Youschkevitch [14] provides a comprehensive discussion of the development of functional concepts from ancient to recent times. See also the introduction to the article by Bos cited in the references to Chapter 9.

Euler's Exponential and Logarithmic Functions

Euler investigates the exponential and logarithmic functions in Chapter VII of the *Introductio* ([5], pp. 122–132). In this section we outline his approach, emphasizing his development of infinite series expansions for these functions.

Euler unhesitatingly accepts the existence of both infinitely small and infinitely large numbers, and uses them to such remarkable advantage that the modern reader's own hesitation must be tinged with envy. His typical argument involves an infinitely small number ω and an infinitely large number i, which in this exposition we will replace by ϵ and N, respectively, thereby reserving i for the imaginary unit $\sqrt{-1}$. Euler did not introduce the now-standard notation $i = \sqrt{-1}$ until late in his career. Finally, we will write x for Euler's usual independent variable z.

In Chapter VI ([5], p. 106) Euler has introduced the logarithm of x with base a, $\log_a x$ (he writes simply lx), as that exponent y such that $a^y = x$. This was the first historical appearance of logarithms interpreted explicitly as exponents. At the beginning of Chapter VII, noting that $a^0 = 1$, he writes

$$a^\epsilon = 1 + k\epsilon \tag{1}$$

for an infinitely small number ϵ. It will turn out that k is a constant depending on a.

EXERCISE 1. Interpreting k as $\lim_{\epsilon \to 0}(1/\epsilon)(a^\epsilon - 1)$, explain why $k = \log_e a$. *Hint*: Interpret this limit as the value of the derivative of a^x when $x = 0$.

Given a (finite) number x, Euler introduces the infinitely large number $N = x/\epsilon$. Then

$$a^x = a^{N\epsilon} = (a^\epsilon)^N$$

$$= (1 + k\epsilon)^N \quad \text{(by Eq. (1))}$$

$$= \left(1 + \frac{kx}{N}\right)^N \tag{2}$$

$$= 1 + N\left(\frac{kx}{N}\right) + \frac{N(N-1)}{2!}\left(\frac{kx}{N}\right)^2$$

$$+ \frac{N(N-1)(N-2)}{3!}\left(\frac{kx}{N}\right)^3 + \cdots \quad \text{(binomial series)}$$

$$a^x = 1 + kx + \frac{1}{2!}\frac{N(N-1)}{N^2}k^2x^2 + \frac{1}{3!}\frac{N(N-1)(N-2)}{N^3}k^3x^3 + \cdots. \tag{3}$$

Because N is infinitely large, he assumes that

$$1 = \frac{N-1}{N} = \frac{N-2}{N} = \cdots .$$

Consequently Equation (3) becomes

$$a^x = 1 + \frac{kx}{1!} + \frac{k^2x^2}{2!} + \frac{k^3x^3}{3!} + \cdots . \tag{4}$$

Substituting $x = 1$, he obtains the relationship between a and k,

$$a = 1 + \frac{k}{1!} + \frac{k^2}{2!} + \frac{k^3}{3!} + \cdots . \tag{5}$$

Euler now introduces his famous number e as the value of a for which $k = 1$,

$$e = 1 + \frac{1}{1!} + \frac{1}{2!} + \frac{1}{3!} + \cdots .$$

He immediately identifies e as the base for natural or hyperbolic logarithms, and writes out its decimal expansion to 23 places,

$$e = 2.71828182845904523536028.$$

Equation (2) then gives

$$e^x = \left(1 + \frac{x}{N}\right)^N, \tag{6}$$

which we may interpret as

$$e^x = \lim_{n \to \infty} \left(1 + \frac{x}{n}\right)^n,$$

so

$$e = \lim_{n \to \infty} \left(1 + \frac{1}{n}\right)^n, \tag{7}$$

the usual modern definition. With $k = 1$, Equation (4) finally gives

$$e^x = 1 + \frac{x}{1!} + \frac{x^2}{2!} + \frac{x^3}{3!} + \cdots . \tag{8}$$

Turning to the logarithm, Euler writes

$$1 + y = a^x = a^{N\epsilon} = (1 + k\epsilon)^N,$$

so $\log_a(1 + y) = N\epsilon$. Then

$$1 + k\epsilon = (1 + y)^{1/N},$$

so $\epsilon = ((1 + y)^{1/N} - 1)/k$, and it follows that

$$\log_a(1 + y) = N\epsilon = \frac{N}{k}\left[(1 + y)^{1/N} - 1\right]. \tag{9}$$

Replacing a with e (so $k = 1$) and y with x, we obtain

$$\log(1 + x) = N\left[(1 + x)^{1/N} - 1\right], \tag{10}$$

which we may interpret as

$$\log(1 + x) = \lim_{n \to \infty} n\left[(1 + x)^{1/n} - 1\right]. \tag{11}$$

EXERCISE 2. With $a = 1 + x$ and $n = 1/h$, Equation (11) becomes

$$\log a = \lim_{h \to 0} \frac{a^h - 1}{h}.$$

Explain why this limit follows from the computation in Exercise 1.

Euler obtain's Mercator's series for $\log(1 + x)$ by using the binomial series to expand the $(1 + x)^{1/N}$ in (10).

$$(1 + x)^{1/N} = 1 + \frac{1}{N}x + \frac{\frac{1}{N}\left(\frac{1}{N} - 1\right)}{2!}x^2 + \frac{\frac{1}{N}\left(\frac{1}{N} - 1\right)\left(\frac{1}{N} - 2\right)}{3!}x^3 + \cdots$$

$$= 1 + \frac{1}{N}x - \frac{1}{2!}\frac{N - 1}{N^2}x^2 + \frac{1}{3!}\frac{(N - 1)(2N - 1)}{N^3}x^3 + \cdots.$$

Setting

$$\frac{N - 1}{N} = 1, \qquad \frac{2N - 1}{N} = 2, \qquad \text{etc.,}$$

because N is infinitely large, he obtains

$$\log(1 + x) = N\left[(1 + x)^{1/N} - 1\right]$$

$$= x - \frac{1}{2}\frac{N - 1}{N}x^2 + \frac{1}{3!}\frac{(N - 1)(2N - 1)}{N^2}x^3 - \cdots$$

$$\log(1 + x) = x - \tfrac{1}{2}x^2 + \tfrac{1}{3}x^3 - \cdots. \tag{12}$$

EXERCISE 3. Replace x with $-x$ in (12), and then subtract logarithms to obtain

$$\log\frac{1 + x}{1 - x} = 2\left(x + \frac{1}{3}x^3 + \frac{1}{5}x^5 + \cdots\right). \tag{13}$$

EXERCISE 4. Note that, if the k in Equation (9) is carried through, the derivation of equation (12) gives

$$\log_a(1 + y) = \frac{1}{k}\left(x - \frac{1}{2}x^2 + \frac{1}{3}x^3 - \frac{1}{4}x^4 + \cdots\right).$$

On page 127 Euler substitutes $a = 10$ and $y = 9$ to obtain the value

$$k = 9 - \frac{9^2}{2} + \frac{9^3}{3} - \frac{9^4}{4} + \cdots$$

for base 10. Do you believe this?

Euler's Trigonometric Functions and Expansions

Prior to Euler, sines and cosines were lengths of line segments relative to a circle of given radius R. The sine of the angle A was half the chord of the circle subtended by a central angle $2A$, and the cosine of A was the length of the perpendicular from the center to this chord. Thus, with $R = 10000$, $\sin 30° = 5000.00$ and $\cos 30° = 8660.25$.

In Chapter VIII of the *Introductio* ([5], pp. 133–152), Euler defined (and standardized) the trigonometric *functions* as follows: $\sin x$ and $\cos x$ denote the sine and cosine of the central angle in a *unit* circle that subtends an arc of length x. This amounts to saying that $\sin x$ and $\cos x$ are the sine and cosine of an angle of x *radians* in a circle of radius *one*. The fundamental identity

$$\sin^2 x + \cos^2 x = 1$$

follows at once. Euler immediately noted the periodicity properties of the sine and cosine, and proceeded to list the standard formulas, e.g.,

$$\sin(y \pm z) = \sin y \cos z \pm \cos y \sin z$$
$$\cos(y \pm z) = \cos y \cos z \mp \sin y \sin z \qquad (14)$$

in precisely the forms that trigonometry textbooks have included them ever since. Next he indicated (on page 140) the inductive derivation of "De Moivre's identity"

$$(\cos z \pm i \sin z)^n = \cos nz \pm i \sin nz, \qquad (15)$$

where $i = \sqrt{-1}$ and n is a positive integer.

EXERCISE 5. Prove De Moivre's identity by induction on n.

Now let ϵ be an infinitely small number and N an infinitely large integer (!). The two choices of signs in (15) give

$$\cos N\epsilon + i \sin N\epsilon = (\cos \epsilon + i \sin \epsilon)^N$$

and

$$\cos N\epsilon - i \sin N\epsilon = (\cos \epsilon - i \sin \epsilon)^N.$$

Addition and subtraction then gives

$$\cos N\epsilon = \tfrac{1}{2}\left[(\cos \epsilon + i \sin \epsilon)^N + (\cos \epsilon - i \sin \epsilon)^N \right]$$

and

$$\sin N\epsilon = \frac{1}{2i}\left[(\cos \epsilon + i \sin \epsilon)^N - (\cos \epsilon - i \sin \epsilon)^N \right]. \qquad (16)$$

Euler expands the right-hand sides of these equations using the binomial series to obtain

$$\cos N\epsilon = \cos^N\epsilon - \frac{N(N-1)}{2!}\cos^{N-2}\epsilon \sin^2\epsilon$$
$$+ \frac{N(N-1)(N-2)(N-3)}{4!}\cos^{N-4}\epsilon \sin^4\epsilon + \cdots$$

and

$$\sin N\epsilon = N\cos^{N-1}\epsilon \sin \epsilon - \frac{N(N-1)(N-2)}{3!}\cos^{N-3}\epsilon \sin^3\epsilon$$
$$+ \frac{N(N-1)(N-2)(N-3)(N-4)}{5!}\cos^{N-5}\epsilon \sin^5\epsilon + \cdots .$$

Finally, writing $N\epsilon = x$, and substituting $\cos \epsilon = 1$, $\sin \epsilon = \epsilon$, $N = N - 1 = N - 2 = \cdots$ because ϵ is infinitely small and N is infinitely large, he obtains the trigonometric series

$$\cos x = 1 - \frac{x^2}{2!} + \frac{x^4}{4!} - \cdots \tag{17}$$

and

$$\sin x = x - \frac{x^3}{3!} + \frac{x^5}{5!} - \cdots . \tag{18}$$

Euler obtains his famous relation between the exponential and trigonometric functions by substituting $\epsilon = x/N$ into Equations (16), obtaining

$$\cos x = \frac{1}{2}\left[\left(1 + \frac{ix}{N}\right)^N + \left(1 - \frac{ix}{N}\right)^N\right]$$

and

$$\sin x = \frac{1}{2i}\left[\left(1 + \frac{ix}{N}\right)^N - \left(1 - \frac{ix}{N}\right)^N\right]$$

because $\cos \epsilon = 1$ and $\sin \epsilon = \epsilon = x/N$. But remember that

$$e^z = \left(1 + \frac{z}{N}\right)^N.$$

These formulas then say

$$\cos x = \frac{e^{ix} + e^{-ix}}{2} \tag{19}$$

and

$$\sin x = \frac{e^{ix} - e^{-ix}}{2i}. \tag{20}$$

EXERCISE 6. Deduce from (19) and (20) that

$$e^{\pm ix} = \cos x \pm i \sin x. \tag{21}$$

The following exercises outline Euler's derivation ([5], pp. 148–150) of Gregory's inverse tangent series.

Exercise 7. Substitute $x = z/N$ (where z is finite) into (20) to obtain

$$\frac{z}{N} = \frac{1}{2i}\left[(e^{iz})^{1/N} - (e^{-iz})^{1/N} \right].$$

Remembering that $\log y = N(y^{1/N} - 1)$, so

$$y^{1/N} = 1 + \frac{1}{N}\log y,$$

conclude that

$$z = \frac{1}{2i}\left[\log(e^{iz}) - \log(e^{-iz})\right].$$

Exercise 8. Substitute Euler's relation (21) into the result of Exercise 7 to deduce the identity

$$z = \frac{1}{2i}\log\frac{1 + i\tan z}{1 - i\tan z}. \tag{22}$$

Exercise 9. Substitute the logarithmic series (13) on the right-hand side of (22), with $x = i\tan z$, to obtain

$$z = \tan z - \frac{\tan^3 z}{3} + \frac{\tan^5 z}{5} - \cdots.$$

With $z = \tan^{-1}t$, this is

$$\tan^{-1}t = t - \frac{t^3}{3} + \frac{t^5}{5} - \cdots.$$

Differentials of Elementary Functions *à la* Euler

Not the least interesting feature of Euler's treatment of elementary transcendental functions in the *Introductio* is the fact that he derives their infinite series expansions without any use of calculus. In the *Calculi Differentialis* he then used these expansions to derive the Leibnizian differentials of the elementary functions.

His approach is simply to delete all higher-order infinitesimals $(dx)^2, (dx)^3, \ldots$, in an appropriate expansion of the differential dy of a given function y of x. For example, if $y = x^n$, then the binomial expansion gives

$$
\begin{aligned}
dy &= (x + dx)^n - x^n \\
&= \left(x^n + nx^{n-1}dx + \tfrac{1}{2}n(n-1)x^{n-2}dx^2 + \cdots \right) - x^n \\
&= nx^{n-1}dx + \tfrac{1}{2}n(n-1)x^{n-2}dx^2 + \cdots \\
dy &= nx^{n-1}dx.
\end{aligned}
$$

The product rule is derived in the usual way,

$$d(pq) = (p + dp)(q + dq) - pq$$
$$= p\, dq + q\, dp + dp\, dq = p\, dq + q\, dp.$$

To derive the quotient rule ([6], p. 109), Euler first expands $1/(q + dq)$ as a geometric series,

$$\frac{1}{q + dq} = \frac{1}{q}\left(1 - \frac{dq}{q} + \frac{dq^2}{q^2} - \cdots\right)$$
$$= \frac{1}{q} - \frac{dq}{q^2} + \frac{dq^2}{q^3} - \cdots$$
$$= \frac{1}{q} - \frac{dq}{q^2}.$$

Then

$$d\left(\frac{p}{q}\right) = \frac{p + dp}{q + dq} - \frac{p}{q} = (p + dp)\left(\frac{1}{q} - \frac{dq}{q^2}\right) - \frac{p}{q}$$
$$= \frac{dp}{q} - \frac{p\, dq}{q^2} - \frac{dp\, dq}{q^2}$$
$$d\left(\frac{p}{q}\right) = \frac{q\, dp - p\, dq}{q^2}.$$

To differentiate the logarithm ([6], p. 122), Euler writes

$$d(\log x) = \log(x + dx) - \log x$$
$$= \log\left(1 + \frac{dx}{x}\right)$$
$$= \frac{dx}{x} - \frac{dx^2}{2x^2} + \frac{dx^3}{3x^3} - \cdots \quad \text{(by Eq. (12))}$$
$$d(\log x) = \frac{dx}{x}. \tag{23}$$

As an example of Euler's occasional flights of fancy, here is an alternative derivation of (23) that he gives. From Equation (11) he writes

$$\log x = \frac{x^{\epsilon} - 1}{\epsilon}$$

where ϵ is an infinitely small number. The power rule then gives

$$d(\log x) = \frac{\epsilon x^{\epsilon - 1}dx}{\epsilon} = \frac{x^{\epsilon}dx}{x} = \frac{dx}{x},$$

because $x^{\epsilon} = x^0 = 1$ since ϵ is infinitely small!

He uses the exponential series (8) to differentiate e^x,

$$d(e^x) = e^{x+dx} - e^x = e^x(e^{dx} - 1)$$

$$= e^x\left(dx + \frac{dx^2}{2!} + \frac{dx^3}{3!} + \cdots\right)$$

$$d(e^x) = e^x dx. \tag{24}$$

EXERCISE 10. Given functions p and q of x, differentiate $y = p^q$ as follows. First expand

$$y + dy = (p + dp)^{q + dq}$$

by the binomial series to obtain

$$dy = p^q(p^{dq} - 1) + (q + dq)p^{q + dq - 1}dp. \tag{$*$}$$

Then expand $p^{dq} - 1 = e^{(\log p)\,dq} - 1$ by the exponential series to obtain

$$p^{dq} - 1 = (\log p)\,dq.$$

Finally substitute this into ($*$) to obtain

$$d(p^q) = p^q(\log p)\,dq + qp^{q-1}dp.$$

To differentiate $\sin x$ ([6], pp. 137–138), Euler uses the trigonometric series (17) and (18).

$$d(\sin x) = \sin(x + dx) - \sin x$$

$$= \sin x \cos dx + \cos x \sin dx - \sin x$$

$$= (\sin x)\left(-\frac{dx^2}{2!} + \frac{dx^4}{4!} - \cdots\right) + (\cos x)\left(dx - \frac{dx^3}{3!} + \cdots\right)$$

$$d(\sin x) = \cos x\,dx. \tag{25}$$

EXERCISE 11. Show similarly that $d(\cos x) = -\sin x\,dx$.

EXERCISE 12. Instead of using the quotient rule to differentiate $\tan x = \sin x / \cos x$, Euler on page 139 starts with

$$\tan(x + dx) = \frac{\tan x + \tan dx}{1 - \tan x \tan dx},$$

using the addition formula for the tangent. Carry through this approach to obtain $d(\tan x) = \sec^2 x\,dx$. Hint: $\tan dx = \sin dx / \cos dx = dx$.

To differentiate the inverse sine, $y = \sin^{-1}x$, Euler ([6], p. 132) starts with

$$e^{iy} = \cos y + i \sin y$$

$$= \sqrt{1 - x^2} + ix, \quad \text{(Eq. (21))}$$

whence

$$y = \frac{1}{i}\log(\sqrt{1 + x^2} + ix).$$

It follows that

$$dy = \frac{1}{i} \frac{-x(1-x^2)^{-1/2}dx + i\,dx}{\sqrt{1-x^2} + ix}$$

$$= \frac{1}{i} \frac{i\sqrt{1-x^2} - x}{\sqrt{1-x^2} + ix} \frac{dx}{\sqrt{1-x^2}}$$

$$d(\sin^{-1}x) = \frac{dx}{\sqrt{1-x^2}}. \tag{26}$$

If $y = \tan^{-1}x$, then $\sin y = x/\sqrt{1+x^2}$, so it follows that

$$d(\tan^{-1}x) = d\left(\sin^{-1}\frac{x}{\sqrt{1+x^2}}\right).$$

EXERCISE 13. Apply Eq. (26) to conclude that

$$d(\tan^{-1}x) = \frac{1}{1+x^2}. \tag{27}$$

This is how Euler does it ([6], pp. 133–134).

EXERCISE 14. Write Equation (22) in the form

$$\tan^{-1}x = \frac{1}{2i}\log\frac{1+ix}{1-ix},$$

and then differentiate the right-hand side to obtain (27).

EXERCISE 15. Derive Equation (27) by termwise differentiation of Gregory's inverse tangent series, using the geometric series to recognize the result.

Euler's admission of complex numbers on an equal footing with real numbers as functional arguments and values was a significant step. In particular, his slick use of complex logarithms in the preceding differential computations deserves comment. During the early eighteenth century there was a dispute between Leibniz and John Bernoulli over the meaning and existence of logarithms of negative and imaginary numbers. Euler sought to settle this dispute in a 1749 paper entitled "De la controverse entre Mrs. Leibniz et Bernoulli sur les logarithmes des nombres negatifs et imaginaires" ([7], pp. 195–232).

On page 210 he asserts that every number x has infinitely many logarithms. His argument is that, since

$$\log x = Nx^{1/N} - N \quad \text{(Eq. (10))}$$

for N infinitely large, the quantity $nx^{1/n} - n$ approaches $\log x$ as the positive integer n increases without bound. But each number x has n

distinct nth roots, or values for $x^{1/n}$. For example, if $\sqrt[n]{x}$ denotes the ordinary positive nth root of the positive number x, then

$$\left(\sqrt[n]{x}\; e^{2\pi ki/n}\right)^n = xe^{2\pi ki} = x$$

for every integer k, because $e^{2k\pi i} = \cos 2k\pi + i \sin 2k\pi = 1$. Thus x has the n distinct nth roots

$$\sqrt[n]{x}\; e^{2\pi ki/n}, \qquad k = 0, 1, 2, \ldots, n-1.$$

Euler concludes that $x^{1/N}$ should have infinitely many values if N is infinitely large, thereby producing infinitely many values of $\log x$. For example, he points out (page 213) that, if $\log a$ denotes the ordinary real logarithm of the positive number a, then each of the infinitely many values

$$(\log a) + 2\pi ki, \qquad k = 0, \pm 1, \pm 2, \ldots,$$

is a logarithm of a, because

$$\exp\left[(\log a) + 2\pi ki\right] = e^{\log a} e^{2\pi ki} = a.$$

EXERCISE 16. Show that $i(\pi/2 + 2k\pi)$ is a logarithm of $i = \sqrt{-1}$ for each integer k. Conclude that

$$i^i = e^{i \log i} = e^{-\pi/2} \cong 0.20788.$$

Interpolation and Numerical Integration

We have seen that infinite power series expansions of particular functions played a central role in the analysis of Euler (as well as in the work of some of his predecessors, especially including Newton). Early in the eighteenth century it was discovered by several people that the series expansions of the various elementary transcendental functions are all special cases of the general expansion that is now called *Taylor's series* (which will be discussed in the following section). The discovery of this quite general approach to infinite series expansions was closely associated with the development of *interpolation methods*.

The construction of mathematical tables (such as tables of logarithms or trigonometric functions) during the seventeenth century focused attention on the problem of accurately interpolating between tabulated values. The goal was to lessen the considerable labor of constructing a table (e.g., of logarithms) by directly computing only a limited number of values, afterwards filling in the remaining entries by interpolation between these directly computed values.

For example, suppose we want to construct a table of 5-place natural logarithms of numbers between 1 and 100, at intervals of 0.1. We saw in Chapter 6 (in the section on Newton's logarithmic computations) that a list of the logarithms of the first several prime integers can be built up on the basis of comparatively few direct logarithmic computations using Mercator's series. Once the logarithms of the 25 primes less than 100 have been found in this way, the logarithms of the remaining integers up to 100 can be obtained by addition (e.g., log 40 = 3 log 2 + log 5). It then remains only to interpolate 9 logarithms between each pair of logarithms of successive integers.

Unfortunately, the familiar process of *linear* interpolation is not sufficiently accurate for this purpose. For example, log 40 = 3.68888 and log 41 = 3.71357, so linear interpolation gives

$$\log 40.4 \cong 3.68888 + (0.4)(3.71357 - 3.68888) = 3.69876,$$

whereas actually log 40.4 = 3.69883. Thus two decimal places of accuracy have been lost in the process of linear interpolation.

It is convenient to describe linear interpolation as follows. Given the values $y_0 = f(x_0)$ and $y_1 = f(x_1)$, where $x_1 - x_0 = \Delta x$, consider the *first difference* $\Delta y_0 = y_1 - y_0$. Then linear interpolation is defined by

$$f(x_0 + s\Delta x) \cong y_0 + s\Delta y_0. \tag{28}$$

Procedures for more accurate interpolations go back at least to Henry Briggs' computations of common logarithms around 1620. Given function values $y_0 = f(x_0)$, $y_1 = f(x_1)$, $y_2 = f(x_2)$ with $x_1 - x_0 = x_2 - x_1 = \Delta x$, he used both the first differences $\Delta y_0 = y_1 - y_0$, $\Delta y_1 = y_2 - y_1$ and the *second difference*

$$\Delta^2 y_0 = \Delta y_1 - \Delta y_0 = y_2 - 2y_1 + y_0.$$

His interpolation for $f(x_0 + s\Delta x)$ was given by

$$f(x_0 + s\Delta x) \cong y_0 + s\Delta y_0 + \tfrac{1}{2}s(s-1)\Delta^2 y_0. \tag{29}$$

For example, the values tabulated below yield

$$\log 40.4 \cong 3.68888 + (0.4)(0.02469) + \tfrac{1}{2}(0.4)(-0.6)(-0.00059)$$
$$= 3.69883$$

which is accurate to five places. An excellent discussion of Briggs' computations can be found in Goldstine's book on the history of numerical analysis ([10], pp. 13–30).

i	x_i	y_i	Δy_i	$\Delta^2 y_i$
0	40	3.68888		
			0.02469	
1	41	3.71357		-0.00059
			0.02410	
2	42	3.73767		

EXERCISE 17. (Cf. [10], p. 27). Show that Formula (29) gives the following interpolation of 9 values between $y_0 = f(x_0)$ and $y_1 = f(x_1)$.

x	y
x_0	y_0
$x_0 + (0.1)\Delta x$	$y_0 + (0.1)y_0 - (0.045)\Delta^2 y_0$
$x_0 + (0.2)\Delta x$	$y_0 + (0.2)y_0 - (0.080)\Delta^2 y_0$
$x_0 + (0.3)\Delta x$	$y_0 + (0.3)y_0 - (0.105)\Delta^2 y_0$
$x_0 + (0.4)\Delta x$	$y_0 + (0.4)y_0 - (0.120)\Delta^2 y_0$
$x_0 + (0.5)\Delta x$	$y_0 + (0.5)y_0 - (0.125)\Delta^2 y_0$
$x_0 + (0.6)\Delta x$	$y_0 + (0.6)y_0 - (0.120)\Delta^2 y_0$
$x_0 + (0.7)\Delta x$	$y_0 + (0.7)y_0 - (0.105)\Delta^2 y_0$
$x_0 + (0.8)\Delta x$	$y_0 + (0.8)y_0 - (0.080)\Delta^2 y_0$
$x_0 + (0.9)\Delta x$	$y_0 + (0.9)y_0 - (0.045)\Delta^2 y_0$
x_1	y_1

A generalization of Formulas (28) and (29) that is now known as *Newton's forward-difference formula* was stated without proof by Newton under Lemma V of Book III of the *Principia Mathematica*. It refers to interpolation between the values $y_0, y_1, \ldots, y_n$ of a function $f(x)$ given at $n+1$ equally spaced points $x_0, x_1, \ldots, x_n$. For a concise statement of this formula, it is convenient to extend the difference notation (which neither Briggs nor Newton used) to higher order differences by defining $\Delta y_j = y_{j+1} - y_j$ and

$$\Delta^{k+1} y_j = \Delta(\Delta^k y_j) = \Delta^k y_{j+1} - \Delta^k y_j$$

recursively. For example,

$$\Delta^2 y_1 = \Delta y_2 - \Delta y_1 = (y_3 - y_2) - (y_2 - y_1)$$
$$= y_3 - 2y_2 + y_1$$

and

$$\Delta^3 y_0 = \Delta^2 y_1 - \Delta^2 y_0$$
$$= (y_3 - 2y_2 + y_1) - (y_2 - 2y_1 + y_0)$$
$$= y_3 - 3y_2 + 3y_1 - y_0.$$

These successive differences are commonly tabulated as in the array shown below. Each entry is the difference of the two entries immediately to its left.

y_0					
	Δy_0				
y_1		$\Delta^2 y_0$			
	Δy_1		$\Delta^3 y_0$		
y_2		$\Delta^2 y_1$		$\Delta^4 y_0$	
	Δy_2		$\Delta^3 y_1$		$\Delta^5 y_0$
y_3		$\Delta^2 y_2$		$\Delta^4 y_1$	
	Δy_3		$\Delta^3 y_2$		
y_4		$\Delta^2 y_3$			
	Δy_4				
y_5					

EXERCISE 18. Show by induction on k that

$$\Delta^k y_j = \sum_{i=0}^{k} (-1)^i \binom{k}{i} y_{k-i+j}$$

$$= y_{k+j} - k y_{k+j-1} + \tfrac{1}{2} k(k-1) y_{k+j-2} \cdots + (-1)^k y_j,$$

where $\binom{k}{i}$ is the usual binomial coefficient.

In this difference notation, Newton's forward-difference formula for the interpolated value $f(x_0 + s\Delta x)$, where $\Delta x = x_{i+1} - x_i$, is

$$f(x_0 + s\Delta x) \cong \sum_{i=0}^{n} \binom{s}{i} \Delta^i y_0$$

$$= y_0 + s\Delta y_0 + \frac{s(s-1)}{2!} \Delta^2 y_0 + \frac{s(s-1)(s-2)}{3!} \Delta^3 y_0$$

$$+ \cdots + \frac{s(s-1) \cdots (s-n+1)}{n!} \Delta^n y_0. \qquad (30)$$

The same formula (except for notation) had been stated (also without proof) by James Gregory in a letter to John Collins dated 23 November 1670 (see Newton's correspondence [NC I], p. 46). The first published derivation of an interpolation formula equivalent to (30) appeared in Newton's *Methodus Differentialis* of 1711 (for an English translation see Newton's works [NW II], pp. 165–173). However, Newton's pioneering investigations of finite difference interpolation, including divided differences and the so-called Newton–Stirling and Newton–Bessel central-difference formulas, were carried out in 1675–1676; this work is presented by Whiteside in Newton's mathematical papers ([NP IV], pp. 3–73).

EXERCISE 19. Given the following five values of the function $y = x^3$, set up a difference array to compute $\Delta y_0 = 61000$, $\Delta^2 y_0 = 30000$, $\Delta^3 y_0 = 6000$, $\Delta^4 y_0 = 0$. Substitute these values with $s = 0.4$ into (30) to calculate $44^3 = 85184$.

i	0	1	2	3	4
x_i	40	50	60	70	80
y_i	64000	125000	216000	343000	512000

In a 1670 manuscript that is discussed in the papers by Gibson ([9], pp. 4–5), Turnbull ([13], pp. 162–164), and Dehn and Hellinger ([4], pp. 152–153), Gregory apparently derived the binomial expansion from his interpolation formula in the manner indicated by the following exercise.

EXERCISE 20. Consider the function $f(x) = (1 + a)^x$. Let $x_j = j = 0, 1, 2, \ldots, n$, so $\Delta x = 1$ and $y_j = (1 + a)^j$. Show by induction on k that

$$\Delta^k y_j = a^k y_j$$

for each $k = 0, 1, 2, \ldots, n$. Substitute $\Delta^k y_0 = a^k$ into (30) to obtain

$$(1 + a)^s \cong 1 + sa + \frac{s(s-1)}{2!} a^2 + \cdots + \frac{s(s-1) \cdots (s-n+1)}{n!} a^n,$$

the first $n + 1$ terms of the binomial expansion of $(1 + a)^s$.

The above-mentioned *Principia* lemma reads, "To find a curved line of the parabolic kind [i.e., a polynomial] which shall pass through any given number of points." This is the basic idea of Gregory–Newton interpolation —to determine a polynomial

$$p(x) = a_0 + a_1 x + a_2 x^2 + \cdots + a_n x^n$$

of degree n which agrees with $f(x)$ at the $n + 1$ equally spaced points $x_0, x_1, \ldots, x_n$, that is, $p(x_i) = f(x_i) = y_i$ for $i = 0, 1, \ldots, n$. Then, for any intermediate point x, the value $p(x)$ can be used as an interpolated approximation to the value $f(x)$.

It is somewhat easier to solve for the coefficients if $p(x)$ is written in the form

$$\begin{aligned} p(x) = A_0 &+ A_1(x - x_0) + A_2(x - x_0)(x - x_1) \\ &+ \cdots + A_n(x - x_0)(x - x_1) \cdots (x - x_{n-1}). \end{aligned} \tag{31}$$

If $x - x_0 = s\Delta x$, then

$$x - x_1 = (s-1)\Delta x, \qquad (x - x_2) = (s-2)\Delta x, \qquad \text{etc.,}$$

so (31) becomes

$$p(x_0 + s\Delta x) = B_0 + B_1 s + B_2 s(s-1) + \cdots + B_n s(s-1) \cdots (s-n+1), \tag{32}$$

where $B_k = A_k(\Delta x)^k$. In order to obtain the interpolation formula (30) it suffices to show that

$$B_k = \frac{\Delta^k y_0}{k!}, \qquad k = 0, 1, \ldots, n. \tag{33}$$

The requirement that $p(x_0 + k\Delta x) = y_k$ for each $k = 0, 1, \ldots, n$ gives the equations

$$\begin{aligned} y_0 &= B_0 \\ y_1 &= B_0 + B_1 \\ y_2 &= B_0 + 2B_1 + 2B_2 \\ y_3 &= B_0 + 3B_1 + 6B_2 + 6B_3 \\ &\vdots \qquad \vdots \\ y_n &= B_0 + nB_1 + n(n-1)B_2 + n(n-1)(n-2)B_3 + \cdots + n!B_n. \end{aligned} \tag{34}$$

EXERCISE 21. Solve the first four equations in system (34) to obtain

$$B_0 = y_0, \qquad B_1 = \Delta y_0, \qquad B_2 = \frac{\Delta^2 y_0}{2!}, \qquad B_3 = \frac{\Delta^3 y_0}{3!}.$$

The solution of system (34) is completed by induction. Assuming that $B_k = \Delta^k y_0 / k!$ for $k < n$, the last equation in (34) gives

$$y_n = y_0 + n\Delta y_0 + \frac{n(n-1)}{2!}\Delta^2 y_0 + \cdots + n\Delta^{n-1}y_0 + n! B_n.$$

Since $y_1 = y_0 + \Delta y_0 = (1+\Delta)y_0$, $y_{k+1} = y_k + \Delta y_k = (1+\Delta)y_k$, it follows easily by induction on n that

$$y_n = (1+\Delta)^n y_0$$

$$= y_0 + n\Delta y_0 + \frac{n(n-1)}{2!}\Delta^2 y_0 + \cdots + n\Delta^{n-1}y_0 + \Delta^n y_0.$$

Finally, comparison of the last two equations shows that $B_n = \Delta^n y_0 / n!$, as desired. This establishes (33), and thereby completes the derivation of the Gregory–Newton interpolation formula.

As a corollary to his interpolation lemma in the *Principia*, Newton wrote

> Hence the areas of all curves may be nearly found; for if some number of points of the curve to be squared are found, and a parabola [polynomial] be supposed to be drawn through those points, the area of this parabola will be nearly the same with the area of the curvilinear figure proposed to be squared: But the parabola can always be squared geometrically by methods generally known.

This remark was the first published allusion to the numerical technique of approximating an integral $\int_a^b f(x)\, dx$ by evaluating the integral $\int_a^b p(x)\, dx$ of an interpolating polynomial for $f(x)$.

In a concluding scholium to the *Methodus Differentialis* ([NW II], p. 172), Newton gave the following example.

> If there are four ordinates at equal intervals, let A be the sum of the first and fourth, B the sum of the second and third, and R the interval between the first and fourth; then ... the area between the first and fourth ordinates will be $\frac{1}{8}(A+3B)R$.

This is the "Newton-Cotes three-eighths rule,"

$$\int_{x_0}^{x_3} f(x)\, dx \simeq \frac{3\Delta x}{8}(y_0 + 3y_1 + 3y_2 + y_3).$$

It is obtained by integration of the interpolating polynomial for four equally spaced points,

$$\int_{x_0}^{x_3} f(x)\,dx = (\Delta x)\int_0^3 f(x_0 + s\Delta x)\,dx$$

$$\cong (\Delta x)\int_0^3 \left[y_0 + s\Delta y_0 + \tfrac{1}{2}s(s-1)\Delta^2 y_0 + \tfrac{1}{6}s(s-1)(s-2)\Delta^3 y_0 \right] ds$$

$$= (\Delta x)\left(3y_0 + \tfrac{9}{2}y_0 + \tfrac{9}{4}\Delta^2 y_0 + \tfrac{3}{8}\Delta^3 y_0\right)$$

$$= (\Delta x)\left[3y_0 + \tfrac{9}{2}(y_1 - y_0) + \tfrac{9}{4}(y_2 - 2y_1 + y_0) + \tfrac{3}{8}(y_3 - 3y_2 + 3y_1 - y_0) \right]$$

$$= \frac{3\Delta x}{8}(y_0 + 3y_1 + 3y_2 + y_3).$$

EXERCISE 22. Use the interpolating polynomial for three equally spaced points to obtain

$$\int_{x_0}^{x_2} f(x)\,dx \cong \frac{\Delta x}{3}(y_0 + 4y_1 + y_2).$$

This approximation which, together with higher-order approximations, was known to Newton's disciples Roger Cotes and James Stirling, was rediscovered by Thomas Simpson in 1743, and is now called "Simpson's rule."

EXERCISE 23. Apply Simpson's rule and the three-eighths rule to approximate

$$\pi = 4\int_0^1 \frac{dx}{1+x^2}.$$

Taylor's Series

The classical Taylor's series is so named because it was first published by Brook Taylor (1685–1731), a disciple of Newton, in his *Methodus incrementorum* of 1715. An English translation of the pertinent passage is included in Struik's source book ([12], pp. 328–333).

Taylor obtained his series by a limit argument based on the Gregory–Newton interpolation formula. In the notation of the previous section, rather than Taylor's own rather cumbersome notation, his derivation may be described as follows. If $x = x_0 + n\Delta x$, then the interpolation formula gives $y = f(x) = f(x_0 + n\Delta x)$ as

$$y = y_0 + n\Delta y_0 + \frac{n(n-1)}{2!}\Delta^2 y_0 + \frac{n(n-1)(n-2)}{3!}\Delta^3 y_0 + \cdots. \quad (35)$$

In essence, Taylor wants to take the limit as $\Delta x \to 0$ and $n \to \infty$, while x_0 and x remain fixed. If we substitute

$$n = \frac{x - x_0}{\Delta x}, \qquad n - 1 = \frac{x - x_1}{\Delta x}, \qquad n - 2 = \frac{x - x_2}{\Delta x}, \qquad \text{etc.},$$

then (35) becomes

$$y = y_0 + (x - x_0)\frac{\Delta y_0}{\Delta x} + \frac{(x - x_0)(x - x_1)}{2!}\frac{\Delta^2 y_0}{(\Delta x)^2}$$

$$+ \frac{(x - x_0)(x - x_1)(x - x_2)}{3!}\frac{\Delta^3 y_0}{(\Delta x)^3} + \cdots . \tag{36}$$

In order to evaluate the limit, Taylor proposes to "substitute for evanescent increments the fluxions proportional to them" ([12], p. 332, Corollary II).

He does this by thinking of x and y as functions of t, with x increasing uniformly (linearly) with $x(0) = x_0$, $x(h) = x_1$, $x(2h) = x_2$, etc. If h is very small, then

$$\Delta x = x_1 - x_0 = x(h) - x(0) \cong \dot{x}_0 h,$$

where $\dot{x}_0 = \dot{x}(0)$ is the fluxion (time derivative) of x when $t = 0$. Similarly,

$$\Delta y_i = y_{i+1} - y_i = y(ih + h) - y(ih) \cong \dot{y}_i h$$

so $\Delta y_0 = \dot{y}_0 h$ where $\dot{y}_0 = \dot{y}(0)$ is the fluxion of y when $t = 0$. Then

$$\Delta^2 y_0 = \Delta y_1 - \Delta y_0 \cong \dot{y}(h)h - \dot{y}(0)h \cong \ddot{y}_0 h^2,$$

$$\Delta^3 y_0 = \Delta^2 y_1 - \Delta^2 y_0 \cong \ddot{y}(h)h^2 - \ddot{y}(0)h^2 \cong \dddot{y}_0 h^3,$$

and so forth. It follows that

$$\frac{\Delta y_0}{\Delta x} \cong \frac{\dot{y}_0}{\dot{x}_0}, \qquad \frac{\Delta^2 y_0}{(\Delta x)^2} \cong \frac{\ddot{y}_0}{(\dot{x}_0)^2}, \qquad \frac{\Delta^3 y_0}{(\Delta x)^3} \cong \frac{\dddot{y}_0}{(\dot{x}_0)^3}, \qquad \text{etc.}$$

It therefore appears that the limiting form of (36) is

$$y = y_0 + (x - x_0)\frac{\dot{y}_0}{\dot{x}_0} + \frac{(x - x_0)^2}{2!}\frac{\ddot{y}_0}{(\dot{x}_0)^2} + \frac{(x - x_0)^3}{3!}\frac{\dddot{y}_0}{(\dot{x}_0)^3} + \cdots \tag{37}$$

because the points x_i all approach x_0 as $\Delta x \to 0$.

Formula (37) is Taylor's original series. Interpreting the fluxional ratios as derivatives, we obtain the standard modern form

$$f(x) = f(x_0) + f'(x_0)(x - x_0) + \frac{f''(x_0)}{2!}(x - x_0)^2 + \frac{f'''(x_0)}{3!}(x - x_0)^3 + \cdots$$

$$\tag{38}$$

of Taylor's series. Taylor's rather audacious leap across the logical gap between (36) and the equivalent of (38) can be partially justified along the lines of the following exercise.

EXERCISE 24. Substitute $\Delta y_0 = f(x_0 + \Delta x) - f(x_0)$, $\Delta^2 y_0 = f(x_0 + 2\Delta x) - 2f(x_0 + \Delta x) + f(x_0)$, $\Delta^3 y_0 = f(x_0 + 3\Delta x) - 3f(x_0 + 2\Delta x) + 3f(x_0 + \Delta x) - f(x_0)$, and then apply l'Hospital's rule to prove that

$$\lim_{\Delta x \to 0} \frac{\Delta y_0}{\Delta x} = f'(x_0), \quad \lim_{\Delta x \to 0} \frac{\Delta^2 y_0}{(\Delta x)^2} = f''(x_0) \quad \text{and} \quad \lim_{\Delta x \to 0} \frac{\Delta^3 y_0}{(\Delta x)^3} = f'''(x_0).$$

Taylor's series was "in the air" when Taylor published it and, indeed, had a certain history that he may not have been aware of. Almost a half-century earlier, in a letter to John Collins dated 15 February 1671 ([NC I], pp. 61–65), James Gregory had listed the first 5 or 6 terms of the power series expansions of the functions

$$\tan x, \quad \tan^{-1} x, \quad \log \sec x, \quad \sec^{-1}(\sqrt{2}\, e^x),$$
$$\log \tan\left(\frac{x}{2} + \frac{\pi}{4}\right), \quad 2 \tan^{-1}\left(\tanh \frac{x}{2}\right).$$

Although he did not include his derivations of these series, there is some indication in Gregory's unpublished papers that he had calculated the derivatives necessary to obtain these series by successive differentiation ([4], pp. 149–150). It remains unclear whether Gregory had a *general* formula for power series expansions, but evidently he could somehow obtain the power series of (almost) any *particular* function. If Gregory's isolation and premature death had not prevented the full development and publication of his research, we might well speak today of the calculus of Newton, Leibniz, *and* Gregory.

The earliest known explicit statement of the general Taylor's series was given by Newton in a 1691–1692 draft of the *De Quadratura*. It was, however, omitted from the version of this paper that eventually appeared in 1704 as an appendix to Newton's *Opticks*. Corollary 3 of the draft (see [NP VII], pp. 97–99) reads as follows.

Hence, indeed, if the series [for y in terms of z] proves to be of this form

$$y = az + bz^2 + cz^3 + dz^4 + ez^5 + \cdots \tag{39}$$

(where any of the terms az, bz^2, cz^3, dz^4, ..., can either be lacking or be negative), the fluxions of y, when z vanishes, are had by setting $\dot{y}/\dot{z} = a$, $\ddot{y}/\dot{z}^2 = 2b$, $\dddot{y}/\dot{z}^3 = 6c$, $\ddddot{y}/\dot{z}^4 = 24d$, $\dddddot{y}/\dot{z}^5 = 120e$, ...

Substituting the value y_0 (which Newton takes as 0) of y when $z = 0$, it follows that

$$y = y_0 + \frac{\dot{y}_0}{\dot{z}_0} z + \frac{1}{2!} \frac{\ddot{y}_0}{\dot{z}_0^2} z^2 + \frac{1}{3!} \frac{\dddot{y}_0}{\dot{z}_0^3} + \cdots .$$

This is the case $x_0 = 0$ of (37). Newton's statement of the general case is his following Corollary 4.

No formal proof is included, but the context indicates that Newton undoubtedly obtained the listed values of the coefficients a, b, c,

$d, e, \ldots,$ by successive differentiations. Thus, fluxional differentiation of (39) gives

$$\dot{y} = a\dot{z} + 2bz\dot{z} + 3cz^2\dot{z} + 4dz^3\dot{z} + \cdots,$$

so substitution of $z = 0$ yields $a = \dot{y}_0/\dot{z}_0$. If it is assumed, as Newton indicates just prior to his statement of Corollary 3, that "z flows uniformly" so $\dot{z} = \dot{z}_0$ is constant, then another differentiation gives

$$\ddot{y} = 2b\dot{z}^2 + 6cz\dot{z}^2 + 12dz^2\dot{z}^2 + \cdots,$$

so substitution of $z = 0$ gives $2b = \ddot{y}_0/\dot{z}_0^2$.

EXERCISE 25. Differentiate once more to obtain $6c = \dddot{y}_0/\dot{z}_0^3$.

In the *Acta Eruditorum* of 1694 John Bernoulli published a series that was sufficiently similar to Taylor's for Bernoulli to accuse Taylor of plagiarism when the *Methodus incrementorum* appeared twenty years later. Whereas Bernoulli's series is sometimes presented as a result of successive integration by parts, he started by writing

$$
n \, dz = n \, dz + \left(z \, dn - z \frac{dn}{dz} \, dz \right) - \left(\frac{z^2}{2!} \frac{d^2n}{dz^2} \, dz - \frac{z^2}{2!} \frac{d^2n}{dz^2} \, dz \right)
$$

$$
+ \left(\frac{z^3}{3!} \frac{d^3n}{dz^3} \, dz - \frac{z^3}{3!} \frac{d^3n}{dz^3} \, dz \right) - \cdots
$$

$$
= (n \, dz + z \, dn) - \left(z \frac{dn}{dz} + \frac{z^2}{2!} \frac{d^2n}{dz^2} \right) dz
$$

$$
+ \left(\frac{z^2}{2!} \frac{d^2n}{dz^2} + \frac{z^3}{3!} \frac{d^3n}{dz^3} \right) dz - \cdots
$$

$$
n \, dz = d(nz) - d\left(\frac{z^2}{2!} \frac{dn}{dz} \right) + d\left(\frac{z^3}{3!} \frac{d^2n}{dz^2} \right) - \cdots.
$$

Termwise integration then gives Bernoulli's series

$$
\int_0^z n \, dz = nz - \frac{z^2}{2!} \frac{dn}{dz} + \frac{z^3}{3!} \frac{d^2n}{dz^2} - \frac{z^4}{4!} \frac{d^3n}{dz^3} + \cdots. \tag{40}
$$

EXERCISE 26. Derive Bernoulli's series by successive integration by parts, starting with

$$
\int n \, dz = nz - \int z \frac{dn}{dz} \, dz = nz - \left(\frac{z^2}{2!} \frac{dn}{dz} - \int \frac{z^2}{2!} \frac{d^2n}{dz^2} \, dz \right).
$$

EXERCISE 27. Derive Taylor's series from Bernoulli's series as follows. Substitute

$n = f'(z)$, $n = f''(z)$, $n = f'''(z)$, $n = f^{(4)}(z)$ in turn into (40) to obtain the formulas

$$f(z) - f(0) = f'(z)z - \frac{f''(z)}{2!}z^2 + \frac{f'''(z)}{3!}z^3 - \frac{f^{(4)}(z)}{4!}z^4 + \cdots$$

$$f'(z) - f'(0) = f''(z)z - \frac{f'''(z)}{2!}z^2 + \frac{f^{(4)}(z)}{3!}z^3 - \cdots$$

$$f''(z) - f''(0) = f'''(z)z - \frac{f^{(4)}(z)}{2!}z^2 + \cdots$$

$$f'''(z) - f'''(0) = f^{(4)}(z)z - \cdots .$$

Then eliminate $f'(z), f''(z), f'''(z)$ successively to obtain

$$f(z) = f(0) + f'(0)z + \frac{f''(0)}{2!}z^2 + \frac{f'''(0)}{3!}z^3 + \frac{f^{(4)}(z)}{4!}z^4 + \cdots .$$

The case of $x_0 = 0$ of Taylor's series,

$$f(x) = f(0) + f'(0)x + \frac{f''(0)}{2!}x^2 + \cdots + \frac{f^{(n)}(0)}{n!}x^n + \cdots ,$$

is often called Maclaurin's series. Colin Maclaurin (1698–1746), perhaps the most successful of Newton's disciples, employed Taylor's series as a fundamental tool in his *Treatise of Fluxions* (1742). The sections in which he introduced Taylor's series by Newton's method of successive differentiation, and then used it to derive sufficient conditions for the existence of local maxima and minima, are included in Struik's source book ([12], pp. 338–341).

If $f'(0) = 0$, then Taylor's series gives

$$f(x) = f(0) + \frac{f''(0)}{2!}x^2 + \frac{f'''(0)}{3!}x^3 + \cdots ;$$

Maclaurin writes this in fluxional rather than derivative notation. Assuming that the terms of degree greater than two are negligible when x is sufficiently small, Maclaurin concludes that $f(x) > f(0)$ on both sides of $x = 0$ if $f''(0) > 0$, while $f(x) < f(0)$ on both sides if $f''(0) < 0$. Thus the conditions $f'(0) = 0$, $f''(0) > 0$ imply a local minimum, while the conditions $f'(0) = 0$, $f''(0) < 0$ imply a local maximum. If $f'(0) = f''(0) = 0$ but $f'''(0) \neq 0$, then

$$f(x) = f(0) + \frac{f'''(0)}{3!}x^3 + \frac{f^{(4)}(0)}{4!}x^4 + \cdots .$$

Hence it appears that, if x is sufficiently small, then $f(x) > f(0)$ on one side of $x = 0$ and $f(x) < f(0)$ on the other side. Thus, if the third derivative is the first non-vanishing one, then neither a maximum nor a minimum occurs.

In general, if the first n derivatives vanish at $x = 0$, then

$$f(x) = f(0) + \frac{f^{(n+1)}(0)}{(n+1)!}x^{n+1} + \cdots .$$

Maclaurin concludes from this that

> If the first fluxion of the ordinate, with its fluxions of several subsequent orders, vanish, the ordinate is a minimum or maximum, when the number of all those fluxions that vanish is 1, 3, 5, or any odd number. The ordinate is a *minimum*, when the fluxion next to those that vanish is positive; but a *maximum* when this fluxion is negative. ... But if the number of all the fluxions of the first and successive orders that vanish be an even number, the ordinate is then neither a *maximum* nor *minimum*.

In his *Institutiones calculi differentialis* ([6], pp. 256–258) Euler takes a characteristically carefree approach to Taylor's formula. Given two values x_0 and x of the independent variable, write $\omega = x - x_0$, and let

$$dx = \frac{\omega}{N}, \quad \text{so} \quad x = x_0 + N\,dx,$$

where N is infinitely large (so dx is infinitely small). Then Euler writes the interpolation series for $f(x) = f(x_0 + N\,dx)$ as

$$f(x) = y + N\,dy + \frac{N(N-1)}{2!}d^2y + \frac{N(N-1)(N-2)}{3!}d^3y + \cdots,$$

$$(41)$$

with the Leibnizian differential d^ky serving as an infinitesimal version of the Newtonian difference Δ^ky_0. But $N = N-1 = N-2 = \cdots$ because N is infinitely large, so (41) simplifies to

$$f(x) = y + N\,dy + \frac{N^2}{2!}d^2y + \frac{N^3}{3!}d^3y + \cdots.$$

Substitution of $N = \omega/dx = (x - x_0)/dx$ then yields Taylor's series

$$f(x) = y + (x - x_0)\frac{dy}{dx} + \frac{(x - x_0)}{2!}\frac{d^2y}{dx^2} + \frac{(x - x_0)^3}{3!}\frac{d^3y}{dx^3} + \cdots,$$

in which Euler takes the quotient $d^ky/(dx)^k$ of higher-order differentials to be the kth order derivative $f^{(k)}(x_0)$.

Fundamental Concepts in the Eighteenth Century

The calculus entered the eighteenth century encumbered with glaring uncertainties regarding the logical foundations of the subject. However, as we have seen in this chapter, these uncertainties, which persisted throughout the century, did little or nothing to impede a rapid development of the now-standard computational tools of differential calculus. During this period integration was generally regarded simply as the inverse of differentiation, so integral calculus also was treated as a formal manipulative subject.

A vigorous exposition of the inconsistencies of the early calculus was presented in a 1734 essay by George Berkeley (1685–1753) entitled *The Analyst, or A Discourse Addressed to an Infidel Mathematician*. The "infidel mathematician" was Edmund Halley (1656–1742), the astronomer and disciple of Newton. The stated purpose of Berkeley, then Anglican Bishop-elect of Cloyne (Ireland), was to question whether the foundations of mathematics are any firmer than those of religion—"He who can digest a second or third fluxion, a second or third difference, need not, methinks, be squeamish about any point in divinity."

The full text (in English) of *The Analyst* is readily available in Berkeley's collected works [1], and selected passages are included in Struik's source book ([12], pp. 333–338).

Bishop Berkeley finds the followers of Newton and Leibniz guilty of using methods that they do not understand, basing even the derivation of valid conclusions on logical inconsistencies and ambiguous concepts.

> And, forasmuch as it may perhaps seem an unaccountable paradox that mathematicians should deduce true propositions from false principles, be right in the conclusion and yet err in the premises; I shall endeavor particularly to explain why this may come to pass, and shew how error may bring forth truth, though it cannot bring forth science ([1], pp. 76–77).

> The Method of Fluxions is the general key by help whereof the modern mathematicians unlock the secrets of Geometry, and consequently of Nature. . . .

> And whereas quantities generated in equal times are greater or lesser according to the greater or lesser velocity wherewith they increase and are generated, a method hath been found to determine quantities from the velocities of their generating motions. And such velocities are called fluxions: and the quantities generated are called flowing quantities. These fluxions are said to be nearly as the increments of the flowing quantities, generated in the least equal particles of time; and to be accurately in the first proportion of the nascent, or in the last of the evanescent increments. . . . and of the aforesaid fluxions there be other fluxions, which fluxions of fluxions are called second fluxions. And the fluxions of these second fluxions are called third fluxions: and so on, fourth, fifth, sixth, &c. ad infinitum. . . . The further the mind analyseth and pursueth these fugitive ideas the more it is lost and bewildered; the objects, at first fleeting and minute, soon vanishing out of sight ([1], pp. 66–67).

> The foreign mathematicians are supposed by some, even of our own, to proceed in a manner less accurate, perhaps, and geometrical, yet more intelligible. Instead of flowing quantities and their fluxions, they consider the variable finite quantities as increasing or diminishing by the continual addition or subduction of infinitely small quantities. Instead of the velocities wherewith increments are generated, they consider the increments or decrements themselves, which they call differences, and which

are supposed to be infinitely small. The difference of a line is an infinitely little line; of a plain an infinitely little plain. They suppose finite quantities to consist of parts infinitely little, and curves to be polygons, whereof the sides are infinitely little, which by the angles they make one with another determine the curvity of the line. Now to conceive a quantity infinitely small, that is, infinitely less than any sensible or imaginable quantity, or than any the least finite magnitude is, I confess, above my capacity. ... And yet in the *calculus differentialis*, which method serves to all the same intents and ends with that of fluxions, our modern analysts are not content to consider only the differences of finite quantities: they also consider the differences of those differences, and the differences of the differences of the first differences. And so on *ad infinitum*. That is, they consider quantities infinitely less than the least discernible quantity; and others infinitely less than those infinitely small ones; and still others infinitely less than the preceding infinitesimals, and so on without end or limit. ([1], pp. 67–68)

In his most memorable passage, Berkeley answered the claim that the difficulties associated with ratios of fluxions or differentials could be circumvented by replacing these "ultimate ratios of evanescent quantities" with proportional ratios of finite line segments.

It must, indeed, be acknowledged that [Newton] used fluxions, like the scaffold of a building, as things to be laid aside or got rid of as soon as finite lines were found proportional to them. But then these finite exponents are found by the help of fluxions. Whatever therefore is got by such exponents and proportions is to be ascribed to fluxions: which must therefore be previously understood. And what are these fluxions? The velocities of evanescent increments? And what are these same evanescent increments? They are neither finite quantities, nor quantities infinitely small, nor yet nothing. May we not call them the ghosts of departed quantities? ([1], pp. 88–89)

Besides pointing out (with some accuracy) the lack of clarity in contemporary conceptions of fluxions and differentials, Berkeley argued that the basic computations of the calculus invariably involved conflicting suppositions, and therefore could arrive at valid results only through compensation of errors. For example, he criticized the extraction of the derivative nx^{n-1} of x^n from the increment

$$(x + o)^n - x^n = nx^{n-1}o + \tfrac{1}{2}n(n-1)x^{n-2}o^2 + \cdots$$

by first dividing by o, supposing that o is non-zero, and then setting o equal to zero.

Hitherto I have supposed that x flows, that x hath a real increment, that o is something. And I have proceeded all along on that supposition, without which I should not have been able to have made so much as one single step. From that supposition it is that I get at the increment of x^n, that I am able to compare it with the increment of x, and that I find the proportion between the two increments. I now beg leave to make a new supposition contrary to the first, i.e. I will suppose that there is no increment of x, or that o is nothing; which second supposition destroys my first, and is inconsistent with it, and therefore with every thing that supposeth it. I do nevertheless beg leave to retain nx^{n-1}, which is an expression obtained in virtue of my first supposition, which necessarily presupposeth such supposition, and which could not be obtained without it: All which seems a most inconsistent way of arguing, and such as would not be allowed of in Divinity ([1], p. 73).

Berkeley's polemic hit close enough to home to inspire a number of spirited rejoinders, most of which proved only that their authors hardly understood Berkeley, much less the calculus (an exception being Maclaurin's profound *Treatise of Fluxions*, which may have been motivated in part by the Berkeley controversy).

The first step towards resolving Berkeley's difficulties by explicitly defining the derivative as a limit of a quotient of increments, in the manner suggested but not stated with sufficient clarity by Newton, was taken by Jean d'Alembert (1717–1783). In the article entitled "Différentiel" in vol. 4 (1754) of the *Encyclopédie* published by the French Academy, d'Alembert wrote

Leibniz was embarassed by the objections he felt to exist against infinitely small quantities, as they appear in the *differential* calculus; thus he preferred to reduce infinitely small to merely incomparable quantities. ... Newton started out from another principle; and one can say that the metaphysics of this great mathematician on the calculus of fluxions is very exact and illuminating, even though he allowed us only an imperfect glimpse of his thoughts. ... He never considered the *differential* calculus as the study of infinitely small quantities, but as the method of first and ultimate ratios, that is to say, the method of finding the limits of ratios. Thus this famous author has never differentiated quantities but only equations; in fact, every equation involves a relation between two variables and the differentiation of equations consists merely in finding the limit of the ratio of the finite differences of the two quantities contained in the equation. ... Once this is well understood, one will feel that the assumption made concerning infinitely small quantities serves only to abbreviate and simplify the reasoning; but that the differential calculus does not necessarily suppose the existence of these quantities; and that moreover this calculus consists in *algebraically determining the limit of a ratio, for which we already have the expression in terms of lines, and in equating those two expressions.* ... We have seen above that in the *differential* calculus there are really no infinitely small quantities of the first order; that actually those quantities [the differentials] are supposed to be divided by other supposedly infinitely small quantities; in this state they do not denote either infinitely small quantities or quotients of infinitely small quantities; they are the limits of the ratio of two finite quantities ([12], pp. 341–345).

Thus d'Alembert presented in this article a view of the derivative that
would now be expressed by simply writing

$$\frac{dy}{dx} = \lim_{\Delta x \to 0} \frac{\Delta y}{\Delta x}.$$

Although he did not describe the limit concept itself with the precision that
would come in the nineteenth century, it was a noteworthy step for him to
clearly identify the derivative as a limit of a ratio of increments, rather
than a ratio of either differentials or fluxions. However, this insight failed
to immediately affect basic expositions of the calculus. Although the
derivative as a limit of a quotient appeared in occasional late eighteenth
century discussions of the "metaphysics" of the calculus, most textbooks of
the time continued to rely mainly on the Leibnizian approach with its
labyrinth of differentials (see [2], p. 250).

In his *Théorie des Fonctions Analytiques* published in 1797, Joseph Louis
Lagrange (1736–1813), the other great figure (with Euler) of eighteenth
century mathematics, presented a comprehensive development of the
calculus that was intended to eliminate all references to differentials,
infinitesimals, and limit concepts from the subject. A brief extract illustrating his approach is included in Struik's source book ([12], pp. 388–391);
the references below are to the second edition of 1813 [11].

Lagrange's new approach was based on a power series expansion of a
given function $f(x)$. If x is replaced by $x + i$ then "by the theory of series",
as he says, we obtain

$$f(x+i) = f(x) + pi + qi^2 + ri^3 + \cdots , \qquad (42)$$

where the coefficients $p, q, r, \ldots$, are new functions of x, derived (in
some way to be determined) from the original function $f(x)$. *Ad hoc*
expansions of this sort are available for particular familiar functions and,
in Chapter I ([11], p. 8–9), Lagrange purports to prove that, except for
particular values of x, every function can be expanded as in (42), with only
positive integral powers of i appearing. Today we would say that, if
$f(x + i)$ is represented by the convergent power series (42) in a neighborhood of $i = 0$, then f is *analytic* at x. In this sense Lagrange's book is
correctly titled—it is a theory of analytic functions (rather than arbitrary
ones). As Cauchy soon pointed out, there are simple functions such as
$f(x) = e^{-1/x^2}$ that are not analytic.

The first coefficient in (42), for which Lagrange introduces the notation
$p(x) = f'(x)$, is called the *first derived function* of $f(x)$. Of course it will turn
out to be the derivative of $f(x)$; indeed, this is the historical origin of the
notation $f'(x)$ for the derivative of $f(x)$, as well as of the term "derivative"
itself. In order to identify the coefficients in (42), Lagrange replaces i by

$i + o$, obtaining

$$f(x+i+o) = f(x) + p \cdot (i+o) + q \cdot (i+o)^2 + r \cdot (i+o)^3 + \cdots$$
$$f(x+i+o) = f(x) + pi + qi^2 + ri^3 + si^4 + \cdots$$
$$+ po + 2pio + 3ri^2o + 4si^3o + \cdots . \tag{43}$$

He then replaces x by $x + o$ in (42), obtaining

$$f(x+i+o) = f(x+o) + p(x+o)i + q(x+o)i^2 + r(x+o)i^3 + \cdots$$
$$= [f(x)+f'(x)o + \cdots] + [p(x)+p'(x)o + \cdots]i$$
$$+ [q(x)+q'(x)o + \cdots]i^2 + \cdots$$
$$f(x+i+o) = f(x) + pi + qi^2 + ri^3 + \cdots$$
$$+ p'io + q'i^2o + r'i^3o + \cdots . \tag{44}$$

Comparison of the coefficients in (43) and (44) then gives

$$q(x) = \tfrac{1}{2}p'(x) = \tfrac{1}{2}f''(x),$$

$$r(x) = \frac{1}{3}q'(x) = \frac{1}{3!}f'''(x)$$

$$s(x) = \frac{1}{4}r'(x) = \frac{1}{4!}f^{(4)}(x), \text{ etc.,}$$

where $f''(x)$ denotes the first derived function of $f'(x)$, etc. Consequently series (42) becomes

$$f(x+i) = f(x) + f'(x)i + \frac{f''(x)}{2!}i^2 + \frac{f'''(x)}{3!}i^3 + \cdots , \tag{45}$$

that is, Taylor's series.

Finally Lagrange remarks ([11], p. 19) that only a little knowledge of the differential calculus is necessary to recognize that the derived functions $f'(x), f''(x), f'''(x), \ldots$, coincide with the successive derivatives of the original function $f(x)$. Actually this verification requires the assumption that termwise differentiation of (42) with respect to i is valid.

EXERCISE 28. Differentiate series (42) termwise and then set $i = 0$ to verify that $f'(x)$ is the first derivative of $f(x)$.

Lagrange's attempt to expunge from the calculus all trace of infinitesimal and limit concepts was inevitably unsuccessful. Nevertheless, his book on calculus includes several contributions of lasting significance. For example, the remainder term for Taylor's formula makes its first appearance in Chapter VII ([11], p. 67). Lagrange's derivation is equivalent to one that is still seen today in calculus textbooks.

He starts by replacing x by $x - i$ in (45), obtaining

$$f(x) = f(x-i) + if'(x-i) + \frac{i^2}{2!}f''(x-i) + \cdots.$$

Substitution of xz for i then gives

$$f(x) = f(x-xz) + xzf'(x-xz) + \frac{x^2z^2}{2!}f''(x-xz) + \cdots$$

$$+ \frac{x^n z^n}{n!}f^{(n)}(x-xz) + \frac{x^{n+1}z^{n+1}}{(n+1)!}f^{(n+1)}(x-xz) + \cdots$$

$$f(x) = f(x-xz) + xzf'(x-xz) + \frac{x^2z^2}{2!}f''(x-xz) + \cdots$$

$$+ \frac{x^n z^n}{n!}f^{(n)}(x-xz) + x^{n+1}R(x, z). \tag{46}$$

Note that $R(x, 0) = 0$. For a modern derivation of Taylor's formula with remainder, we would *define* $R(x, z)$ for $z \neq 0$ by (46), instead of assuming (45) to start with.

Next Lagrange differentiates Equation (46) with respect to z, obtaining

$$0 = -xf'(x-xz) + xf'(x-xz) - x^2zf''(x-xz) + x^2zf''(x-xz)$$

$$- \cdots - \frac{x^{n+1}z^n}{n!}f^{(n+1)}(x-xz) + x^{n+1}R'(x, z),$$

so pairwise cancellation of all but the last two terms yields

$$R'(x, z) = \frac{z^n}{n!}f^{(n+1)}(x-xz) \tag{47}$$

for the partial derivative of $R(x, z)$ with respect to z. Now let M and N be the minimum and maximum values, respectively, of $f^{(n+1)}(x-xz)$ for $z \in [0, 1]$. Then (47) gives

$$\frac{Mz^n}{n!} \leqslant R'(x, z) \leqslant \frac{Nz^n}{n!}$$

for $z \in [0, 1]$. Because $R(x, 0) = 0$, antidifferentiation of this inequality yields

$$\frac{Mz^{n+1}}{(n+1)!} \leqslant R(x, z) \leqslant \frac{Nz^{n+1}}{(n+1)!}$$

for $z \in [0, 1]$. Taking $z = 1$ in particular, we obtain

$$\frac{M}{(n+1)!} \leqslant R(x, 1) \leqslant \frac{N}{(n+1)!}.$$

At this point Lagrange assumes what is now called the intermediate value theorem to conclude that

$$R(x, 1) = \frac{f^{(n+1)}(x - x\bar{z})}{(n+1)!}$$

for some $\bar{z} \in [0, 1]$. Substitution of this value with $z = 1$ into Equation (46) finally gives

$$f(x) = f(0) + f'(0)x + \frac{f''(0)}{2!}x^2 + \cdots + \frac{f^{(n)}(0)}{n!}x^n + \frac{f^{(n+1)}(u)}{(n+1)!}x^{n+1},$$

(48)

where $u = x - x\bar{z} \in [0, x]$.

Of course the final term of (48) is the "Lagrange form" of the remainder term. Lagrange treats explicitly only the particular cases $n = 0, 1, 2$, indicating that the general result follows in an analogous manner. He quite accurately identifies his Taylor's formula with remainder as a "new theorem remarkable for its simplicity and generality."

This work of Lagrange was a fitting climax for the eighteenth century development of the calculus. It provided a reasonably firm foundation for the Taylor series that had typified the central role of infinite series expansions in eighteenth century investigations, and at the same time pointed up the need for the studies of basic properties of continuous functions (e.g., maximum-minimum and intermediate value properties) that were soon to follow in the nineteenth century.

References

[1] *The Works of George Berkeley*, vol. 4. London: Nelson, 1951.

[2] C. B. Boyer, *The History of the Calculus and its Conceptual Development*. New York: Dover (reprint), 1959, Chapter VI.

[3] F. Cajorie, Indivisibles and 'ghosts of departed quantities' in the history of mathematics. *Scientia* 37, 303–306, 1925.

[4] M. Dehn and E. D. Hellinger, Certain mathematical achievements of James Gregory. *Am Math Mon* 50, 149–163, 1943.

[5] L. Euler, *Introductio in Analysin Infinitorum*, vol. 1. *Opera Omnia*, Ser. 1, vol. 8. Leipzig, 1922.

[6] L. Euler, *Institutiones Calculi Differentialis*. *Opera Omnia*, Ser. 1, vol. 10. Leipzig, 1913.

[7] L. Euler, De la controverse entre Mrs. Leibniz et Bernoulli sur les logarithmes des nombres negatifs et imaginaires. *Opera Omnia*, Ser. 1, vol. 17. Leipzig, 1914.

[8] G. A. Gibson, Taylor's theorem and Bernoulli's theorem: A historical note. *Proc Edinburgh Math Soc* 39 (1) 25–33, 1921–22.

[9] G. A. Gibson, James Gregory's mathematical work. *Proc Edinburgh Math Soc* 41 (1) 2–25, 1922–23.

[10] H. H. Goldstine, *A History of Numerical Analysis from the 16th through the 19th Century*. New York: Springer-Verlag, 1977.

[11] J. L. Lagrange, *Theorie des Fonctions Analytiques*. Paris: Courcier, 1813, 2nd ed.

[12] D. J. Struik, *A Source Book in Mathematics, 1200–1800*. Cambridge, MA: Harvard University Press, 1969.

[13] H. W. Turnbull, James Gregory: A study in the early history of interpolation. *Proc Edinburgh Math Soc* **3**(2), 151–178, 1932–33.

[14] A. P. Youschkevitch, The concept of function up to the middle of the 19th century. *Arch Hist Exact Sci.* **16**, 37–85, 1976.

The Calculus According to Cauchy, Riemann, and Weierstrass

11

Functions and Continuity at the Turn of the Century

We have seen that, at the beginning of his *Introductio*, Euler defined a function of a variable quantity as "an *analytical expression* composed in any way from this variable quantity and from numbers or constant quantities." Thus a particular function was defined in terms of a specific formula or equation; Euler assumed, moreover, that a given function is prescribed throughout its "domain of definition" by one and the same "analytical expression."

This narrowly analytical conception of functions generally prevailed in eighteenth century calculus, but was called into question by discussions between Euler, d'Alembert, and Daniel Bernoulli (1700–1782, son of John Bernoulli) concerning the nature of the "arbitrary functions" that arise in the integration of partial differential equations, such as the one that represents the motion of a vibrating string. For some extracts from this discussion see Struik's source book ([17], pp. 351–368). An exhaustive exposition of the whole matter is given by C. A. Truesdell in his introduction to Volume 11 (part 2), Series II of Euler's collected works [18].

In a 1747 paper d'Alembert investigated the motion of a vibrating string that is stretched between the points $x = 0$ and $x = L$ on the x-axis. He introduced a condition equivalent to the partial differential equation

$$\frac{\partial^2 y}{\partial t^2} = a^2 \frac{\partial^2 y}{\partial x^2}, \tag{1}$$

$y(x, t)$ being the (transverse) displacement at time t of the point x on the string. (Actually, d'Alembert used the arclength s in place of the variable x, and the partial differential notation was supplied by Euler several years later.) He then observed that (1) is satisfied by any function of the form

$$y(x, t) = \phi(x + at) + \psi(x - at) \qquad (2)$$

where ϕ and ψ are "arbitrary functions" of a single variable. Assuming that the string is initially set into motion by first deforming it into the shape $y = f(x)$, and then releasing it at time $t = 0$, it follows that its subsequent motion is described by

$$y(x, t) = \tfrac{1}{2}f(x + at) + \tfrac{1}{2}f(x - at). \qquad (3)$$

EXERCISE 1. If ϕ and ψ are twice differentiable functions of a single variable, show that the function $y(x, t)$ defined by (2) satisfies Equation (1).

EXERCISE 2. Derive the solution (3) by imposing the initial conditions $y(x, 0) = f(x)$ and $D_t y(x, 0) = 0$ on the function defined by (2).

Now the disagreement that arose concerned the type of arbitrary function $y = f(x)$ that could be assumed to represent the initial shape of the string. D'Alembert argued that, in order to apply legitimately the operations of the calculus, it was necessary that each such function be expressed everywhere in terms of one and the same algebraic or transcendental equation. This requirement was described at the time by saying that "the function is subject to the law of continuity of form."

Euler countered that this requirement is physically unrealistic—a string may well be plucked in such a way that its initial shape is described by different analytical expressions in different intervals. For example, if a stretched string is set into motion by displacing its midpoint a unit distance before releasing it at time $t = 0$, then its initial shape is described by

$$y = \begin{cases} \dfrac{2x}{L} & \text{if } x \in \left[0, \dfrac{L}{2}\right], \\[2ex] \dfrac{2}{L}(L - x) & \text{if } x \in \left[\dfrac{L}{2}, L\right]. \end{cases}$$

Euler therefore argued for the admission into mathematical analysis of functions that today would be called piecewise-smooth, but which he referred to as "mixed" or "irregular and discontinuous," because they corresponded to different "continuous" (i.e. smooth) functions on different intervals. He also included as a "discontinuous" function one whose graph can be traced with a free motion of the hand, such a function not being subject to any "law of continuity" whatever.

Thus, in the late eighteenth century view of functions that stemmed largely from the vibrating string controversy, "continuity" referred to a constancy of the analytical expression of a function, rather than to connectedness of its graph (the modern idea of continuity). Indeed, essentially all functions treated in eighteenth century analysis were continuous in the modern sense; "discontinuity" then referred either to failure at isolated points (where the analytical expression changed) of a function to be smooth (in the modern sense), or to the lack of any analytical expression at all (as in the case of freehand curves).

By contrast, the modern sense of discontinuity is more nearly that of *discontiguity*. This distinction was first made explicit by Louis Arbogast (1759–1803). In 1787 the Academy of St. Petersburg offered a prize for the best answer to the question:

> Whether the arbitrary functions that one obtains by the integration of an equation in three or more variables represent any curves or surfaces whatsoever, either algebraic or transcendental, either mechanical, discontinuous, or produced by a free motion of the hand; or whether these functions include only continuous curves represented by an algebraic or transcendental equation.

In his 1791 memoir which won this prize, Arbogast wrote

> The law of continuity consists in that a quantity cannot pass from one state to another without passing through all the intermediate states which are subject to the same law. Algebraic functions are regarded as continuous because the different values of these functions depend in the same manner on those of the variable; and, supposing that the variable increases continually, the function will receive corresponding variations; but it will not pass from one value to another without also passing through all the intermediate values. Thus the ordinate y of an algebraic curve, when the abscissa x varies, cannot pass brusquely from one value to another; there cannot be a saltus from one ordinate to another which differs from it by an assignable quantity; but all the successive values of y must be linked together by one and the same law which makes the extremities of these ordinates make up a regular and continuous curve.

Here he singles out the "intermediate value property" that was soon to play an important role in calculus.

> This continuity may be destroyed in two manners: 1. The function may change its form, that is to say, the law by which the function depends on the variable may change all at once. A curve formed by the assemblage of many portions of different curves is of this kind ... It is not even necessary that the function y should be expressed by an equation for a certain interval of the variable; it may continually change its form, and the line representing it, instead of being an assemblage of regular curves, may be such that at each of its points it becomes a different curve; that is to say, it may be entirely irregular and not follow any law for any interval however small.
>
> Such would be a curve traced at hazard by the free movement of the hand. These kinds of curves can neither be represented by one nor by many algebraic or transcendental equations.

2. The law of continuity is again broken when the different parts of a curve do not join to one another ... We will call curves of this kind discontiguous curves, because all their parts are not contiguous, and similarly for discontiguous functions (cited by Jourdain [14], pp. 675–676).

Arbogast decided that the arbitrary functions that appear in the solutions of partial differential equations may be neither continuous nor contiguous. However, as we will see in the next section, the pivotal and clinching argument for the necessity of considering discontinuous functions in mathematical analysis was provided by Joseph Fourier (1768–1830) in the first decade of the nineteenth century.

Fourier and Discontinuity

Fourier's celebrated book *Theorie analytique de la chaleur* (The Analytical Theory of Heat) was published in 1822, but much that it contains dates back to a memoir that he presented to the Paris Academy of Sciences in 1807. In it he developed into a comprehensive general theory the method of trigonometric series that Euler and Bernoulli had applied to isolated special cases in their work on the vibrating string a half-century earlier (see [17], pp. 360–367).

A typical problem in the theory of heat of the sort that Fourier considers asks for a steady-state temperature function $u(x, y)$ in the region $0 \leqslant x \leqslant \pi$, $y \geqslant 0$ satisfying the conditions

$$\frac{\partial^2 u}{\partial x^2} + \frac{\partial^2 u}{\partial y^2} = 0, \tag{4}$$

$$u(0, y) = u(\pi, y) = 0, \tag{5}$$

$$u(x, 0) = \phi(x), \tag{6}$$

where $\phi(x)$ is a given function prescribing the temperature on the base of the region. Fourier observes (e.g., by separation of variables) that each of the simple functions

$$e^{-ny}\sin nx, \quad n = 1, 2, 3, \ldots,$$

satisfies conditions (4) and (5), and concludes by superposition (assuming convergence, termwise differentiability, etc.) that the more general function

$$u(x, y) = \sum_{n=1}^{\infty} b_n e^{-ny}\sin nx$$

does also, $b_1, b_2, b_3, \ldots$, being arbitrary constants. The final condition (6) can then be satisfied as well if these constants can be chosen so that

$$\phi(x) = \sum_{n=1}^{\infty} b_n \sin nx, \quad x \in (0, \pi). \tag{7}$$

Fourier attacks the "development of an arbitrary function in trigonometric series" in Chapter III, Section VI of [8]. His first heuristic procedure for the evaluation of the constants $\{b_n\}$ may be outlined as follows. He assumes that $\phi(x)$ is an odd function with Taylor series

$$\phi(x) = x\phi'(0) + \frac{x^3}{3!}\phi'''(0) + \frac{x^5}{5!}\phi^{(5)}(0) + \cdots . \tag{8}$$

By first substituting the Taylor series for $\sin nx$ on the right-hand side of (7), and then equating coefficients of like powers of x in (7) and (8), Fourier obtains the infinite set of linear equations

$$\phi'(0) = b_1 + 2b_2 + 3b_3 + 4b_4 + \cdots$$
$$-\phi'''(0) = b_1 + 2^3b_2 + 3^3b_3 + 4^3b_4 + \cdots$$
$$\phi^{(5)}(0) = b_1 + 2^5b_2 + 3^5b_3 + 4^5b_4 + \cdots$$
$$-\phi^{(7)}(0) = b_1 + 2^7b_2 + 3^7b_3 + 4^7b_4 + \cdots$$
$$\vdots$$

To solve this system, he first approximates $b_1, \ldots, b_m$ by deleting all terms in the first m equations with subscripts greater than m (hence m linear equations in m unknowns), and finally computes the limits of these approximations as $m\to\infty$. The result of this complicated elimination process ([8], pp. 169–183) is

$$a_n = (-1)^{n+1}\frac{\pi}{2}b_n$$

$$a_n = \phi(\pi) - \frac{1}{n^2}\phi''(\pi) + \frac{1}{n^4}\phi^{(4)}(\pi) - \frac{1}{n^6}\phi^{(6)}(\pi) + \cdots . \tag{9}$$

Considering a_n as a function of π, differentiating twice, comparing the results, and finally replacing π by x, he obtains

$$\frac{1}{n^2}\frac{d^2a_n}{dx^2} + a_n(x) = \phi(x).$$

The general solution of this second order ordinary differential equation is

$$a_n(x) = A\cos nx + B\sin nx$$
$$+ (n\sin nx)\int_0^x \phi(t)\cos nt\,dt - (n\cos nx)\int_0^x \phi(t)\sin nt\,dt.$$

Because $A = a_n(0) = 0$ (why?), it finally follows that

$$b_n = (-1)^{n+1}\frac{2}{\pi}a_n(\pi) = \frac{2}{\pi}\int_0^\pi \phi(x)\sin nx\,dx, \tag{10}$$

the now-familiar formula for the Fourier coefficients! Only after carrying through this technical tour de force does Fourier point out ([8], pp. 187–188) that formula (10) can be "verified" by the now standard

device of multiplying both sides of equation (7) by sin mx and then
integrating termwise from 0 to π, making use of the orthogonality of the
sine functions on the interval $[0, \pi]$.

EXERCISE 3. Assuming (7) and the validity of termwise integration, verify (10) in
this way. The needed "orthogonality" is the fact that

$$\int_0^{\pi} \sin mx \, \sin nx \, dx = 0 \quad \text{if } m \neq n.$$

We see by this that the coefficients $a, b, c, d, e, f, \ldots$, which enter
into the equation

$$\tfrac{1}{2}\pi\phi(x) = a \sin x + b \sin 2x + c \sin 3x + d \sin 4x + \cdots,$$

and which we found formerly by way of successive eliminations, are the
values of definite integrals expressed by the general term $\int \sin ix \, \phi(x) \, dx$,
i being the number of the term whose coefficient is required. This remark
is important, because it shows how even entirely arbitrary functions may
be developed in series of sines of multiple arcs. In fact, if the function
$\phi(x)$ be represented by the variable ordinate of any curve whatever whose
abscissa extends from $x = 0$ to $x = \pi$, and if on the same part of the axis
the known trigonometric curve, whose ordinate is $y = \sin x$, be constructed, it is easy to represent the value of any integral term. We must
suppose that for each abscissa x, to which corresponds one value of $\phi(x)$,
and one value of $\sin x$, we multiply the latter value by the first, and at the
same point of the axis raise an ordinate equal to the product $\phi(x) \sin x$.
By this continuous operation a third curve is formed, whose ordinates are
those of the trigonometric curve, reduced in proportion to the ordinates of
the arbitrary curve which represents $\phi(x)$. This done, the area of the
reduced curve taken from $x = 0$ to $x = \pi$ gives the exact value of the
coefficient of sin x; and whatever the given curve may be which corresponds to $\phi(x)$, whether we can assign to it an analytical equation, or
whether it depends on no regular law, it is evident that it always serves to
reduce in any manner whatever the trigonometric curve; so that the area
of the reduced curve has, in all possible cases, a definite value which is the
value of the coefficient of sin x in the development of the function. The
same is the case with the following coefficient b, or $\int \phi(x) \sin 2x \, dx$. ([8],
p. 186).

Here Fourier makes the important observation that, to permit the
calculation of the coefficients in the Fourier series of $\phi(x)$, it suffices for
the region under $y = \phi(x) \sin nx$ to have an area (for each n) that can be
interpreted as the value of the integral $\int_0^{\pi}\phi(x) \sin nx \, dx$. It is not necessary
that $\phi(x) \sin nx$ be continuous and therefore have an integral that can be
calculated by antidifferentiation. Moreover, Fourier observed that even if
$\phi(x)$ is continuous on $[0, \pi]$, but $\phi(\pi) \neq 0$, then the extended function to
which its Fourier series converges (presumably) on the whole real line will
necessarily be discontinuous (i.e., discontiguous) at points x that are odd

multiples of π, because this extended function is odd with period 2π. This is the case with so simple a function as $\phi(x) = x$. Consequently, the introduction of Fourier series techniques essentially forced the consideration of discontinuous functions on an equal footing with continuous ones, and called for the development of a theory of integration of discontinuous functions (soon provided, as we will see, by Cauchy and Riemann).

> It is remarkable that we can express by convergent series, and, as we shall see in the sequel, by definite integrals, the ordinates of lines and surfaces which are not subject to a continuous law. We see by this that we must admit into analysis functions which have equal values, whenever the variable receives any values whatever included between two given limits, even though on substituting in these two functions, instead of the variable, a number included in another interval, the results of the two substitutions are not the same. The functions which enjoy this property are represented by different lines, which coincide in a definite portion only of their course, and offer a singular species of finite osculation. These considerations arise in the calculus of partial differential equations; they throw a new light on this calculus, and serve to facilitate its employment in physical theories ([8], p. 199).

In the final chapter of his book Fourier presented an outline of a proof of the convergence of his trigonometric series, and included the following quite general formulation of the functional concept.

> Above all, it must be remarked that the function $f(x)$, to which this proof applies, is entirely arbitrary, and not subject to a continuous law. . . . In general the function $f(x)$ represents a succession of values being given to the abscissa x, there are an equal number of ordinates $f(x)$. . . . We do not suppose these ordinates to be subject to a common law; they succeed each other in any manner whatever, and each of them is given as if it were a single quantity.
>
> It may follow from the very nature of the problem, and from the analysis which is applicable to it, that the passage from one ordinate to the following is effected in a continuous manner. But special conditions are then concerned, and the general equation, considered by itself, is independent of these conditions. It is rigourously applicable to discontinuous functions ([8], p. 430).

Although Fourier approaches here the modern concept of a function, his working definition of discontinuity in actual practice was that of the eighteenth century (discontinuity of analytic form)—his functions (like everyone else's at that time) were at worst piecewise-smooth, with only a finite number of "discontiguities" in each finite interval.

The first example of a "genuinely discontinuous" function, one that exhibited the full potential of the concept of a function as an arbitrary pairing, was provided by Peter Lejeune-Dirichlet (1805–1859). In an 1829 paper Dirichlet formulated sufficient conditions for the convergence of a Fourier series, and gave the first complete rigorous proof of such convergence (see [9], Chapter 5). At the end of this paper he gave an example of a function not satisfying the "Dirichlet conditions": "$f(x)$ equals a determined constant c when the variable x takes a rational value, and equals another constant d when this variable is irrational." This famous function is of course discontinuous everywhere.

Bolzano, Cauchy, and Continuity

A precise formulation of the modern concept of continuity first appeared in a pamphlet published privately in Prague by the isolated Bohemian scholar and priest Bernard Bolzano (1781–1848). Its title stated its purpose: *Purely analytical proof of the theorem, that between each two roots which guarantee an opposing result* [in sign], *at least one real root of the equation lies.* Thus he proposed to give a "purely analytical proof" of the intermediate value theorem for continuous functions.

Bolzano argued that the intuitive geometric proof—a continuous curve must somewhere cross any straight line that separates its endpoints—is based on an inadequate conception of continuity. To correctly explain the concept of continuity, he said, one must understand that the meaning of the phrase "A function $f(x)$ varies according to the law of continuity for all values of x which lie inside or outside certain limits, is nothing other than this: If x is any such value, the difference $f(x+\omega)-f(x)$ can be made smaller than any given quantity, if one makes ω as small as one wishes." In other words, $f(x)$ is continuous on an interval provided that $\lim_{\omega \to 0} f(x+\omega) = f(x)$ for each point x of the interval.

As a crucial lemma, Bolzano asserted that, if M is a property of real numbers that does not hold for all x, and there exists a number u such that all numbers $x < u$ have property M, then there exists a *largest* U such that all numbers $x < U$ have property M. In his attempted proof by the now-familiar bisection method, he produced a "Cauchy sequence" $\{u_n\}_1^\infty$ intended to converge to the desired U. Although he (and later Cauchy) correctly stated what is now called the "Cauchy convergence criterion" [$\{u_n\}_1^\infty$ converges if and only if, given $\epsilon > 0$, $|u_m - u_n| < \epsilon$ for m and n sufficiently large], he could not (nor could Cauchy) supply a complete proof, for lack of a completeness property of the real number system.

EXERCISE 4. Use Bolzano's lemma to prove that, if $\{a_n\}_1^\infty$ is a bounded, monotone increasing sequence,

$$a_1 \leqslant a_2 \leqslant \cdots \leqslant a_n \leqslant a_{n+1} \leqslant \cdots \leqslant A,$$

then there is a number U such that $\lim_{n \to \infty} a_n = U$. *Hint*: Let the number x have property M provided that $x > a_n$ for some n.

Bolzano applied the lemma above to prove the following generalization of the theorem of his title: If $f(x)$ and $g(x)$ are continuous functions on the interval $[a, b]$ with $f(a) < g(a)$ and $f(b) > g(b)$, then $f(\bar{x}) = g(\bar{x})$ for some number $\bar{x}$ between a and b.

EXERCISE 5. (Bolzano's Proof). Let the number r have property M if either $r \leqslant 0$ or $r > 0$ and $f(a + r) < g(a + r)$. If U is the largest number such that all numbers $r < U$ have property M, show that $a < a + U < b$ and $f(a + U) = g(a + U)$.

Bolzano's little pamphlet was not widely circulated among contemporary mathematicians, and the extent of its influence is unclear. However, in his *Cours d'analyse* of 1821, Cauchy essentially duplicated Bolzano's definition and immediate applications of continuity. An interesting (if perhaps controversial) discussion of the question as to whether Cauchy knew of the earlier work of Bolzano is included in an article by Grattan-Guinness [10].

Augustin-Louis Cauchy (1789–1857) was the dominant mathematical figure in a Paris that still regarded itself as the center of the mathematical world (despite the fact that Gauss never left Germany). Today Cauchy is often credited with the founding of the modern age of rigor in mathematics. In this tradition he may be a beneficiary by default on the part of Gauss, whose personal standards of rigor were equally high, but whose publication policy—"few but ripe"—was the opposite of Cauchy's. Moreover, it may be noted that Cauchy occasionally stumbled conspicuously, as in failing to distinguish between continuity and uniform continuity or between convergence and uniform convergence.

Nevertheless, it was Cauchy whose expositions of analysis first stamped elementary calculus with the general character that it retains today. Continuing the pedagogical tradition of the Ecole Polytechnique (Paris), he wrote three great textbooks—the *Cours d'analyse* (1821), *Resume des leçons sur le calcul infinitesimal* (1822), and *Leçons sur le calcul differentiel* (1829) —which were the first to set forth the establishment of complete rigor in mathematical analysis as a principal goal. These books are reprinted in volumes 3 and 4 (Series 2) of Cauchy's collected works, and English translations of substantial portions of them are provided by R. Iacobacci in her 1965 dissertation [12]. In the forward to his *Resume* ([4], pp. ii–iii; [12], p. 188) Cauchy wrote:

> ... The methods which I have followed differ in many respects from those which are expounded in other works of the same type. My principal aim has been to reconcile rigor, which I have made a law to myself in my *Cours d'analyse*, with the simplicity which the direct consideration of infinitely small quantities produces. For this reason, I believed it to be my duty to reject the development of functions into infinite series each time that the series obtained is not convergent ... In the integral calculus, it has appeared to me necessary to demonstrate generally the existence of the *integrals* or *primitive functions* before making known their diverse

properties. In order to attain this object, it was found necessary to establish at the outset the notion of *integrals taken between given limits* or *definite integrals*.

Cauchy's was the first comprehensive treatment of mathematical analysis to be based from the outset on a reasonably clear definition of the limit concept:

> When the successive values attributed to a variable approach indefinitely a fixed value so as to end by differing from it by as little as one wishes, this last [fixed value] is called the *limit* of all the others. Thus, for example, an irrational number is the limit of diverse fractions which furnish more and more approximate values of it ([3], p. 19; [12], p. 191).

The device that enabled him to "reconcile rigor with infinitesimals" was a new definition of infinitesimals that avoided the infinitely small *fixed numbers* of earlier mathematicians. Cauchy defined an infinitesimal ("un *infiniment petit*") or *infinitely small* quantity ("quantite *infiniment petite*") to be simply a variable with zero as its limit ([3], p. 19). Again,

> One says that a variable quantity becomes *infinitely small* when its numerical value decreases indefinitely in such a way as to converge towards the value zero ([3], p. 37; [12], p. 194).

> Let α be an infinitesimal ("une quantite infiniment petites"), that is to say a variable whose numerical value decreases indefinitely ([3], p. 38; [12], p. 196).

Although Cauchy complicates his exposition by discussing infinitely small quantities in the language of variables rather than that of functions, it is clear that by the phrases "un infiniment petit" and "une quantite infiniment petite"—both of which are frequently rendered into English as "an infinitesimal" (e.g., by Iacobacci)—he means a dependent variable or function $\alpha(h)$ that approaches zero as $h \to 0$. In particular, his "infinitesimals" are no longer the infinitely small fixed numbers that earlier had been the source of so much confusion and controversy.

In Chapter II of the *Cours d'analyse* Cauchy introduces the concept of continuity for a function defined on an interval, with essentially the same definition that Bolzano had given. If, given a value of x within the interval where the function f is defined,

> one assigns to the variable x an infinitely small increment α, the function itself will take on for an increment the difference
>
> $$f(x + \alpha) - f(x),$$
>
> which will depend at the same time on the new variable α and on the value of x. This granted, the function $f(x)$ will be, between the two limits assigned to the variable x, a *continuous* function of the variable if, for

each value of x intermediate between these limits, the numerical value of the difference

$$f(x + \alpha) - f(x)$$

decreases indefinitely with that of α. In other words, *the function $f(x)$ will remain continuous with respect to x between the given limits, if, between these limits, an infinitely small increment of the variable always produces an infinitely small increment of the function itself* ([3], p. 43; [12], p. 201).

Note that the final italicized statement does *not* refer to fixed infinitesimals; it says simply that the *variable $f(x + \alpha) - f(x)$* is an infinitely small quantity (as previously defined) whenever the variable α is, that is, that $f(x + \alpha) - f(x)$ approaches zero as α does. Thus Cauchy uses the statement that "$f(x + \alpha) - f(x)$ is infinitely small when α is" in the same way that one often does today—as a convenient abbreviation for a more complicated statement involving limits.

He points out that the continuity of the familiar elementary functions is easily verified (on intervals containing no singular points corresponding to zero denominators). For example, the function $\sin x$ is continuous on every interval because "the numerical value of $\sin\left(\frac{1}{2}\alpha\right)$, and consequently that of the difference

$$\sin(x + \alpha) - \sin x = 2 \sin\left(\tfrac{1}{2}\alpha\right) \cos\left(x + \tfrac{1}{2}\alpha\right)$$

decreases indefinitely with that of α" ([3], p. 44).

EXERCISE 6. Show directly that e^x is continuous everywhere. What must you assume?

Cauchy next discusses the continuity of a composition of continuous functions. However, he errs in his attempted proof that a continuous function of *several* continuous functions is continuous, erroneously thinking that he has proved that $f(x, y, z)$ is continuous, that is,

$$\lim_{\alpha, \, \beta, \, \gamma \to 0} f(x + \alpha, y + \beta, z + \gamma) = f(x, y, z)$$

provided that f is continuous in each of the independent variables x, y, z separately (see Theorem I on page 47 of [3]).

Cauchy states the intermediate value theorem as Theorem IV on page 50: If $f(x)$ is continuous on the interval $[x_0, X]$, and b is a number between $f(x_0)$ and $f(X)$, then there exists at least one point x of the interval such that $f(x) = b$. He provides an intuitive geometric proof, but in a note on the numerical solution of equations ([3], pp. 378–425) includes an alternative proof by "une methode directe et purement analytique" (shades of Bolzano?).

Taking $b = 0$ and m an integer larger than one, he first subdivides $[x_0, X]$ into m equal subintervals. Since $f(x)$ changes sign on $[x_0, X]$ it must change sign on some subinterval $[x_1, X_1]$. Next $[x_1, X_1]$ is subdivided into

m equal subintervals, on one of which, say $[x_2, X_2]$, $f(x)$ again changes sign. Continuing in this way Cauchy constructs an increasing sequence $\{x_n\}_1^\infty$ of points of $[x_0, X]$ such that each value $f(x_n)$ has the same sign as $f(x_0)$, and a decreasing sequence $\{X_n\}_1^\infty$ such that each $f(X_n)$ has the same sign as $f(X)$. Because $X_n - x_n = (X - x_0)/m^n \to 0$ as $n \to \infty$, he concludes that these two sequences converge to a common limit point $a \in (x_0, X)$. By continuity, $f(x_n) > 0$ (say) for each n implies $f(a) = \lim f(x_n) \geq 0$, while $f(X_n) < 0$ implies $f(a) \leq 0$. It therefore follows that $f(a) = 0$, as desired. This proof of the intermediate value theorem is the one that is perhaps most frequently found in modern textbooks.

In Chapter VI of his *Cours d'analyse* Cauchy presents the first systematic study of convergence of infinite series, including statements and proofs of the ratio and root tests. In Theorem I on page 120 he makes his famous incorrect assertion that the sum of a convergent infinite series of continuous functions is itself a continuous function. Under Problem I on page 146 he makes an incomplete but quite interesting attempt to establish Newton's binomial series

$$(1+x)^\mu = 1 + \frac{\mu}{1}x + \frac{\mu(\mu-1)}{2!}x^2 + \frac{\mu(\mu-1)(\mu-2)}{3!}x^3 + \cdots . \quad (11)$$

First he points out that (by the ratio test) this infinite series converges if $|x| < 1$. With x fixed, he denotes its sum by $\phi(\mu)$, and concludes (by his incorrect assertion) that ϕ is a continuous function of μ. Direct computation (by the "Cauchy product" of two infinite series) then shows that

$$\phi(\mu)\phi(\mu') = \phi(\mu + \mu'). \quad (12)$$

But he has shown in Problem 2 of Chapter V that this functional equation for a *continuous* function implies that

$$\phi(\mu) = [\phi(1)]^\mu,$$

that is, $\phi(\mu)$ is an exponential function of μ. But $\phi(1) = 1 + x$, so this means that $\phi(\mu) = (1+x)^\mu$ for $|x| < 1$ and μ arbitrary, as desired. The first complete verification of the binomial series was given by Abel in 1826.

EXERCISE 7. (a) Assuming (12) for all μ and μ', show that $\phi(m) = [\phi(1)]^m$ and $\phi(1/m) = [\phi(1)]^{1/m}$ if m is a positive integer.
  (b) Deduce that $\phi(m/n) = [\phi(1)]^{m/n}$ if m and n are integers.
  (c) Conclude by continuity that $\phi(\mu) = [\phi(1)]^\mu$ if μ is irrational.

Cauchy's Differential Calculus

Previous expositions of the calculus (with the exception of Lagrange and his attempted calculus as algebra without limits) had generally taken the differential in some form as the fundamental concept. The derivative of

$y = f(x)$ was then introduced as the "differential coefficient" in the expression $dy = f'(x) \, dx$. Cauchy, by contrast, took as his starting point a clearcut definition of the derivative as the limit of a difference quotient:

> When a function $y = f(x)$ remains continuous between two given limits of the variable x, and when one assigns to such a variable a value enclosed between the two limits at issue, then an infinitely small increment assigned to the variable produces an infinitely small increment in the function itself. Consequently, if one puts $\Delta x = i$, the two terms *of the ratio of differences*
>
> $$\frac{\Delta y}{\Delta x} = \frac{f(x+i) - f(x)}{i}$$
>
> will be infinitely small quantities. But though these two terms will approach the limit zero indefinitely and simultaneously, the ratio itself can converge towards another limit, be it positive or be it negative. This limit, when it exists, has a definite value for each particular value of x; but it varies with x . . . The form of the new function which serves as the limit of the ratio $[f(x+i) - f(x)]/i$ will depend on the form of the proposed function $y = f(x)$. In order to indicate this dependence, one gives the new function the name of *derived function*, and designates it with the aid of an accent by the notation, y' or $f'(x)$ ([4], pp. 22–23; [12], p. 240).

After some typical computations of derivatives of elementary functions, Cauchy introduces what is now called the *chain rule* for computing the derivative of the composition of two functions ([4], p. 25; [12], p. 243):

> Now let z be a second function of x, bound to the first $y = f(x)$ by the formula
>
> $$z = F(y).$$
>
> z or $F[f(x)]$ will be that which one calls a function of a function of a variable x; and, if one designates the infinitely small and simultaneous increments of x, y, and z by $\Delta x, \Delta y, \Delta z$, one will find
>
> $$\frac{\Delta z}{\Delta x} = \frac{F(y + \Delta y) - F(y)}{\Delta x} = \frac{F(y + \Delta y) - F(y)}{\Delta y} \cdot \frac{\Delta y}{\Delta x},$$
>
> then, on passing to the limits,
>
> $$z' = y' F'(y) = f'(x) F'[f(x)].$$

Here (as sometimes happens in elementary calculus texts today) the possibility that $\Delta y = f(x + \Delta x) - f(x) = 0$ for small non-zero values of Δx is overlooked.

His explicit formulation of continuity and differentiability in terms of limits was one of three features of Cauchy's calculus that set the pattern for subsequent expositions of the subject. The second was the central role he accorded to the mean value theorem (which was known previously to Lagrange, but not extensively used by him). The third—his definition of the integral and proof of the fundamental theorem of calculus—will be discussed in the following section.

In modern textbooks the mean value theorem—if f is differentiable on $[a, b]$ then $f(b) - f(a) = f'(\xi)(b - a)$ for some $\xi \in (a, b)$—is generally proved by applying Rolle's theorem to a suitably contrived auxiliary function; this approach apparently was first discovered by Ossian Bonnet (1819–1892). The fact, that a positive (negative) derivative on an interval corresponds to an increasing (decreasing) function on that interval, is then deduced as an immediate corollary to the mean value theorem.

By contrast, in the fourth of his *Leçons sur le calcul differentiel*, Cauchy begins his approach to the mean value theorem by first investigating the significance of the sign of the derivative. Because

$$y' = \frac{dy}{dx} = \lim \frac{\Delta y}{\Delta x},$$

he observes that, if $y' > 0$ at x_0, then Δy and Δx must have the same sign for Δx sufficiently small (and different signs if $y' < 0$). Hence $y = f(x)$ increases as x increases through x_0. Therefore, he says, if one increases x "by insensible degrees" from $x = x_0$ to $x = X$, the function $f(x)$ will be increasing at all times that its derivative is positive, and decreasing when it is negative ([5], p. 308, Corollary I). In particular, $f(X) > f(x_0)$ if $f'(x) > 0$ on $[x_0, X]$.

With this preparation, Cauchy is ready for his "generalized mean value theorem": Let $f(x)$ and $F(x)$ have continuous derivatives on the interval $[x_0, X]$, and suppose $F'(x)$ is non-zero on this interval. Then

$$\frac{f(X) - f(x_0)}{F(X) - F(x_0)} = \frac{f'(\xi)}{F'(\xi)} \tag{13}$$

for some point $\xi \in (x_0, X)$. (See Theorem II and its first corollary ([5], pp. 308–310)).

Cauchy's proof is more clearly motivated than the auxiliary function proof (which, however, does not require continuity of the derivatives). Consider the case $F'(x) > 0$, and assume without loss of generality that $f(x_0) = F(x_0) = 0$. Let A and B be the minimum and maximum values, respectively, of the quotient $f'(x)/F'(x)$ on the interval $[x_0, X]$ (it may be noted that Cauchy overlooks the need to prove that a continuous function on a closed interval attains minimum and maximum values). Then

$$f'(x) - AF'(x) \geqslant 0 \quad \text{and} \quad f'(x) - BF'(x) \leqslant 0.$$

Thus the derivative of $f(x) - AF(x)$ is non-negative on $[x_0, X]$, while that of $f(x) - BF(x)$ is non-positive. Hence the first of these latter functions is non-decreasing on the interval, while the second is non-increasing. Since both vanish at x_0, it follows that

$$f(X) - AF(X) \geqslant 0 \quad \text{and} \quad f(X) - BF(X) \leqslant 0,$$

and hence that

$$A \leqslant \frac{f(X)}{F(X)} \leqslant B.$$

An application of the intermediate value theorem to the quotient $f'(x)/F'(x)$ then gives a point $\xi \in [x_0, X]$ such that Equation (13) holds (remembering that $f(x_0) = F(x_0) = 0$).

EXERCISE 8. Obtain the ordinary mean value theorem by substituting $F(x) = x$ in Cauchy's generalized mean value theorem.

With $X = x_0 + h$ and $f(x_0) = F(x_0) = 0$, Equation (13) becomes

$$\frac{f(x_0 + h)}{F(x_0 + h)} = \frac{f'(x_0 + \theta h)}{F'(x_0 + \theta h)}$$

for some $\theta \in (0, 1)$. If also $f'(x_0) = F'(x_0) = 0$, then a second application of the generalized mean value theorem gives

$$\frac{f'(x_0 + \theta h)}{F'(x_0 + \theta h)} = \frac{f''(x_0 + \theta_1 h)}{F''(x_0 + \theta_1 h)}$$

for some $\theta_1 \in (0, 1)$. Continuing in this way, after n applications of the generalized mean value theorem Cauchy obtains the result that

$$\frac{f(x_0 + h)}{F(x_0 + h)} = \frac{f^{(n)}(x_0 + \theta h)}{F^{(n)}(x_0 + \theta h)} \tag{14}$$

for some $\theta \in (0, 1)$, under the assumptions that f and F and their first $n - 1$ derivatives vanish at x_0, the first n derivatives of f are continuous between x_0 and $x_0 + h$, and that the first n derivatives of F are continuous and non-zero between x_0 and $x_0 + h$ ([5], p. 310, Corollary II). In his fifth *leçon* he takes the limit in (14) as $h \to 0$ to obtain l'Hospital's rule

$$\lim_{x \to x_0} \frac{f(x)}{F(x)} = \lim_{x \to x_0} \frac{f^{(n)}(x)}{F^{(n)}(x)}$$

for higher order $0/0$ indeterminate forms (with the first n derivatives of f and F continuous in a neighborhood of x_0, and the first $n - 1$ of them having x_0 as an isolated zero).

In his seventh *leçon* Cauchy applies (14) to rigorously establish the higher derivative test for local maxima and minima that had been known to Euler and Maclaurin in the eighteenth century. Suppose that f and its first $n - 1$ derivatives vanish at x_0. If $F(x) = (x - x_0)^n$, then F and its first $n - 1$ derivatives also vanish at x_0, while $F^{(n)}(x) \equiv n!$. Therefore Equation (14) yields

$$f(x_0 + h) = \frac{h^n}{n!} f^{(n)}(x_0 + \theta h) \tag{15}$$

for some $\theta \in (0, 1)$.

EXERCISE 9. Deduce from (15) that f has (a) a local maximum at x_0 if n is even and $f^{(n)}(x_0) < 0$; (b) a local minimum at x_0 if n is even and $f^{(n)}(x_0) > 0$; (c) neither if n is odd and $f^{(n)}(x_0) \neq 0$.

In his eighth *leçon* Cauchy gives what is in some ways still the most appealing derivation of Taylor's formula with remainder. If $f(x)$ is a function whose first n derivatives are continuous on $[x_0, x_0 + h]$ then the function

$$F(x) = f(x) - f(x_0) - f'(x_0)(x - x_0) - \frac{f''(x_0)}{2!}(x - x_0)^2$$
$$- \cdots - \frac{f^{(n-1)}(x_0)}{(n-1)!}(x - x_0)^{n-1}$$

vanishes together with its first $n - 1$ derivatives at x_0, so

$$F(x_0 + h) = \frac{h^n}{n!} F^{(n)}(x_0 + \theta h)$$

by Equation (15). But obviously $F^{(n)}(x) = f^{(n)}(x)$, so this yields

$$f(x) = f(x_0) + f'(x_0)h + \cdots + \frac{f^{(n-1)}(x_0)}{(n-1)!}h^{n-1} + \frac{f^{(n)}(x_0 + \theta h)}{n!}h^n.$$

With $x_0 = a$ and $h = x - a$ this becomes

$$f(x) = f(a) + f'(a)(x - a)$$
$$+ \cdots + \frac{f^{(n-1)}(a)}{(n-1)!}(x - a)^{n-1} + \frac{f^{(n)}(a + \theta(x - a))}{n!}(x - a)^n,$$

that is, Taylor's formula with the Lagrange form of the remainder.

On pages 360–361 of the *leçons* we find the "Cauchy form" of the remainder. Regarding x as a constant and a as a variable, Cauchy defines the function $\phi(a)$ by the equation

$$f(x) = f(a) + f'(a)(x - a) + \cdots + \frac{f^{(n-1)}(a)}{(n-1)!}(x - a)^{n-1} + \phi(a).$$
$$(16)$$

Then $\phi(x) = 0$, and a simple computation yields

$$\phi'(a) = -\frac{f^{(n)}(a)}{(n-1)!}(x - a)^{n-1}. \qquad (17)$$

Application of the (ordinary) mean value theorem to the function $\phi(a)$ on the interval with end points a and x then gives

$$\phi(a) = \phi(x) - (x - a)\phi'(a + \theta(x - a))$$
$$\phi(a) = \frac{(1 - \theta)^{n-1}(x - a)^n}{(n-1)!} f^{(n)}(a + \theta(x - a)),$$

the Cauchy form of the remainder. In succeeding lessons Cauchy applies
Taylor's formula with remainder to rigorously establish the convergence of
the Taylor series of the elementary transcendental functions, e.g.

$$e^x = 1 + x + \frac{x^2}{2!} + \cdots + \frac{x^n}{n!} + \cdots$$

and

$$\sin x = x - \frac{x^3}{3!} + \frac{x^5}{5!} + \cdots .$$

EXERCISE 10. Differentiate Equation (16) with respect to a to verify (17).

The Cauchy Integral

During the eighteenth century the integral was generally regarded simply
as the inverse of the derivative. That is, a function $f(x)$ was integrated by
finding an antiderivative or primitive function $F(x)$ such that $F'(x)=f(x)$.
The integral of $f(x)$ over the interval $[a, b]$ was then given, according to
the still only heuristically understood fundamental theorem of calculus, by

$$\int_a^b f(x) \, dx = F(b) - F(a).$$

At the same time, the idea of the integral as some sort of limit of a sum,
or as the area of an ordinate set under a curve, was familiar, but was
generally relied upon only in approximating integrals when it was incon-
venient or impossible to find the antiderivative needed in order to apply
the fundamental theorem of calculus. Neither limits of sums nor areas of
plane sets were sufficiently well understood to provide a solid basis for a
logical treatment of the integral. In particular, the notion of area was still
wholly intuitive—it was regarded as a self-evident concept, and no need
for a precise definition had yet been perceived. Indeed, the analytical
integral in the antiderivative sense of Newton was adequate in practice, so
long as the only functions to be integrated were continuous in the sense of
Euler, that is, each such function was defined by a single explicit analytical
expression.

But in the early nineteenth century, as we have seen, the work of Fourier
brought to light the need to make integration meaningful for functions that
are discontinuous (at least in the sense of Euler). Such functions appeared
naturally in applied problems, and the coefficients of their Fourier series
were expressed as integrals that did not fit the narrowly analytical pattern
of eighteenth century integration.

It was Cauchy who first addressed this necessity "to demonstrate the
existence of the integrals or primitive functions before making known their
diverse properties"—that is, to first provide a *general* definition and proof

of the existence of the integral for a broad class of functions that could
then provide a basis for the discussion of *particular* integrals and their
properties. In his *Resume des leçons donnees a l'Ecole Royale Polytechnique
sur le calcul infinitesimal* of 1823 he formulated the definition of the
integral which (as later completed by Riemann) appears in modern elementary treatments of the integral calculus.

In his twenty-first *leçon* Cauchy starts with a function $f(x)$ that is
continuous (in the modern sense) on the interval $[x_0, X]$, and subdivides
this interval into n subintervals by means of the points $x_0, x_1, \ldots, x_n = X$.
With this subdivision or partition P of $[x_0, X]$ he associates the approximating sum

$$S = \sum_{i=1}^{n} f(x_{i-1})(x_i - x_{i-1}) \tag{18}$$

obtained by adding the areas of rectangles based on the subintervals of the
partition, the rectangle with base $[x_{i-1}, x_i]$ having height $f(x_{i-1})$. He wants
to define the integral $\int_{x_0}^{X} f(x)\, dx$ as the limit of the sum (18) as the
maximum of the lengths $x_i - x_{i-1}$ of the subintervals approaches zero. Of
course the existence of this limit must be established. To this end, he says,

> It is important to remark that if the numerical values [lengths] of these
> elements [subintervals] become very small and the number n very large
> the mode of subdivision will have only an imperceptible influence on the
> value of S ([4], p. 122; [12], p. 261).

To prove this he applies the following elementary arithmetical result
from his *Cours d'analyse* ([3], p. 28): If $\alpha_1, \ldots, \alpha_n$ are positive numbers,
and $a_1, \ldots, a_n$ are arbitrary numbers, then

$$\sum_{i=1}^{n} \alpha_i a_i = \bar{a}(\alpha_1 + \cdots + \alpha_n) \tag{19}$$

where $\bar{a}$ is a "mean" of the numbers $a_1, \ldots, a_n$, that is, $\bar{a}$ lies between the
largest and smallest of them. With $\alpha_i = x_i - x_{i-1}$ and $a_i = f(x_{i-1})$, (19)
yields

$$S = f(x_0 + \theta(X - x_0))(X - x_0) \tag{20}$$

for some $\theta \in (0, 1)$, because by the intermediate value theorem any mean
of the numbers $f(x_0), \ldots, f(x_{n-1})$ is a value of the continuous function f
at some point of the interval.

Now Cauchy considers a refinement P' of the above partition P, that is,
each subinterval of the partition P' lies in some subinterval of P. Then the
sum S' of the form (18) associated with this new partition can be written as

$$S' = S_1' + S_2' + \cdots + S_n'$$

where S_i' is the sum of those terms of S' that correspond to subintervals of

P' that lie in the ith subinterval of P. Then (20), applied on this ith subinterval, gives

$$S_i' = f(x_{i-1} + \theta_i(x_i - x_{i-1}))(x_i - x_{i-1})$$

for some $\theta_i \in (0, 1)$, $i = 1, \ldots, n$, so

$$S' = \sum_{i=1}^{n} f(x_{i-1} + \theta_i(x_i - x_{i-1}))(x_i - x_{i-1}). \tag{21}$$

If we write

$$\epsilon_i = f(x_{i-1} + \theta_i(x_i - x_{i-1})) - f(x_{i-1}) \tag{22}$$

for each $i = 1, \ldots, n$, then comparison of (18) and (21) yields

$$S' - S = \sum_{i=1}^{n} \epsilon_i(x_i - x_{i-1}) = \bar{\epsilon}(X - x_0) \tag{23}$$

for some mean $\bar{\epsilon}$ of $\epsilon_1, \ldots, \epsilon_n$.

Cauchy concludes from (23) that "one will not alter perceptibly the value of S calculated by a mode of division [partition] in which the elements [subintervals] of the difference $X - x_0$ have very small numerical values, if one passes to a second mode in which each of these elements is subdivided into many others" ([4], p. 125). This is where he overlooks the need to prove that the continuous function f is *uniformly* continuous on $[x_0, X]$, that is, that given $\epsilon > 0$ there exists $\delta > 0$ such that $|f(x') - f(x'')| < \epsilon$ for any two points $x', x'' \in [x_0, X]$ with $|x' - x''| < \delta$. Knowing this, the numbers ϵ_i defined by (22) could be made as small as desired by choosing P with sufficiently short subintervals.

Now let P_1 and P_2 be arbitrary partitions of $[x_0, X]$, and let P' be the common refinement obtained by amalgamating the points of subdivision of P_1 and P_2. If S_1, S_2, S' are the associated approximating sums, then (23) gives

$$S' - S_1 = \bar{\epsilon}_1(X - x_0) \quad \text{and} \quad S' - S_2 = \bar{\epsilon}_2(X - x_0),$$

so

$$S_1 - S_2 = (\bar{\epsilon}_2 - \bar{\epsilon}_1)(X - x_0).$$

Hence the difference between S_1 and S_2 can be made arbitrarily small by choosing P_1 and P_2 with sufficiently short subintervals.

Cauchy summarizes this situation as follows ([4], p. 125; [12], p. 265):

> Conceive for the present that one considers at the same time two modes of division of the difference $X - x_0$, in each of which the elements of the difference have very small numerical values. One will be able to compare these two modes with a third in such a way that each element, be it from the first, or from the second mode is formed by the union of several elements of the third. For this condition to be satisfied, it will suffice that each of the values of x interplaced in the first two modes between the

limits x_0 and X be employed in the third, and one will prove that one alters the value of S very little in passing from the first or the second mode to the third, and consequently, in passing from the first to the second. Therefore, when the elements of the difference $X - x_0$ become infinitely small, the mode of division has no more than an imperceptible influence on the value of S; and, if one makes the numerical values of these elements decrease indefinitely, by increasing their number, the value of S will end by being perceptibly constant or, in other words, it will end by attaining a certain limit which will depend solely on the form of the function $f(x)$ and on the extreme values x_0 and X attributed to the variable x. This limit is that which one calls a *definite integral*.

To clinch his final argument—the actual existence of the limit—Cauchy would have needed a completeness property of the real numbers (just as in the earlier problem of the existence of the limit of a Cauchy sequence of numbers).

EXERCISE 11. Let f be a continuous function on $[a, b]$, and denote by S_n the Cauchy sum (18) associated with the partition of $[a, b]$ into 2^n equal subintervals. Then use Cauchy's results to prove that $\{S_n\}_1^\infty$ is a Cauchy sequence, that is, given $\epsilon > 0$ there exists an integer N such that $|S_m - S_n| < \epsilon$ if $m, n \geq N$.

EXERCISE 12. Let $\{P_n\}_1^\infty$ and $\{P_n'\}_1^\infty$ be two sequences of partitions of the interval $[a, b]$ into subintervals whose lengths approach zero as $n \to \infty$. Let $\{S_n\}_1^\infty$ and $\{S_n'\}_1^\infty$ be the associated Cauchy sums (18) for a continuous function f on $[a, b]$. Assuming that $\lim_{n \to \infty} S_n = I$, use Cauchy's results to show that $\lim_{n \to \infty} S_n' = I$ also.

In the twenty-second *leçon* ([4], p. 131), Cauchy argues that Equation (20) for approximating sums carries over to the integral itself, that is,

$$\int_{x_0}^X f(x)\, dx = f(x_0 + \theta(X - x_0))(X - x_0)$$

$$= f(\bar{x})(X - x_0) \tag{24}$$

for some $\theta \in [0, 1]$ or $\bar{x} \in [x_0, X]$. This is the "mean value property of integrals". In the twenty-third *leçon* ([4], pp. 134–136) he carefully observes that the simple properties

$$\int_{x_0}^X [af(x) + bg(x)]\, dx = a \int_{x_0}^X f(x)\, dx + b \int_{x_0}^X g(x)\, dx \tag{25}$$

and

$$\int_{x_0}^X f(x)\, dx = \int_{x_0}^{\bar{x}} f(x)\, dx + \int_{\bar{x}}^X f(x)\, dx \tag{26}$$

follow from the definition of the integral as a limit of a sum.

In his twenty-sixth *leçon* ([4], pp. 151–155) Cauchy presents the rigorous formulation of the "fundamental theorem of calculus" that is duplicated in almost every modern calculus text. Having given an arithmetical definition of the integral, he could now establish the inverse relationship between differentiation and integration without relying on intuitive area concepts. Given a continuous function $f(x)$ on the interval $[x_0, X]$, he wants to prove that the new function $\tilde{F}(x)$ defined for $x \in [x_0, X]$ by

$$\tilde{F}(x) = \int_{x_0}^{x} f(x)\, dx \tag{27}$$

is a primitive function or antiderivative of $f(x)$, that is, $\tilde{F}'(x)=f(x)$ on $[a, b]$. The only change of detail in his exposition that one might make today would be to write $f(t)\, dt$ instead of $f(x)\, dx$ in the integrand of (27), so as to distinguish the dummy variable of integration from the variable upper limit.

Applying property (26) and then the mean value property (24), Cauchy notes that

$$\tilde{F}(x+\alpha) - \tilde{F}(x) = \int_{x_0}^{x+\alpha} f(x)\, dx - \int_{x_0}^{x} f(x)\, dx$$
$$= \int_{x}^{x+\alpha} f(x)\, dx$$
$$\tilde{F}(x+\alpha) - \tilde{F}(x) = \alpha f(x+\theta\alpha) \tag{28}$$

for some $\theta \in [0, 1]$. Dividing both sides of (28) by α and then taking limits as $\alpha \to 0$, he concludes from the definition of the derivative and the continuity of f that

$$\tilde{F}'(x) = f(x) \tag{29}$$

as desired. Thus

$$\frac{d}{dx}\left(\int_{x_0}^{x} f(t)\, dt \right) = f(x) \tag{30}$$

if f is continuous.

To deduce from this first form the second familiar form of the fundamental theorem, Cauchy considers an arbitrary function $F(x)$ such that $F'(x)=f(x)$ on $[x_0, X]$. If

$$\omega(x) = \tilde{F}(x) - F(x),$$

then $\omega'(x)= \tilde{F}'(x)- F'(x)=f(x)-f(x)=0$, so the mean value theorem gives

$$\omega(x) = \omega(x_0) + (x - x_0)\omega'(\bar{x}) = \omega(x_0)$$

for all $x \in [x_0, X]$. Therefore

$$\tilde{F}(x) - F(x) = \tilde{F}(x_0) - F(x_0) = -F(x_0),$$
$$\tilde{F}(x) = F(x) - F(x_0),$$
$$\int_{x_0}^{x} f(x)\, dx = F(X) - F(x_0) \tag{31}$$

for any antiderivative F of f.

Thus we see that Cauchy essentially completed the general theory of integration of *continuous* functions on closed intervals. Although we have seen that quite general definitions of functions were formulated in the early nineteenth century, it does not appear that anyone then took seriously the importance for analysis (or perhaps even the existence) of functions having more than a *finite* number of discontinuities in each finite interval. It may be observed that Cauchy's theory of integration for continuous functions also suffices for *piecewise* continuous functions. For if the interval $[x_0, X]$ is partitioned into subintervals $[x_{i-1}, x_i]$, $i = 1, \ldots, n$, such that f agrees on (x_{i-1}, x_i) with a function f_i that is continuous on $[x_{i-1}, x_i]$, then the integral of f is satisfactorily defined by

$$\int_{x_0}^{X} f(x)\, dx = \sum_{i=1}^{n} \int_{x_{i-1}}^{x_i} f_i(x)\, dx.$$

In addition, Cauchy considered integrals of functions having isolated infinite discontinuities, that is, improper integrals. For example, if $\lim_{x \to X} f(x) = \pm \infty$ but f is continuous on $[x_0, X - \epsilon]$ for each $\epsilon > 0$, he defined the integral of f on $[x_0, X]$ by

$$\int_{x_0}^{X} f(x)\, dx = \lim_{\epsilon \to 0} \int_{x_0}^{X-\epsilon} f(x)\, dx$$

provided that this limit exists.

The Riemann Integral and Its Reformulations

Genuinely discontinuous functions entered the mainstream of mathematics through the work of G. B. F. Riemann (1826–1866) on the convergence of Fourier series. In the course of extending the applicability of Dirichlet's convergence proof to a wider class of functions, Riemann formulated the generalization of Cauchy's integral that to this day remains the most convenient and useful one for elementary applications of the calculus.

In his 1854 "Habilitationschrift" Riemann, who in 1859 would succeed Dirichlet in the Göttingen chair that Gauss had occupied, took under Dirichlet's influence a fresh look at the representability of functions by means of trigonometric series. Although willing to concede that the more general functions he proposed to consider probably do not occur in nature, he felt an investigation of their Fourier series would be worthwhile because "this subject is closely related to the principles of the infinitesimal calculus and can serve to bring greater clarity and precision to these principles" ([16], p. 238). This 1854 investigation, which did not appear in print until 1867, is reprinted in Riemann's collected works ([16] pp. 227–265), and selected portions are translated in Birkhoff's source book ([1], pp. 16–23).

The first three sections of the paper are devoted to a summary of the history of Fourier series. In his 1829 paper Dirichlet had shown that, if the

period 2π function $f(x)$
(a) is piecewise continuous and
(b) has only finitely many local maxima and minima in the interval $[-\pi, \pi]$,

then its Fourier series converges pointwise to $\frac{1}{2}[f(x+0)+f(x-0)]$, the average of its righthand and lefthand limits (assuming that these exist at every point). However, it appeared that the only need for the piecewise continuity assumption (a) was to ensure the integrability of the function $f(x)$ and the meaningfulness of the integrals appearing as its Fourier coefficients.

Riemann therefore poses at the beginning of Section 4 the question "What is one to understand by $\int_a^b f(x)\,dx$?", to which he immediately supplies the following answer ([16], p. 239; [1], p. 22):

> In order to establish this, we take a sequence of values $x_1, x_2, \ldots, x_{n-1}$ lying between a and b and ordered by size, and for brevity, denote $x_1 - a$ by δ_1, $x_2 - x_1$ by δ_2, $\ldots$, $b - x_{n-1}$ by δ_n, and proper positive fractions by ϵ_i. Then the value of the sum
>
> $$S = \delta_1 f(a+\epsilon_1\delta_1) + \delta_2 f(x_1+\epsilon_2\delta_2) + \delta_3 f(x_2+\epsilon_3\delta_3)$$
> $$+ \cdots + \delta_n f(x_{n-1}+\epsilon_n\delta_n)$$
>
> will depend on the choice of the intervals δ_i and the quantities ϵ_i. If it has the property that, however the δ_i and the ϵ_i may be chosen, it tends to a fixed limit A as soon as all the δ_i become infinitely small, then this value is called $\int_a^b f(x)\,dx$. If it does not have this property, then $\int_a^b f(x)\,dx$ is meaningless.

Thus Riemann chooses an *arbitrary* point $\bar{x}_i = x_{i-1} + \epsilon_i\delta_i$ in the ith subinterval $[x_{i-1}, x_i]$ of his partition, $i = 1, \ldots, n$, and defines the integral by

$$\int_a^b f(x)\,dx = \lim_{\delta \to 0} \sum_{i=1}^n f(\bar{x}_i)(x_i - x_{i-1}), \tag{32}$$

where δ denotes the maximum of the lengths δ_i of the subintervals of the partition of $[a, b]$. This is a direct generalization of Cauchy's definition,

$$\int_a^b f(x)\,dx = \lim_{\delta \to 0} \sum_{i=1}^n f(x_{i-1})(x_i - x_{i-1}).$$

Riemann has simply replaced Cauchy's initial point x_{i-1} with an arbitrary point $\bar{x}_i$ of $[x_{i-1}, x_i]$, and insists (if the integral is to exist) that the approximating sums thereby associated with a partition approach a fixed value (the integral) as the "norm" δ approaches zero, independently of the choice of the points $\bar{x}_i$.

Now he says, "Let us determine the extent of the validity of this concept, and ask: in what cases is a function integrable and in what cases is it not?" ([16], p. 240; [1], p. 22). Starting with a *bounded* function f and a partition

P of $[a, b]$, he considers the "total oscillation" $D(P) = D_1\delta_1 + D_2\delta_2 + \cdots + D_n\delta_n$ of $f(x)$ with respect to P, where $\delta_i = x_i - x_{i-1}$ and D_i denotes the difference between the largest and smallest values of $f(x)$ on the ith subinterval $[x_{i-1}, x_i]$. Sidestepping (or overlooking) the question of the completeness of the real numbers, he takes it for granted that the integral (32) exists if and only if $D(P) \to 0$ as the norm $\delta \to 0$,

$$\lim_{\delta \to 0} (D_1\delta_1 + D_2\delta_2 + \cdots + D_n\delta_n) = 0. \tag{33}$$

EXERCISE 13. Let $\{S_n\}_1^\infty$ and $\{S_n'\}_1^\infty$ be two different sequences of Riemann sums (corresponding to different choices of the points $\bar{x}_i$) associated with a sequence of partitions whose norms approach zero as $n \to \infty$. Assuming that (33) holds, show that $\lim_{n\to\infty} S_n = \lim_{n\to\infty} S_n'$ provided that one of these limits exist.

EXERCISE 14. If $f(x)$ is a monotone non-decreasing function on $[a, b]$, show that $D(P) \leqslant D\delta$ where D is the oscillation of $f(x)$ on $[a, b]$. Hence conclude that f satisfies (33) and is therefore integrable.

EXERCISE 15. Let ϕ be Dirichlet's discontinuous function such that $\phi(x) = 0$ if x is rational, but $\phi(x) = 1$ if x is irrational, $x \in [0, 1]$. Show that $D(P) = 1$ for any partition P of $[0, 1]$, and therefore conclude that ϕ is *not* Riemann integrable.

Next Riemann defines $\Delta = \Delta(d)$ as the maximum value of the total oscillation $D(P)$ for all partitions P with norm $\delta \leqslant d$. Then $\Delta(d)$ is obviously a decreasing function of d, and f is integrable on $[a, b]$ if and only if $\lim_{d\to 0}\Delta(d) = 0$. Given $\sigma > 0$ and a partition P he denotes by $s = S(\sigma, P)$ the sum of the lengths δ_i of those subintervals of P for which the oscillation $D_i > \sigma$. He now establishes the following necessary and sufficient condition for the existence of the integral of a *bounded* function.

If $f(x)$ is bounded for $x \in [a, b]$, then $\int_a^b f(x)\, dx$ exists if and only if, given $\sigma > 0$, it follows that $s(\sigma, P)$ approaches zero as the norm of the partition P approaches zero.

That is, given $\sigma > 0$ and $\epsilon > 0$, there exists $d > 0$ such that for any partition P with norm $\delta < d$, the sum s of the lengths of those subintervals of P, on which the oscillation of $f(x)$ is greater than σ, is less than ϵ.

To see that this condition is necessary for the integrability of f, note that

$$\sigma s < D_1\delta_1 + D_2\delta_2 + \cdots + D_n\delta_n \leqslant \Delta(d)$$

if $\delta < d$, because $D_i > \sigma$ on subintervals having a total length of s. Therefore $s(\sigma, P) < \Delta(d)/\sigma$, which approaches zero as $d \to 0$ with σ fixed, assuming that $\int_a^b f(x)\, dx$ exists.

To see that the above condition is sufficient for integrability, let $\sigma > 0$ and $\epsilon > 0$ be given, and choose $d > 0$ as above. If the norm of the partition P is less than d, then those subintervals on which the oscillation of $f(x)$ is greater than σ contribute to $D(P)$ an amount less than $D\epsilon$ (because their lengths add up to $s < \epsilon$), where D is the oscillation of $f(x)$ on $[a, b]$. The remaining subintervals contribute to $D(P)$ an amount at most $\sigma(b-a)$. Hence

$$D(P) < D\epsilon + \sigma(b-a),$$

so $D(P)$ can be made as small as desired by taking ϵ and σ sufficiently small. Thus condition (33) is satisfied, so $\int_a^b f(x)\, dx$ exists.

Riemann's theorem immediately implies that $\int_a^b f(x)\, dx$ exists if f is *uniformly* continuous on $[a, b]$. For in this case, given $\sigma > 0$, there exists $d > 0$ such that the oscillation of $f(x)$ is less than σ on any subinterval of length less than d. Therefore $s(\sigma, P) = 0$ if the norm of the partition P is less than d.

EXERCISE 16. Apply condition (33) to show that any uniformly continuous function is integrable.

However, Riemann refrained from concluding that every continuous function is integrable. The uniform continuity of a continuous function on a closed interval was not rigorously established until the early 1870's when the Bolzano-Weierstrass theorem (stated by Bolzano but proved by Weierstrass in his lectures) was available. According to this theorem, every infinite sequence of points in an interval has a subsequence that converges to some point of the interval. If the function f is *not* uniformly continuous on $[a, b]$, then for some $\delta > 0$ and each positive integer n there exist points $a_n, b_n \in [a, b]$ such that $|a_n - b_n| < 1/n$ but $|f(a_n) - f(b_n)| \geqslant \delta$. By the Bolzano-Weierstrass theorem it may be assumed without loss of generality that the sequences $\{a_n\}_1^\infty$ and $\{b_n\}_1^\infty$ both converge to some point $c \in [a, b]$. It follows that f is *not* continuous at c (why?).

In the opposite direction, Riemann pointed out that a function can be discontinuous at a dense set of points but nevertheless be integrable. "Since these functions are as yet nowhere considered, it will be good to start with a specific example" ([16], p. 242). His example is described as follows. For each real number x, let $(x) = x - i(x)$ where $i(x)$ is the integer nearest to x, unless x is an odd multiple of $\frac{1}{2}$, in which case $(x) = 0$. Then $-\frac{1}{2} < (x) < \frac{1}{2}$ for all x, and (x) is continuous at x unless x is an odd multiple of $\frac{1}{2}$, in which case (x) has a "jump", or difference of lefthand and righthand limits, of $(x-0) - (x+0) = 1$. Riemann's exotic function is then defined by

$$f(x) = \frac{(x)}{1^2} + \frac{(2x)}{2^2} + \frac{(3x)}{3^2} + \cdots = \sum_{k=1}^{\infty} \frac{(kx)}{k^2}. \tag{34}$$

If x is *not* a rational number of the form $m/2n$, where m and n are relatively prime integers with m odd, then kx is not an odd multiple of $\frac{1}{2}$

for any k. Consequently each term of (34) is continuous at x and it can be proved, as expected, that f is continuous at x.

But if x *is* of the indicated form $m/2n$, then kx is an odd multiple of $\frac{1}{2}$ when k is an odd multiple of n, $k = n(2p + 1)$. In this case the term $(kx)/k^2$ of (34) has a (negative) jump of $1/k^2 = 1/n^2(2p+1)^2$ at $x = m/2n$. This indicates (and it can be verified) that f is discontinuous at each such point $x = m/2n$, having there a jump

$$ f(x - 0) - f(x + 0) = \frac{1}{n^2} \sum_{p=1}^{\infty} \frac{1}{(2p+1)^2} = \frac{\pi^2}{8n^2}, $$

using one of Euler's summations. Of course these points of discontinuity are dense in every interval.

On the other hand, if an interval $[a, b]$ and a number $\sigma > 0$ are given, there are only a finite number of these points $x = m/2n$ in $[a, b]$ such that $\pi^2/8n^2 > \sigma$. Consequently the sum $s(\sigma, P)$ of the lengths of the subintervals containing these latter points can be made arbitrarily small by choosing the norm of the partition P small enough. Thus Riemann's function (34) satisfies his sufficient condition for integrability, so $\int_a^b f(x)\, dx$ exists despite the denseness of the set of discontinuities of f.

Riemann's definition (32) of the integral was the most general one that could be based directly on Cauchy's original device of approximating sums associated with partitions of the interval of integration into subintervals. Nevertheless, during the last three decades of the nineteenth century, this definition was reformulated in several ways that further illuminated the concept of the integral and paved the way for important additional generalizations in the early twentieth century.

In the middle 1870's several authors independently introduced the so-called upper and lower Riemann sums for the bounded function f on the interval $[a, b]$,

$$ U(P) = \sum_{i=1}^{n} M_i(x_i - x_{i-1}) \quad \text{and} \quad L(P) = \sum_{i=1}^{n} m_i(x_i - x_{i-1}), \quad (35) $$

where P is a partition of $[a, b]$ into n subintervals, and M_i and m_i are the maximum and minimum values (actually, the least upper and greatest lower bounds) of $f(x)$ on the ith subinterval $[x_{i-1}, x_i]$. Today these are often called "Darboux sums" after Gaston Darboux (1842–1917)—see [6].

EXERCISE 17. If P' is a refinement of the partition P, show that

$$ L(P) \leqslant L(P') \leqslant U(P') \leqslant U(P). $$

Using this observation, it is easily verfied that the upper and lower sums $U(P)$ and $L(P)$ approach limits U and L, respectively, as the norm δ of the partition P approaches zero, whether or not the bounded function f is integrable. In the 1880's Vito Volterra (1860–1940) introduced the terms "upper integral" and "lower integral" for U and L, together with the descriptive notation

$$U = \overline{\int_a^b} f(x)\, dx \quad \text{and} \quad L = \underline{\int_a^b} f(x)\, dx,$$

and Giuseppe Peano (1858–1932) noted that these upper and lower integrals could be defined conveniently as the greatest lower and least upper bounds of the upper and lower Riemann sums, respectively, for all partitions P of the interval $[a, b]$,

$$\overline{\int_a^b} f(x)\, dx = glb\{U(P)\} \quad \text{and} \quad \underline{\int_a^b} f(x)\, dx = lub\{L(P)\}. \tag{36}$$

The function f is the integrable if and only if its upper and lower integrals are equal, $\underline{\int_a^b} f(x)\, dx = \overline{\int_a^b} f(x)\, dx$.

EXERCISE 18. Use (36) and Exercise 17 to show that $\underline{\int_a^b} f(x)\, dx \leqslant \overline{\int_a^b} f(x)\, dx$.

EXERCISE 19. If ϕ is Dirichlet's function of Exercise 15, show that $\underline{\int_0^1} \phi(x)\, dx = 0$ while $\overline{\int_0^1} \phi(x)\, dx = 1$.

Since the seventeenth century the idea of the integral had always been motivated by the concept of area. In particular, if O_f denotes the ordinate set of the non-negative function f on the interval—the set of all points (x, y) with $a \leqslant x \leqslant b$ and $0 \leqslant y \leqslant f(x)$—the idea was that the value of the integral $\int_a^b f(x)\, dx$ should be the area $a(O_f)$. Yet, prior to the late nineteenth century, the concept of area itself had been wholly intuitive and not based on any precise definition.

The first formal mathematical definition of area apparently was given by Peano in a book published in 1887 [15]. Beginning with Eudoxus and his method of exhaustion, it had always been taken as obvious that the area of a plane set S is the upper bound of the areas of all polygons that are contained in S, and the lower bound of the areas of all polygons that contain S (of course the area of a polygon is obtained by dissecting it into triangles).

Peano took this ancient idea as the starting point for an actual definition of area. He defined the *inner area* $a_i(S)$ of S as the least upper bound of the areas of all polygons that are contained in S, and the *outer area* $a_o(S)$ as the greatest lower bound of the areas of all polygons that contain S. It is clear that $a_i(S) \leqslant a_o(S)$, but the two may not be equal. For example, if S is the set of all points (x, y) in the unit square $0 \leqslant x,\ y \leqslant 1$ such that the numbers x and y are both irrational, then $a_o(S) = 1$, the area of the square, but $a_i(S) = 0$ because only degenerate polygons are contained in S.

With Peano's definition of inner and outer area, it was easy to establish that

$$\underline{\int_a^b} f(x)\, dx = a_i(O_f) \quad \text{and} \quad \overline{\int_a^b} f(x)\, dx = a_o(O_f) \tag{37}$$

for any non-negative function f on $[a, b]$. In case $a_i(S) = a_o(S)$, the common value is *the area* $a(S)$ of S. If f is integrable, then (37) reduces to

$$\int_a^b f(x)\, dx = a(O_f). \tag{38}$$

At this point the concept of the integral had come full circle, back to its original motivation.

Peano's area is now generally referred to as "Jordan content" because of its definitive treatment in the second edition (1893) of Camille Jordan's influential *Cours d'analyse* [13]. With the minor difference that he uses only polygons that are made up of small squares with horizontal and vertical sides, Jordan defines what he called the *inner content* $c_i(S)$ and *outer content* $c_o(S)$ of a plane set S, equivalent to Peano's inner and outer areas. However, his approach works equally well in all dimensions, so his concept of content ("etendue") simultaneously generalizes the concepts of length, area, and volume ([13], pp. 28–31). He calls the set S *measurable* with content $c(S)$ if $c_i(S) = c_o(S)$.

Jordan proceeds to define the Riemann integral of bounded function f of n real variables defined on a measurable set E in Euclidean n-space ([13], pp. 32–37). Let P be a partition of E into measurable sets $E_1, \ldots, E_m$ with non-overlapping interiors, and let $\bar{p}_i$ be an arbitrary point of E_i, $i = 1, \ldots, m$. Then

$$S(P) = \sum_{i=1}^m f(\bar{p}_i) c(E_i)$$

is a Riemann sum for f on E.

The function f is *integrable* on the set E provided that the limit

$$\int_E f = \lim_{\delta \to 0} \sum_{i=1}^m f(\bar{p}_i) c(E_i) \tag{39}$$

exists, δ being the maximum of the diameters of the sets $E_1, \ldots, E_m$. In the one-dimensional case with E being an interval $[a, b]$, this is

$$\int_a^b f(x)\, dx = \lim_{\delta \to 0} \sum_{i=1}^m f(\bar{x}_i) c(E_i) \tag{40}$$

where, in comparison with Riemann's definition (32), the partition of $[a, b]$ into subintervals has been generalized to a partition of the interval into measurable sets.

Both (38) and (40) were reformulations of the definition of the Riemann integral that, in contrast to (32), are directly susceptible of significant generalization. A detailed discussion of the role of these reformulations as forerunners to the Lebesgue integral (see Chapter 12) is given by Hawkins in Chapter 2–4 of his book on the origins and development of Lebesgue's theory of integration [11].

The Arithmetization of Analysis

The calculus of Newton and Leibniz was a calculus of geometric variables, of quantities explicitly associated with geometric curves, and much of their analysis depended upon intuitive geometric concepts. Euler, Lagrange, and Cauchy attempted to substitute the principles of arithmetic for geometric intuition in the foundations of analysis; not a single geometrical diagram appears in their books on the infinitesimal calculus. However, these first attempts at an arithmetization of the subject were only partially successful because, prior to the late nineteenth century, the real numbers themselves were understood only in an intuitive fashion.

Since the seventeenth century mathematicians had pragmatically used irrational numbers (such as $\sqrt{2}$) in an uncritical way without seriously questioning their precise meaning or nature, relying for computational purposes upon the assumption that any irrational number can be arbitrarily closely approximated by rational numbers (e.g., $\sqrt{2} = 1.41421\ldots$). In particular, it was assumed not only that irrational numbers exist as needed for the ordinary purposes of analysis, but that they obey the same laws of algebraic operation as the familiar rational numbers. Nevertheless, as Richard Dedekind (1831–1916) remarked in an 1872 essay ([7], p. 22), such a simple fact as $\sqrt{2} \cdot \sqrt{3} = \sqrt{6}$ had never been rigorously established.

In the absence of a full understanding of the real number system it was impossible to provide firm foundations for the calculus. For example, the Bolzano-Cauchy proof of the intermediate value theorem for continuous functions required the "bounded monotone sequence property" of the real numbers—to the effect that every bounded sequence $\{a_n\}$ of numbers, that is either increasing ($a_n \leqslant a_{n+1}$ for all n) or decreasing ($a_n \geqslant a_{n+1}$ for all n), is convergent. This same property of the real numbers is needed to establish the sufficiency of the Cauchy convergence criterion, and it was implicitly assumed by both Cauchy and Riemann in their proofs of the existence of the integral under appropriate hypotheses. However, the validity of this basic property of the real numbers was not verified, but merely assumed to be evident on geometrical grounds.

This vagueness in the foundations of analysis resulted not only in logical gaps but also in actual errors on occasion. For example, it was generally thought during the early nineteenth century that every continuous function is differentiable except perhaps at isolated singular points (such as $x = 0$ for the function $f(x) = |x|$); indeed, several calculus texts of this period purported to prove this false proposition. It therefore came as a healthy shock when Karl Weierstrass (1815–1897) exhibited, in his Berlin lectures as early as 1861, a function that is continuous everywhere but differentiable nowhere. His example was the function

$$f(x) = \sum_{n=0}^{\infty} b^n \cos(a^n \pi x)$$

where a is an odd integer and $b \in (0, 1)$ a constant such that $ab > 1 + 3\pi/2$. This infinite series converges uniformly on the real line, so f is continuous everywhere. However, it turns out that given any point x_0 and any positive number M, there exist points x_1 and x_2 arbitrarily close to x_0 such that

$$\frac{f(x_1) - f(x_0)}{x_1 - x_0} > M \quad \text{while} \quad \frac{f(x_2) - f(x_0)}{x_2 - x_0} < -M.$$

It obviously follows that f is *not* differentiable at x_0.

Actually, Bolzano had described an example of a non-differentiable continuous function in 1834, but it had gone unnoticed. It was Weierstrass' example whose impact made clear the necessity of a re-examination of the foundations of analysis. In particular, instead of taking the real number system for granted, as "given", it was necessary that the real numbers be constructed or defined in such a way that the existence and properties of irrational numbers could be rigorously proved. Dedekind later wrote that, when he first lectured on the calculus in 1858, he

> felt more keenly than ever before the lack of a really scientific foundation for arithmetic. In discussing the notion of the approach of a variable magnitude to a fixed limiting value, and especially in proving the theorem that every magnitude which grows continually, but not beyond all limits, must certainly approach a limiting value, I had recourse to geometric evidences. Even now such resort to geometric intuition in a first presentation of the differential calculus, I regard as exceedingly useful, from the didactic standpoint, and indeed indispensable, if one does not wish to lose too much time. But that this form of introduction into the differential calculus can make no claim to being scientific, no one will deny. For myself this feeling of dissatisfaction was so overpowering that I made the fixed resolve to keep meditating on the question till I should find a purely arithmetic and perfectly rigorous foundation for the principles of infinitesimal analysis. The statement is so frequently made that the differential calculus deals with continuous magnitude, and yet an explanation of this continuity is nowhere given; even the most rigorous expositions of the differential calculus do not base their proofs upon continuity but, with more or less consciousness of the fact, they either appeal to geometric notions or those suggested by geometry, or depend upon theorems which are never established in a purely arithmetic manner. Among these, for example, belongs the above-mentioned theorem, and a more careful investigation convinced me that this theorem, or any one equivalent to it, can be regarded in some way as a sufficient basis for infinitesimal analysis. It then only remained to discover its true origin in the elements of arithmetic and thus at the same time to secure a real definition of the essence of continuity ([7], pp. 1–2).

The year 1872 was marked by the almost simultaneous publication of constructions of the real numbers by Dedekind, Georg Cantor (1845–1918), Charles Meray (1835–1911), and Edward Heine (1821–1881); Weierstrass had given an earlier construction in his Berlin University lectures. The constructions of Dedekind and Cantor are those that are generally employed today.

Dedekind's approach was closely related to Eudoxus' definition of proportionality for ratios of geometric magnitudes. We saw in Chapter 1 that, given incommensurable magnitudes a and b, this definition of proportionality serves to separate the set Q of all rational numbers into two disjoint subsets L and U such that every element of L is less than every element of U—the rational number m/n is in L if $m : n < a : b$, and otherwise is in U. Dedekind noted that, similarly, every rational number r partitions Q into two sets A_1 and A_2 such that every element of A_1 is less than every element of A_2. There are actually two possibilities, according to whether r is the largest element of A_1 or the smallest element of A_2, but these two corresponding partitions of Q may be regarded as essentially equivalent.

Dedekind defined a *cut* of the rational numbers as a partition (A_1, A_2) of Q into two non-empty disjoint subsets such that every element of A_1 is less than every element of A_2. Whereas some cuts are generated by rational numbers, others are not. For example, if A_2 consists of all positive rational numbers x such that $x^2 > 2$, while A_1 consists of all other rational numbers, then there is neither a largest element of A_1 nor a smallest element of A_2, because there is no rational number x such that $x^2 = 2$. Intuitively, this cut (A_1, A_2) may be regarded as generated by the irrational number $\sqrt{2}$.

"Whenever", Dedekind says, "we have to do with a cut (A_1, A_2) produced by no rational number, we create a new, an *irrational* number α, which we regard as completely defined by this cut (A_1, A_2); we shall say that the number α corresponds to this cut, or that it produces this cut" ([7], p. 15). In more modern language, the set R of all real numbers is *defined* to be the set of all cuts of the rational numbers, except that the two essentially equivalent cuts produced by a given rational number are identified with each other. Thus a *real number* is a cut $\alpha = (A_1, A_2)$; α is a rational real number if this cut is generated by a rational number, otherwise α is an irrational real number.

Order and algebraic operations are easily defined for real numbers regarded as cuts. If $\alpha = (A_1, A_2)$ and $\beta = (B_1, B_2)$ are different real numbers, we say that $\alpha < \beta$ provided that A_1 is a proper subset of B_1. Dedekind proves that, for any two real numbers α and β, either $\alpha < \beta$ or $\alpha = \beta$ or $\alpha > \beta$. He defines the sum $\gamma = (C_1, C_2) = \alpha + \beta$ as follows: the rational number c is in C_1 if there exist $a_1 \in A_1$ and $b_1 \in B_1$ such that $a_1 + b_1 > c$; otherwise $c \in C_2$ ([7], p. 21).

EXERCISE 20. Define the product of two real numbers α and β. It suffices to consider the case in which α and β are both positive (that is, are greater than the real number generated by the rational number 0).

Dedekind proves what he calls the continuity property of the real number system—every cut of the set of real numbers is generated by some real number. Thus the real numbers are *complete* in that, whereas the real numbers are generated by taking cuts of the rational numbers, no new numbers are generated by taking cuts of the real numbers.

The crucial bounded monotone sequence property is equivalent to this completeness property, but is easily established directly. Let $\{\alpha_n\}_1^\infty = \{(A_n, B_n)\}_1^\infty$ be a sequence of real numbers such that $\alpha_n \leqslant \alpha_{n+1}$ and $\alpha_n \leqslant \mu$ for every n, where $\mu = (M, N)$ is a fixed real number. If A_0 is the union of the increasing sequence of sets $\{A_n\}_1^\infty$ and B_0 is the intersection of the decreasing sequence $\{B_n\}_1^\infty$, then it turns out that $\alpha_0 = (A_0, B_0)$ is a real number such that $\lim_{n\to\infty} \alpha_n = \alpha_0$.

EXERCISE 21. Verify that $\alpha_0 = (A_0, B_0)$ is a cut of the rational numbers. Why is the set B_0 non-empty?

Cantor's approach was based on the idea of a real number α as the limit of a sequence $\{a_n\}_1^\infty$ of rational numbers. In this case $\{a_n\}_1^\infty$ will be a *fundamental sequence* satisfying Cauchy's convergence criterion that $a_m - a_n$ approach 0 as $m, n \to \infty$. Cantor wants to identify the real number α with this fundamental sequence of rational numbers. However, two fundamental sequences $\{a_n\}_1^\infty$ and $\{b_n\}_1^\infty$ will have the same limit if $\lim_{n\to\infty}(a_n - b_n) = 0$, in which case these two sequences are called *equivalent*.

In modern language, Cantor's construction amounts to defining the set of real numbers to be the set of all equivalence classes of fundamental sequences of rational numbers. If r is a rational number, then the sequence $\{r, r, r, \ldots, \}$ represents the real number that corresponds to r. However, the most obvious representative of the real number $\sqrt{2}$ is the sequence $\{1, 1.4, 1.41, 1.414, \ldots, \}$ consisting of its finite decimal approximations.

Cantor's real number is perhaps a more complicated object than Dedekind's real number, but the sequential definitions of the algebraic operations are simpler. Let the real numbers α and β be represented by the fundamental sequences $\{a_n\}_1^\infty$ and $\{b_n\}_1^\infty$. Then the sum $\alpha + \beta$ and product $\alpha\beta$ are represented by the fundamental sequences $\{a_n + b_n\}_1^\infty$ and $\{a_n b_n\}_1^\infty$, respectively. We say that $\alpha > \beta$ provided there is a positive d such that $a_n \geqslant b_n + d$ for n sufficiently large. Cantor proves that his real numbers are complete in the sense that every fundamental sequence of real numbers converges to a real number. In particular, every bounded monotone sequence of real numbers, being a fundamental sequence, converges.

EXERCISE 22. Show that $\{a_n b_n\}_1^\infty$ is a fundamental sequence if $\{a_n\}_1^\infty$ and $\{b_n\}_1^\infty$ are. *Hint*: $a_m b_m - a_n b_n = a_m(b_m - b_n) + b_n(a_m - a_n)$. Use the fact that every fundamental sequence is bounded.

In either approach, Dedekind's or Cantor's, the set Q of rational numbers is taken as the starting point for the construction of the set R of all real numbers. Then Q is enlarged by the addition of irrational numbers which continue to obey the familiar algebraic laws. Most important, the construction of R permits the rigorous verification of the completeness property that, in its various forms, plays a crucial role in infinitesimal analysis.

The construction of the real number system was the principal step in the arithmetization of analysis during the closing third of the nineteenth century. The final loose end was tied by Weierstrass in his purely arithmetical formulation of the limit concept, which previously had been tinged with connotations of continuous motion—it was said that $\lim_{x \to a} f(x) = L$ provided that $f(x)$ approaches L as x approaches a. Weierstrass objected to this "dynamic" description of limits, and replaced it with a "static" formulation involving only real numbers, with no appeal to motion or geometry: $\lim_{x \to a} f(x) = L$ provided that, given $\epsilon > 0$, there exists a number $\delta > 0$ such that $|f(x) - L| < \epsilon$ if $0 < |x - a| < \delta$. With the various types of limits appearing in the calculus reformulated in this way, the arithmetization of analysis was complete, and the calculus had assumed precisely the form in which it appears in twentieth century expositions.

References

[1] G. Birkhoff, *A Source Book in Classical Analysis*. Cambridge, MA: Harvard University Press, 1973.

[2] C. B. Boyer, *The History of the Calculus and its Conceptual Development*. New York: Dover (reprint), 1959, Chapter VII.

[3] A.-L. Cauchy, *Cours D'analyse de l'Ecole Royale Polytechnique, Oeuvres*, Ser. 2, Vol. 3. Paris: Gauthier-Villars, 1897.

[4] A.-L. Cauchy, *Resume des leçons donnees a l'Ecole Royale Polytechnique sur le calcul infinitesimal, Oeuvres*, Ser. 2, Vol. 4. Paris: Gauthier-Villars, 1899.

[5] A.-L. Cauchy, *Leçons sur les calcul differentiel, Oeuvres*, Ser. 2, Vol. 4. Paris: Gauthier-Villars, 1899.

[6] G. Darboux, Memoire sur la theorie des functions discontinues. *Ann Sci Ecole Norm Sup* **4** (2) 57–112, 1875.

[7] R. Dedekind, *Essays on the Theory of Numbers*, New York: Dover (reprint), 1963.

[8] J. Fourier, *The Analytical Theory of Heat* (translated by A. Freeman). New York: Dover (reprint), 1955.

[9] I. Grattan-Guinness, *The Development of the Foundations of Analysis from Euler to Riemann*. MIT Press, 1970.

[10] I. Grattan-Guinness, Bolzano, Cauchy and the 'New analysis' of the early nineteenth century. *Arch Hist Exact Sci* **6**, 372–400, 1969-70.

[11] T. Hawkins, *Lebesgue's Theory of Integration: Its Origins and Development*. New York: Chelsea, 1975, 2nd ed.

[12] R. Iacobacci, *Augustin-Louis Cauchy and the Development of Mathematical Analysis*, Ph. D. Dissertation, New York University, 1965.

[13] C. Jordan, *Cours d' analyse*. Vol. 1. Paris: Gauthier-Villars, 1893, 2nd. ed.

[14] P. E. B. Jourdain, The origin of Cauchy's conceptions of a definite integral and of the continuity of a function. *Isis* **1**, 661–703, 1913.

[15] G. Peano, *Applicazione Geometriche del Calcolo Infinitesimale*. Torino: Bocca, 1887.

[16] B. Riemann, Uber die Darstellbarkeit einer Function durch eine trigono-metrische Reine. *Mathematische Werke*. Leipzig:Teubner, 1892, 2nd ed.

[17] D. J. Struik, *A Source Book in Mathematics 1200–1800*. Cambridge, MA: Harvard University Press, 1969.

[18] C. A. Truesdell, The rational mechanics of flexible or elastic bodies, 1638–1788. Euler's *Opera Omina*, Ser. 2, Vol. 11, Part 2, 1960.

[19] A. P. Youschkevitch, The concept of function up to the middle of the 19th century. *Arch Hist Exact Sci* **16**, 37–85, 1976.

Postscript: The Twentieth Century

12

This brief closing chapter is devoted to two twentieth century develop-ments that have in very different ways served to complete the historical development of the calculus. The comprehensive theory of integration that stems from the work of Henri Lebesgue (1875–1941) is (in a certain technical sense) the ultimate generalization of the concept of the integral for real-valued functions of a real variable. The non-standard analysis of Abraham Robinson (1918–1974) provides at long last a logical foundation for infinitesimals as they were frequently used in the seventeenth and eighteenth centuries.

The Lebesgue Integral and the Fundamental Theorem of Calculus

The "fundamental theorem of calculus" provides a generic formulation of the inverse relationship between differentiation and integration. The de-rivative of the (indefinite) integral of a function is that function,

$$\frac{d}{dx} \int_a^x f(t)\, dt = f(x), \tag{1}$$

and the integral of the derivative of a function is (to within a constant) that function,

$$\int_a^x f'(t)\, dt = f(x) - f(a). \tag{2}$$

The status of the fundamental theorem in Cauchy's theory of integration was quite satisfying—Formulas (1) and (2) held so long as the functions being integrated were continuous, and Cauchy only defined the integral for continuous functions.

With Riemann's more general definition of the integral for discontinuous functions, however, the fundamental theorem lost its completeness of scope. Formula (2) is not meaningful for a differentiable function whose derivative is not Riemann integrable, and the classical proof establishes Formula (1) only at points of continuity of the function f. The restoration of the fundamental theorem to a satisfactory status was one of the first fruits of the new theory of integration that Lebesgue introduced in his 1902 doctoral thesis [4] and expanded in his 1904 book [5] and subsequent papers. An excellent outline of Lebesgue's work is given by Hawkins ([1], Chapter 5).

In Jordan's approach to Riemann integration, the lower and upper Riemann integrals were defined by

$$R\underline{\int}_a^b f(x)\ dx\ =\ lub \sum_{i=1}^n m_i c(E_i) \tag{3}$$

and

$$R\overline{\int}_a^b f(x)\ dx\ =\ glb \sum_{i=1}^n M_i c(E_i), \tag{4}$$

where $\{E_1, \ldots, E_n\}$ denotes a partition of $[a, b]$ into Jordan measurable sets with Jordan contents $c(E_i)$, and m_i and M_i denote the greatest lower bound and least upper bound, respectively, of $f(x)$ for $x \in E_i$. Lebesgue's basic idea was to enlarge the class of functions for which the integral is defined by enlarging the class of sets that are measurable. If the measure $m(E)$ is defined for a class of sets that properly includes the Jordan measurable sets, and $m(E) = c(E)$ if E is Jordan measurable, then the integral can be generalized by replacing the sets E_i in (3) and (4) by these new measurable sets.

Lebesgue generalized the concept of measure by basing it on countably infinite rather than finite coverings. Given a subset E of the real line, he defined its *outer measure* $m_o(E)$ as the greatest lower bound of the sums

$$\sum_{n=1}^\infty l(I_i)$$

where $\{I_n\}_1^\infty$ is a sequence of intervals whose union contains E. If $E \subset [a, b]$, then the *inner measure* of E is defined by

$$m_i(E) = (b - a) - m_o([a, b] - E).$$

The bounded set E is called (*Lebesgue*) *measurable* with *measure* $m(E)$ provided that $m_o(E) = m_i(E) = m(E)$. Since it is clear that

$$c_i(E) \leqslant m_i(E) \leqslant m_o(E) \leqslant c_o(E),$$

it follows that the set E is Lebesgue measurable if it is Jordan measurable, in which case $m(E) = c(E)$. Although for convenience we have defined Lebesgue measure only for bounded sets, this restriction is actually unnecessary.

The principal advantage of Lebesgue measure over previous measure concepts is that it is countably additive: If $\{E_n\}_1^\infty$ is a sequence of mutually disjoint measurable sets, then their union is measurable with

$$m\left(\bigcup_{n=1}^\infty E_n \right) = \sum_{n=1}^\infty m(E_n).$$

For example, if Q is the (countable) set of rational numbers in the unit interval, it follows immediately that Q is measurable with $m(Q) = 0$. It is a simple consequence of countable additivity that

$$m\left(\bigcup_{n=1}^\infty E_n \right) = \lim_{n \to \infty} m(E_n) \quad \text{if } E_n \subset E_{n+1} \tag{5}$$

for each n, while

$$m\left(\bigcap_{n=1}^\infty E_n \right) = \lim_{n \to \infty} m(E_n) \quad \text{if } E_{n+1} \subset E_n \tag{6}$$

for each n. Also, if E and F are measurable sets with $F \subset E$, then the difference $E - F$ is measurable with

$$m(E - F) = m(E) - m(F).$$

The lower and upper Lebesgue integrals $L \underline{\int_a^b} f(x)\, dx$ and $L \overline{\int_a^b} f(x)\, dx$ of a bounded function f are defined just as the lower and upper Riemann integrals (3) and (4), except that now $\{E_1, \ldots, E_n\}$ is a partition of $[a, b]$ into Lebesgue measurable sets and $c(E_i)$ is replaced by $m(E_i)$. Then f is Lebesgue integrable on $[a, b]$ provided that these lower and upper integrals are equal. Since it is clear that

$$R\underline{\int_a^b} f(x)\, dx \leqslant L \underline{\int_a^b} f(x)\, dx \leqslant L \overline{\int_a^b} f(x)\, dx \leqslant R \overline{\int_a^b} f(x)\, dx,$$

it follows that the function f is Lebesgue integrable on $[a, b]$ if it is Riemann integrable on $[a, b]$, in which case its Lebesgue and Riemann integrals are equal.

Lebesgue proved that a bounded function on a closed interval is Riemann integrable if and only if its set of discontinuities has measure zero, whereas every bounded measurable function on a closed interval is Lebesgue integrable. A function is called *measurable* if the inverse image of each open interval is a measurable set. For example the Dirichlet function on $[0, 1]$ is not Riemann integrable because its set of discontinuities is the whole interval (of measure $1 \neq 0$), but it is Lebesgue integrable because it is obviously measurable (why?).

Lebesgue's proof that every bounded measurable function is (Lebesgue) integrable was based on the new idea of partitioning the range of a function (rather than its domain, à la Cauchy and Riemann) into subintervals. If $m \leqslant f(x) \leqslant M$ for $x \in [a, b]$, let the points

$$m = y_0 < y_1 < \cdots < y_n = M$$

partition the interval $[m, M]$ into n equal subintervals each having length $(M - m)/n$. If E_i denotes the set of points $x \in [a, b]$ such that $y_{i-1} \leqslant f(x) < y_i$, then the sets $\{E_i\}$ are measurable if f is measurable. The lower and upper Lebesgue sums corresponding to the partition $\{E_1, \ldots, E_n\}$ of $[a, b]$ are

$$\sum_{i=1}^{n} y_{i-1} m(E_i) \quad \text{and} \quad \sum_{i=1}^{n} y_i m(E_i).$$

Because the difference

$$\sum_{i=1}^{n} (y_i - y_{i-1}) m(E_i) = \frac{M-m}{n} \sum_{i=1}^{n} m(E_i) = \frac{1}{n}(M-m)(b-a)$$

of these sums can be made arbitrarily small by taking n sufficiently large, it follows that the lower and upper Lebesgue integrals of f are equal, so the bounded measurable function f is Lebesgue integrable on $[a, b]$.

In addition to the fact that the class of integrable functions has been enlarged considerably, the power of the Lebesgue integral results from the ease with which it handles limits of functions. Suppose that $\lim_{n \to \infty} f_n(x) = f(x)$ for each $x \in [a, b]$, and we ask whether

$$\lim_{n \to \infty} \int_a^b f_n(x) \, dx = \int_a^b f(x) \, dx. \tag{7}$$

The only easy result for the Riemann integral is that (7) holds if the functions $\{f_n\}_1^\infty$ are continuous and the convergence is uniform (otherwise f may not even be Riemann integrable). But these conditions are too strong for many applications; indeed, we have seen in previous chapters numerous examples of "termwise integration" under weaker conditions. For the Lebesgue integral it turns out that (7) holds under very weak conditions; this is the main reason for the modern prominence of Lebesgue's theory of integration.

It is a consequence of elementary properties of measurable sets that the pointwise limit f of a convergent sequence $\{f_n\}_1^\infty$ of measurable functions is itself a measurable function (see Royden [7], p. 56). If in addition f is bounded on $[a, b]$, then it follows that f is Lebesgue integrable on $[a, b]$. Lebesgue's *bounded convergence theorem* asserts that if $\{f_n\}_1^\infty$ is a convergent sequence of measurable functions such that $|f_n(x)| \leqslant K$ for some fixed

K and each $x \in [a, b]$, and $f(x) = \lim_{n \to \infty} f_n(x)$, then

$$\lim_{n \to \infty} \int_a^b f_n(x)\, dx = \int_a^b f(x)\, dx \tag{7}$$

(the integrals here being Lebesgue integrals). Thus the limit of the sequence of integrals is "what it ought to be" if, in addition to being measurable, the functions $\{f_n\}^\infty$, are "uniformly bounded" on $[a, b]$. This illustrates the nice convergence properties (some of them even stronger) that the Lebesgue integral enjoys.

For example, let $\{r_n\}_1^\infty$ be the set of rational numbers in the interval $[0, 1]$ and define

$$\phi_n(x) = \begin{cases} 0 & \text{if } x \in \{r_1, r_2, \ldots, r_n\}, \\ 1 & \text{otherwise.} \end{cases}$$

Then it is clear that each ϕ_n is measurable with $\int_0^1 \phi_n(x)\, dx = 1$. But $\phi(x) = \lim_{n \to \infty} \phi_n(x)$ is the Dirichlet function, and Lebesgue's bounded convergence theorem now gives

$$\int_0^1 \phi(x)\, dx = \lim_{n \to \infty} \int_0^1 \phi_n(x)\, dx = 1.$$

The proof of the bounded convergence theorem is notable for its simplicity. Let $\{f_n\}^\infty$ be a convergent sequence of measurable functions such that $|f_n(x)| \leqslant K$ for all n and $x \in [a, b]$. Given $\epsilon > 0$, denote by E_n the set of all points $x \in [a, b]$ such that $|f_m(x) - f(x)| \geqslant \epsilon$ for some $m \geqslant n$. Then $\{E_n\}_1^\infty$ is a decreasing sequence of measurable sets, and $\cap_{n=1}^\infty E_n$ is empty because $f_n(x)$ converges to $f(x)$ for all $x \in [a, b]$. Therefore (6) implies that $\lim_{n \to \infty} m(E_n) = 0$. Since $|f_n(x) - f(x)| < \epsilon$ unless $x \in E_n$, and $|f_n(x) - f(x)| \leqslant 2K$ everywhere, it follows that

$$\left| \int_a^b f_n(x)\, dx - \int_a^b f(x)\, dx \right| \leqslant \int_a^b |f_n(x) - f(x)|\, dx$$
$$< 2Km(E_n) + \epsilon(b - a).$$

Since $\lim_{n \to \infty} m(E_n) = 0$ and $\epsilon > 0$ can be taken arbitrarily small, it follows that $\lim_{n \to \infty} \int_a^b f_n(x)\, dx = \int_a^b f(x)\, dx$ as desired.

The bounded convergence theorem is the tool that is needed to establish the fundamental theorem of calculus for the Lebesgue integral.

Theorem A. *If f is differentiable and f' is bounded on $[a, b]$, then f' is Lebesgue integrable, and*

$$\int_a^b f'(x)\, dx = f(b) - f(a). \tag{2}$$

PROOF. Let $g_n(x) = [f(x + h_n) - f(x)]/h_n$ with $h_n = 1/n$. Then $\{g_n\}_1^\infty$ is a uniformly bounded sequence of measurable functions converging to f' on

$[a, b]$, so the bounded convergence theorem gives

$$\int_a^b f'(x)\, dx = \lim_{n\to\infty} \int_a^b g_n(x)\, dx$$

$$= \lim_{n\to\infty} \frac{1}{h_n} \int_a^b [f(x+h_n)-f(x)]\, dx$$

$$= \lim_{n\to\infty} \frac{1}{h_n} \int_b^{b+h_n} f(x)\, dx - \lim_{n\to\infty} \frac{1}{h_n} \int_a^{a+h_n} f(x)\, dx$$

$$\int_a^b f'(x)\, dx = f(b) - f(a)$$

because the function f, being differentiable, is continuous.   □

Theorem B. *Let f be bounded and measurable on $[a, b]$, and define*

$$F(x) = \int_a^x f(t)\, dt.$$

Then there exists a set $E \subset [a, b]$ having measure zero such that

$$F'(x) = f(x) \qquad\qquad (1)$$

for all x not in E. That is, (1) holds "almost everywhere."

PROOF. Let $f_+(x) = \max\{f(x), 0\}$ and $f_-(x) = \max\{-f(x), 0\}$. Then f_+ and f_- are non-negative functions such that $f = f_+ - f_-$. Then

$$F(x) = \int_a^x f_+(t)\, dt - \int_a^x f_-(t)\, dt = F_1(x) - F_2(x),$$

where F_1 and F_2 are monotone non-decreasing functions. But it is a fact independent of integration that every monotone function is differentiable almost everywhere (see Royden [7], p. 82). Hence $F'(x)$ exists almost everywhere. A slight generalization of the proof of Theorem A now yields

$$\int_a^c F'(x)\, dx = F(c) - F(a) = \int_a^c f(x)\, dx,$$

so

$$\int_a^c [F'(x) - f(x)]\, dx = 0$$

for all $c \in [a, b]$. By an elementary property of the Lebesgue integral ([7], p. 87, Lemma 7), it therefore follows that $F'(x) = f(x)$ except possibly on a set of measure zero.   □

Theorems A and B together say that differentiation and Lebesgue integration are inverse operations on very large classes of functions. If f is a bounded measurable function on $[a, b]$, then

$$\frac{d}{dx} \int_a^x f(t)\, dt = f(x) \qquad\qquad (1)$$

except possibly on a set of measure zero. If f is a bounded measurable function with $f(a) = 0$, and its derivative f' exists and is bounded on $[a, b]$, then

$$\int_a^x f'(t)\, dt = f(x). \tag{2}$$

These two versions of the fundamental theorem of calculus provide a definitive rigorous formulation of the inverse relationship between differentiation and integration that Newton and Leibniz discovered on conceptual grounds and exploited in the seventeenth century.

Non-standard Analysis—The Vindication of Euler?

For most of the past century, since Weierstrass, students of the calculus have been assiduously taught that infinitesimals do not exist, and must not be mentioned in formal mathematical discourse. But in 1960 Abraham Robinson proved that infinitesimals *do* exist as genuine mathematical objects, and can serve as the basis for an alternative rigorous development of the calculus! His 1966 book *Non-standard Analysis* [6] showed how to develop much of modern analysis in terms of infinitesimals, and in 1976 a "non-standard" introductory calculus textbook [2] by H. J. Keisler appeared. Keisler's monograph *Foundations of Infinitesimal Calculus* [3] is the best introduction to infinitesimals and non-standard calculus at an intermediate level.

Recall that the ordered field R of real numbers is complete—it satisfies the least upper bound axiom. In fact, every complete ordered field is isomorphic to the field R of real numbers. Non-standard analysis is based on the fact that there exists a (non-complete) field R^* of "hyperreal" numbers that contains R as a proper subfield (Axiom 1 below), such that every function $f(x_1, x_2, \ldots, x_n)$ of n real variables has a natural extension f^* which is a function of n "hyperreal" variables (Axiom 2), and such that if two systems of formulas have the same real solutions then they have the same "hyperreal" solutions (Axiom 3). Axioms 1–3 are established in Section 1E of [3].

Axiom 1. There exists a proper ordered field extension R^* of the field R of real numbers.

The elements of R^* are called *hyperreal numbers*. An element $x \in R^*$ is called *infinitesimal* if $|x| < r$ for every positive real number r; *finite* if $|x| < r$ for some positive real number r; *infinite* if $|x| > r$ for all real numbers r. Both infinitesimal and infinite hyperreal numbers actually exist ([3], p. 7, Theorem 8). Sums, differences, and products of infinitesimals are infinitesimal, as is the reciprocal of an infinite number, whereas the product of an infinitesimal and a finite number is infinitesimal ([3], p. 4, Theorem 3).

Two elements $x, y \in R^*$ are said to be *infinitely close*, written $x \approx y$, provided their difference $x - y$ is infinitesimal. According to the "standard part theorem" ([3], p. 5) every *finite* hyperreal number x is infinitely close to a unique real number r. This unique $r \approx x$ is called the *standard part* of x, written $r = \operatorname{st}(x)$. If x and y are finite then $x \approx y$ if and only if $\operatorname{st}(x) = \operatorname{st}(y)$, and

(i) $\operatorname{st}(x \pm y) = \operatorname{st}(x) \pm \operatorname{st}(y)$,
(ii) $\operatorname{st}(xy) = \operatorname{st}(x)\operatorname{st}(y)$,
(iii) $\operatorname{st}(x/y) = \operatorname{st}(x)/\operatorname{st}(y)$ if $\operatorname{st}(y) \neq 0$,
(iv) $\operatorname{st}(\sqrt[n]{x}) = \sqrt[n]{\operatorname{st}(x)}$.

> **Axiom 2** (Function Axiom). Let f be a real-valued function defined on some subset of the set R^n of all n-tuples of real numbers. Then to f there corresponds a hyperreal-valued function f^* on n hyperreal variables, called the *natural extension* of f. The field operations of R^* are the natural extensions of the field operations of R.

The domain definition of f^* is the natural extension of the domain of definition of f. Given a set $X \subset R$, the *natural extension* of X is defined as follows. Consider a (finite) system F of formulas that has X as its set of real solutions (intuitively, a formula is simply an equality or inequality of functions—see ([3], p. 10) for the precise definition)). If F^* is the system of natural extensions of the formulas in F, then the *natural extension* of X is the set of hyperreal solutions of the system F^*. By Axiom 3 below, X^* is independent of the particular choice F of a system of formulas having the set X as its set of solutions. Natural extension of sets preserves the usual set operations,

$$(X \cup Y)^* = X^* \cup Y^*, \qquad (X \cap Y)^* = X^* \cap Y^*,$$
$$X \subset Y \quad \text{if and only if} \quad X^* \subset Y^*.$$

If X is a *bounded* set of real numbers, then X^* consists of *finite* hyperreal numbers. If f is a real-valued function of one real variable, then

$$(\text{domain } f)^* = \text{domain}(f^*) \quad \text{and} \quad (\text{range } f)^* = \text{range}(f^*).$$

> **Axiom 3** (Solution Axiom). If two systems of formulas have exactly the same real solutions, then their natural extensions have exactly the same hyperreal solutions.

For example, let f and g be real-valued functions defined on the set $D \subset R$, with $f(x) \leqslant g(x)$ for all $x \in D$. Then the two formulas $f(x) = g(x)$ and $f(x) \leqslant g(x)$ have the same set of real solutions, namely D. Hence the two formulas $f^*(x) = g^*(x)$ and $f^*(x) \leqslant g^*(x)$ have the same set D^* of hyperreal solutions. Thus it follows from the solution axiom that $f(x) \leqslant g(x)$ for all $x \in D$ implies $f^*(x) \leqslant g^*(x)$ for all $x \in D^*$.

When no confusion will result, we write f in place of f^*. The function f of a real variable is *differentiable* at $a \in R$ provided that the quotient

$$\frac{f(a+\Delta x)-f(a)}{\Delta x}$$

is finite and has the same standard part for every non-zero infinitesimal $\Delta x \approx 0$. Its *derivative* at a is then

$$f'(a) = \text{st}\left(\frac{f(a+\Delta x)-f(a)}{\Delta x}\right). \tag{8}$$

Of course this definition turns out to be equivalent to the usual one in terms of limits of real numbers. But (8) can be applied directly to calculate derivatives by taking standard parts instead of using limits. For example,

$$\frac{d(\sqrt{x})}{dx} = \text{st}\left(\frac{\sqrt{x+\Delta x} - \sqrt{x}}{\Delta x}\right) = \text{st}\left(\frac{x+\Delta x - x}{\Delta x(\sqrt{x+\Delta x} + \sqrt{x})}\right)$$

$$= \text{st}\frac{1}{\sqrt{x+\Delta x} + \sqrt{x}} = \frac{1}{\text{st}(\sqrt{x+\Delta x} + \sqrt{x})}$$

$$= \frac{1}{\text{st}(\sqrt{x+\Delta x}) + \sqrt{x}} = \frac{1}{2\sqrt{x}}.$$

If we write $y = f(x)$, $\Delta y = f(x+\Delta x) - f(x)$, then

$$\Delta y = f'(x)\Delta x + \epsilon\Delta x$$

where $\epsilon = (\Delta y/\Delta x) - f'(x)$ is an infinitesimal if Δx is, so $\Delta y \approx 0$ also. The non-standard derivation of the product rule for $y = uv$ is

$$\frac{\Delta y}{\Delta x} = \frac{(u+\Delta u)(v+\Delta v) - uv}{\Delta x} = u\frac{\Delta v}{\Delta x} + v\frac{\Delta u}{\Delta x} + \Delta u\frac{\Delta v}{\Delta x},$$

$$\frac{dy}{dx} = u\,\text{st}\left(\frac{\Delta v}{\Delta x}\right) + v\,\text{st}\left(\frac{\Delta u}{\Delta x}\right) + 0\,\text{st}\left(\frac{\Delta v}{\Delta x}\right)$$

$$= u\frac{dv}{dx} + v\frac{du}{dx}$$

because $\Delta u \approx 0$. To derive the chain rule for $y = g(x)$, $x = f(t)$, first write

$$\Delta y = g'(x)\Delta x + \epsilon\Delta x$$

where $\epsilon \approx 0$. Then

$$\frac{\Delta y}{\Delta t} = g'(x)\frac{\Delta x}{\Delta t} + \epsilon\frac{\Delta x}{\Delta t}$$

so, taking standard parts, we obtain

$$\frac{dy}{dt} = g'(x) \, \text{st}\left(\frac{\Delta x}{\Delta t}\right) + 0 \, \text{st}\left(\frac{\Delta x}{\Delta t}\right) = g'(f(t))f'(t).$$

Thus the non-standard proof is simpler than the standard one.

To see that $f'(a)=0$ at a local maximum, let Δx be a positive infinitesimal. Then

$$f(a+\Delta x) \leqslant f(a), \qquad f(a-\Delta x) \leqslant f(a)$$

so

$$\frac{f(a+\Delta x)-f(a)}{\Delta x} \leqslant 0 \leqslant \frac{f(a-\Delta x)-f(a)}{-\Delta x}.$$

Taking standard parts then gives $f'(a) \leqslant 0 \leqslant f'(a)$.

We say the real function f is *continuous* at $a \in R$ if $f(x) \approx f(a)$ for all $x \approx a$, that is, $f(x)$ is infinitely close to $f(a)$ when x is infinitely close to a (just as Cauchy said it!). The function f is *uniformly continuous* on the set $S \subset R$ if $f(x) \approx f(y)$ for all $x, y \in S^*$ such that $x \approx y$. The non-standard proof, that f is uniformly continuous on the closed and bounded set S if it is continuous at each point of S, is almost trivial (in contrast with the standard proof). Consider $x, y \in S^*$ with $x \approx y$. Then x and y are finite because S is bounded. From the fact that S is closed it follows easily that $a = \text{st}(x) = \text{st}(y)$ is a point of S. Then $f(x) \approx f(a)$ and $f(y) \approx f(a)$ because f is continuous at a, so $f(x) \approx f(y)$ as desired. Of course it turns out that the above definitions of continuity and uniform continuity are equivalent to the standard ones ([3], Chapter 5).

The non-standard definition of the integral requires the notion of a *hyperinteger*. The set of hyperintegers is the natural extension Z^* of the set Z of (real) integers. A real number is an integer if and only if it is a hyperinteger, and every finite hyperinteger is an integer.

Given a *bounded* function f on the (real) interval $[a, b]$, let $\Delta x = (b-a)/n$ where n is a (real) positive integer. The usual lower and upper Riemann sums for f,

$$L(n) = \sum_{i=1}^{n} m_i \, \Delta x \quad \text{and} \quad U(n) = \sum_{i=1}^{n} M_i \, \Delta x,$$

corresponding to the partition of $[a, b]$ into n equal subintervals each having length Δx, may be regarded as functions defined on the set Z_+ of positive integers. Their natural extensions L^* and U^* are defined on the set Z_+^* of positive hyperintegers, and are finite-valued because $f(x)$ is bounded on $[a, b]$. Then f is *Riemann integrable* on $[a, b]$ if, for any *infinite* positive hyperinteger N, the values $L^*(N)$ and $U^*(N)$ are infinitely close, $L^*(N) \approx U^*(N)$, and their common standard part

$$\text{st}(L^*(N)) = \text{st}(U^*(N))$$

is independent of the infinite hyperinteger N.

A suggestive notation for $L^*(N)$ and $U^*(N)$ is

$$L^*(N) = \sum_a^b m_f\, dx \quad \text{and} \quad U^*(N) = \sum_a^b M_f\, dx$$

where $dx = (b-a)/N \approx 0$. These are called the *infinite lower and upper Riemann sums* of f with respect to the infinitesimal dx; each may be visualized as the sum of the areas of infinitely many vertical strips having infinitesimal width dx. The value of the integral is then

$$\int_a^b f(x)\, dx = \text{st}\left(\sum_a^b m_f\, dx \right) = \text{st}\left(\sum_a^b M_f\, dx \right).$$

There is a non-standard proof that the function f is integrable if it is continuous ([3], p. 96, Lemma 1).

The following proof of the fundamental theorem of calculus, in the form that says that

$$\int_a^b f(x)\, dx = F(b) - F(a)$$

if $F' = f$ and f is integrable on $[a, b]$, is the non-standard version of a standard proof that was given by Darboux in 1875 (see the references to Chapter 11). Let $\Delta x = (b-a)/n$ where n is a (real) positive integer, and consider the partition of $[a, b]$ into n equal subintervals. Application of the mean value theorem to F on the ith subinterval $[x_{i-1}, x_i]$ gives

$$F(x_i) - F(x_{i-1}) = F'(\bar{x}_i)\, \Delta x = f(\bar{x}_i)\, \Delta x$$

for some $\bar{x}_i \in [x_{i-1}, x_i]$. Hence

$$m_i\, \Delta x \leqslant F(x_i) - F(x_{i-1}) \leqslant M_i\, \Delta x.$$

Summation of these inequalities for $i = 1, 2, \ldots, n$ yields

$$\sum_{i=1}^n m_i\, \Delta x \leqslant \sum_{i=1}^n \left[F(x_i) - F(x_{i-1}) \right] \leqslant \sum_{i=1}^n M_i\, \Delta x,$$
$$L(n) \leqslant F(b) - F(a) \leqslant U(n)$$

for every $n \in Z_+$. It then follows from the solution axiom (see the remark following it above) that

$$L^*(N) \leqslant F(b) - F(a) \leqslant U^*(N)$$

for every $N \in Z_+^*$. Taking standard parts with N an infinite hyperinteger, we therefore obtain

$$\int_a^b f(x)\, dx \leqslant F(b) - F(a) \leqslant \int_a^b f(x)\, dx,$$

because the hypothesis that f is integrable means that $\text{st}(L^*(N)) = \text{st}(U^*(N)) = \int_a^b f(x)\, dx$.

The title of this final section of our long account of the history of the calculus should not be taken too literally. It is true, as the above discussion suggests, that non-standard analysis can be employed to convert most of the intuitive infinitesimal arguments of the seventeenth and eighteenth centuries into logically precise arguments. But this is an *a posteriori* interpretation in terms of twentieth century mathematical thought rather than a "vindication" of the seventeenth and eighteenth centuries on their own terms. My own view is that non-standard analysis is a significant development of contemporary mathematics with more implications for the future than for the past. Nevertheless, the non-standard analysis of our time has given clearcut answers to some of the most venerable questions of our subject, and it cannot be denied that this success provides both a palpable satisfaction and a perfect ending for a book on the historical development of the calculus.

References

[1] T. Hawkins, *Lebesgue's Theory of Integration: Its Origins and Development.* New York: Chelsea, 1975, 2nd. ed.
[2] H. J. Keisler, *Elementary Calculus.* Boston: Prindle, Weber & Schmidt, 1976.
[3] H. J. Keisler, *Foundations of Infinitesimal Calculus.* Boston: Prindle, Weber & Schmidt, 1976.
[4] H. Lebesgue, Integrale, longueur, aire. *Ann Mat* 7 (3) 231–359, 1902.
[5] H. Lebesgue, *Leçons sur l'integration et la recherche des fonctions primitives.* Paris: Gauthier-Villars, 1904.
[6] A. Robinson, *Non-standard Analysis.* Amsterdam, London: North-Holland, 1966.
[7] H. L. Royden, *Real Analysis.* New York: Macmillan, 1964.

Index

Adelard of Bath 85
al-Haitham (Alhazen) 83—84
al-Khowarizmi 82—83
Apollonius of Perga 77
analytic geometry 95—97
application of areas 11—12
Arabian mathematics 81—85
Arbogast, Louis 303—304
Archimedes 8, 14, 19, 26, 29—76, 77, 86,
    91, 98, 101, 103, 109, 122
  approximation to π 31—35
  area of ellipse 40—42
  axiom of 14, 16
  convexity axioms 44—46
  discovery method 8, 44, 68—74
  method of compression 31, 54
  quadrature of parabola 35—39
  solids of revolution 62—67
  the spiral 54—62
  volume and area of sphere 42—54
Archimedes-Eudoxus axiom, principle 14,
    16
arclength computations 217—222,
    242—243, 257
area, concept of 2, 15—16, 327—328,
    336—337
area and volume methods of
  Alhazen 83—85
  Archimedes 29—76

Babylonians and Egyptians 2—4
Cavalieri 104—109
Democritus 8—9
Euclid-Eudoxus 16—27
Fermat 116—117
Hippocrates 7—8
Kepler 101—103
Leibniz 231—267
Newton 189—230
Pascal 113, 239—241
Torricelli 116—117
Wallis 113—116
Aristotle 86

Babylonian mathematics 3—5
Barrow, Isaac 132—134, 138—140, 190,
    244
Berkeley, George 293—295
Bernoulli, Daniel 301
Bernoulli, James 268
Bernoulli, John 263, 268—269, 280, 290
Bernoulli's series 290
binomial series 167—168, 178—186, 222,
    284—285, 312
Boethius 81
Bolzano, Bernard 308—309, 325, 330
Briggs, Henry 153—154, 282
Bürgi, Jost 152

Cantor, Georg 330−332
Cauchy, A. L. 296, 309−322
   derivative 313
   fundamental theorem of calculus
      320−321
   integral 317−320
   intermediate value theorem 311−312
   limits and continuity 310−311
   mean value theorem 314−315
Cauchy sequence 308
Cavalieri, Bonaventura 104−109
Cavalieri's principle 9, 104−105
centroid of
   paraboloid 73
   triangle 69
chain rule 196−199, 313, 343
characteristic triangle 218, 239−244
circle
   area of 7−8, 17−19, 31−35, 170
   circumference of 31−35
circle method, Descartes' 125−127
cissoid 176−178, 212, 219−220
common logarithms 153−154, 161
composition of instantaneous motions 55,
   134−137
compression, method of 31, 54
cone
   surface area of 46−49
   volume of 8−10, 23−24, 105
continuity 302−304, 307−308, 310−311,
   344
convexity axioms 44−46
cosine series 205−206, 276
Cotes, Roger 287
cubic and quartic equations, solution of
   93−94
cycloid 135−136, 207, 239−240,
   250−251

d'Alembert, Jean 295−296, 301−302
Darboux, Gaston 326, 345
dark ages, mathematics in 80−81
Dedekind, Richard 14, 329−331
Democritus 8−9
derivative, the 190, 295−297, 313, 343
Descartes, Rene 95−97, 125−127, 168,
   190
differential 253−254, 260−263,
   293−294, 295−296
Diophantus 82

Dirichlet, P. G. Lejeune 307−308,
   322−323
discovery method of Archimedes 8, 44,
   68−74
dissection methods 2

e, the number 273
Egyptian geometry 1−2
ellipse
   area of 40−42
   arclength of 220−221
   tangent to 137
epistola posterior 179−180, 183−185,
   222−223, 226−227
epistola prior 178−179, 222−223, 245
Euclid, the Elements 10−12, 14−15, 81,
   83, 85, 190
Eudemus 7
Eudoxus 10, 13−15, 16−18, 327, 331
   geometric proportions 13−15
   method of exhaustion 15−18
Euler, Leonhard 268−281, 292, 301−302
   complex logarithms 280−281
   function concept 270−271
   differential formulas 277−280
   exponential and logarithmic functions
      272−274
   infinite series expansions 275−277
   Taylor's series 292
Euler's formula 276
Eutocius 68, 78
exhaustion, method of 7, 16
exponential function 272−273, 279
exponential series 205, 273, 317
exponents 115, 143−144, 164, 168, 179

Fermat, Pierre de 95−97, 110−111, 116,
   122−125, 189−190
figurate numbers 111, 172, 237−238
fluxions 191−194, 210, 293−294
Fourier, Joseph 304−307
folium of Descartes 130, 133
Franco of Liege 81
function, concept of 143, 270−271,
   301−304, 306−308
fundamental theorem of calculus 120,
   138−140, 190, 194−196,
   257−258, 317, 320−321,
   335−336, 339−341, 345

Galilei, Galileo 90, 138
Gauss, C. F. 309, 322
geometric algebra 11−12, 79−80
geometric series 22, 39, 91−92, 186
Gerard of Cremona 85
Gerbert (Pope Sylvester II) 81
Greek mathematics, the decline of 78−80
Gregory, James 140−141, 284−289
Gregory's series 277
Gregory St. Vincent 154−156, 159

harmonic series 92
harmonic triangle 237−238
Heine, Edward 330
Herodotus 1
Hipparchus 78
Hippocrates of Chios 7−8
Hobbes, Thomas 176
Hudde, Johann 127−129
Hudde's rule 127−129
Huygens, Christiaan 98, 236−239, 242

incommensurable magnitudes 10−13
indivisibles, concept and use of 8−9, 31,
    69, 74, 86, 89, 102−104, 110, 113,
    246
infinite series techniques and the calculus
    166−167, 187
infinitely large or small magnitudes 28, 86,
    132, 226, 245, 264−265, 272, 275,
    293−294, 310
infinitesimals 226, 252, 256−257,
    264−265, 310, 341
integral, the 253, 257, 260, 306, 309−310,
    317−320, 323−328, 337−338,
    344−345
  definition of
  Cauchy 317−320
  Jordan 328
  Lebesgue 337
  Leibniz 257
  Riemann 323−324
integration by substitution 199−200,
    210−212
intermediate value theorem 298, 303,
    308−309, 311−312
interpolation 147−148, 281−287
inverse sine series 205−207
inverse tangent series 277

Jordan, Camille 328

Kepler, Johann 99−103
  laws of planetary motion, 99−101

Lagrange, J. L. 296−299
Lagrange remainder formula 298−299, 316
latitude of forms 87
Lebesgue, Henri 335−340
Leibniz, Gottfried Wilhelm 141, 166,
    178−179, 189−190, 222−223,
    231−267, 271, 280, 295
  characteristic triangle 239−244
  harmonic triangle 237−238
  higher-order differentials 260−262
  infinitesimals 264−265
  invention of calculus 252−258
  series 247−248
  sums and differences 234−238
  transmutation method 245−251
Leibniz-Newton correspondence 178−186,
    222−224
Leibniz' series 223, 247−248
lever, law of 69−70
L'Hospital, G. F. A. de 269
L'Hospital's rule 269, 315
limit concept 7, 16, 27, 75, 105, 124,
    225−226, 295−296, 310, 333
logarithmic function 272, 278, 280−281
logarithms 142−165
  common 153−154, 161
  hyperbolic 154−158
  laws of 149, 153
  Mercator's computations 161−162
  Napier's definition 148−150
  Napier's motivation 143−144
  Napier's tables 144−148
  natural 144, 148−149, 152−153, 164
  Newton's computations 158−161

Maclaurin, Colin 291−292, 295
maximum-minimum methods
  Cauchy 315
  Fermat 122−125
  Hudde 129
  Leibniz 258−259
  Maclaurin 291−292
  Newton 209
maximum-minimum value theorem 298,
    314

mean value theorem 314−315
medieval infinite series summations 91−93
medieval speculations on motion and
    variability 86−90
Meray, Charles 330
Mercator, Nicolas 161−164, 200
Mercator's series 158, 162−163, 167, 187,
    200, 250, 274
Merton rule 87−89

Napier, John 142−153
natural logarithms 144, 148−149,
    152−153, 164
Neil, William 118−120
Newton, Isaac 113, 132, 139, 141,
    158−160, 162, 166−169,
    178−187, 189−230, 270−271,
    285−286, 289−290
  arclength computations 217−222
  binomial series 167−168, 178−186, 222
  fundamental theorem of calculus
      194−196
  integral tables 199−200, 212−215
  interpolation 283−286
  introduction of fluxions 191−194
  logarithmic computations 158−161
  Newton's method 201−203
  numerical integration 286
  Principia Mathematica, 224−226
  reversion of series 204−205
  sine and cosine series 205−207
  Taylor's series 289−290
  the "prime theorem" 226−229
Newton-Leibniz correspondence 178−186,
    222−224
Newton's method 201−203
Niomachus 81
non-standard analysis 341−346
number, concept of
  of Cantor 332
  of Dedekind 329−332
  Greek 2, 10, 12, 79
  non-standard 341
numerical integration 286−287

Oldenburg, Henry 178, 222
Oresme, Nicole 88−93

Pappus of Alexandria 78
parabola
  quadrature of 35−39, 110

tangent to 136−137
  semi-cubical 118−120, 218−219, 243
Pascal, Blaise 110, 112−113, 239−240
Pascal's triangle 112, 168−169, 174, 186,
    237−238
Peano, Giuseppe 327−328
pi, the number 2, 4, 19, 26, 31, 34, 170,
    176, 247−248
Principia Mathematica 191, 224−226
Proclus 5, 78
proportionality 6, 13−15
pseudo-equality method, Fermat's
    122−125
Ptolemy 78
pyramid, volume of 8−10, 19−22, 105
Pythagoras 5−6

quadratic equations, solution of 3, 11−12,
    82−83
quadratrix 208, 221−222
quadrature of
  Archimedean spiral 56−62
  circle 2, 7, 17−19, 31−32, 81,
      170−176, 247−248
  cissoid 176−178
  cycloid 207, 250−251
  ellipse 40−42
  generalized parabolas 109−117, 244,
      247
  hyperbola 154−160, 249
  kappa curve 216−217
  lune 8
  parabola 35−39
  quadratrix 208−209
  segment of circle 180
  versiera 215−216

radius of curvature 263−264
rectification, methods of 118−120,
    217−222, 242−243, 257
rectification of
  circle 31−35
  cissoid 219−220
  ellipse 220−221
  parabola 120
  quadratrix 221−222
  semi-cubical parabola 118−120,
      218−219, 243
reversion of series 204−205
Rheticus, G. J. 142

Rhind Papyrus 1
Riemann G. B. F. 322–326
Robert of Chester 85
Roberval, G. P. de 110, 134–137, 189, 191, 239
Robinson, Abraham 335, 341–345
roots, extraction of 183–185

Sarasa, A. A. de 156, 164
semi-cubical parabola 118–120, 218–219, 243
Simplicius 7
Simpson, Thomas 287
Simpson's rule 287
Sine series 205–206, 276, 317
Sluse, R. F. de 127, 129–131
Sluse's rule 129–131, 134, 254–255
sphere
    surface area of 42–44, 50–53, 102, 241
    volume of 24–26, 43, 53, 102
Stifel, Michael 143–144
Stirling, James 287
Stirling's formula 169–170
sums of powers of integers 57, 83–84, 109–113
sums of powers of lines 106–109
surface area of
    cone 46–49
    paraboloid 242
    sphere 42–44, 50–53, 102, 241
Swineshead, Richard 87, 91

tangent constructions of
    Archimedes 55
    Barrow 132–134
    Descartes 125–127
    Fermat 124–125
    Hudde 127–129
    Leibniz 245, 258, 261
    Newton 191–194
    Roberval 134–137
    Sluse 129–131

tangent to
    Archimedean spiral 55, 135
    cycloid 135–136
    ellipse 137
    folium of Descartes 130, 133
    parabola 125–127, 136–137
Taylor, Brook 287–289
Taylor's series 281, 287–292, 297–299, 316
Thales 5
Torricelli, Evangelista 116–117, 138–139
torus, volume of 102
transmutation 223, 245–251
trigonometric functions 275–279

ultimate ratios 225, 295

versiera 215–216
vibrating string 301–302
Viète, François 94, 168, 184–185
volume of
    cone 8, 23–24, 105
    cylinder 19
    cylindrical wedge 73–74
    ellipsoid 65–67
    paraboloid 62–65
    pyramid 8–9, 19–22, 105
    segment of sphere 54, 72
    solids of revolution 62–67, 83–85
    sphere 24–26, 42–44, 50, 71–72, 102, 105
    torus 102–103
Volterra, Vito 326

Wallis, John 113–116, 162–164, 168–176, 179–181
Wallis' infinite product 169, 176
Weierstrass, Karl 325, 329, 333
William of Moerbeke 86